Frequency Variations in Power Systems

Frequency Variations in Power Systems

Modeling, State Estimation, and Control

Federico Milano
University College Dublin
Dublin, Ireland

Álvaro Ortega Manjavacas
Universidad Loyola Andalucía
Seville, Spain

Registered Offices
John Wiley & Sons, Inc., 111 River Street, Hoboken, NJ 07030, USA
John Wiley & Sons Ltd, The Atrium, Southern Gate, Chichester, West Sussex, PO19 8SQ, UK

Editorial Office
The Atrium, Southern Gate, Chichester, West Sussex, PO19 8SQ, UK

For details of our global editorial offices, customer services, and more information about Wiley products visit us at www.wiley.com.

Wiley also publishes its books in a variety of electronic formats and by print-on-demand. Some content that appears in standard print versions of this book may not be available in other formats.

Library of Congress Cataloging-in-Publication Data

Names: Milano, Federico, author. — Ortega Manjavacas, Álvaro,
    author.
Title: Frequency variations in power systems : modeling, state estimation
    and control / Federico Milano, Álvaro Ortega Manjavacas.
Description: Hoboken, NJ : Wiley-IEEE Press, 2020. — Includes
    bibliographical references and index.
Identifiers: LCCN 2019053575 (print) — LCCN 2019053576 (ebook) — ISBN
    9781119551843 (hardback) — ISBN 9781119551898 (adobe pdf) — ISBN
    9781119551874 (epub)
Subjects: LCSH: Electric power systems–Mathematical models. — Electric
    power systems–Control. — Frequencies of oscillating
    systems–Measurement. — Frequency dividers.
Classification: LCC TK1005 .M4975 2020 (print) — LCC TK1005 (ebook) — DDC
    621.31–dc23
LC record available at https://lccn.loc.gov/2019053575
LC ebook record available at https://lccn.loc.gov/2019053576

Cover Design: Wiley
Cover Image: © ThomasVogel/Getty Images

Set in 11/13.5pt Computer Modern Roman by SPi Global, Chennai, India

Printed and bound by CPI Group (UK) Ltd, Croydon, CR0 4YY

10 9 8 7 6 5 4 3 2 1

To my grandfather, Alessandro Cesare.

F.M.

To Alicia.

Á.O.M.

Space itself wobbles and rumbles like a drum...

Janna Levin

I should point out, however, that at first some difficulty was experienced in observing the phenomena predicted by the theory, owing to the extreme smallness of the variations in the period of oscillation.

Pieter Zeeman

If you want to find the secrets of the universe, think in terms of energy, frequency and vibration.

Nikola Tesla

The definition, measurement, and interpretation of the frequency of a power system is a challenging problem without (so far) a satisfying solution.

Harold Kirkham, William Dickerson, and Arun Phadke

Contents

List of Figures

List of Tables

Preface

This book originated from the following question: is it possible, using the conventional transient stability model of power systems, which is based on phasors, to know the frequency variations at load buses? The answer appeared since the beginning not trivial, even possibly inconsistent as the construction of the admittance matrix assumes that the frequency is constant and equal to the synchronous reference in every point of the network.

At the beginning, this question was aimed at solving a mere simulation problem, without any particular theoretical implication. The approximated numerical derivative of the phase angle of bus voltage phasors and the frequency of the center of inertia are well-known simple ways to determine frequency variations in transient stability analysis. Existing solutions, however, were inadequate and unsatisfactory either because prone to numerical issues or because unable to capture local modes.

A breakthrough appeared all of a sudden in the form of a simple example. Consider a transmission line connecting two synchronous machines. The frequency at the air gap between the stator and the rotor of each machine is known as it is the rotor angular speed of the machine itself multiplied by the number of pairs of poles. During a transient following a contingency, the two generators will oscillate with different instantaneous rotor speeds. Hence, the frequency takes different values at the beginning and at the end of the line. *Somehow*, such a frequency has to change along the transmission line to satisfy the boundary conditions imposed by the synchronous machines.

The hypothesis that originates the *frequency divider formula* presented in Chapter 5 dictates that the frequency varies *linearly* and *instantaneously* along the transmission line. For each instant, thus, the frequency along the line will change so that it linearly interpolates the rotor speeds of the two machines. A graphical representation on how the frequency varies in space and in time after a large perturbation for a two-machine system is shown in the figure below.

This simple reasoning led, little by little, to a variety of deductions and conclusions that went well beyond the original question and solved many more problems than just a modeling inconsistency. Very often, while developing the examples included in this book, the deductions anticipated experimental results and simulations. The latter were used just to confirm the theory.

This is, ultimately, a book on a linear expression born from the simple concept above and based on well-assessed approximations, which are common

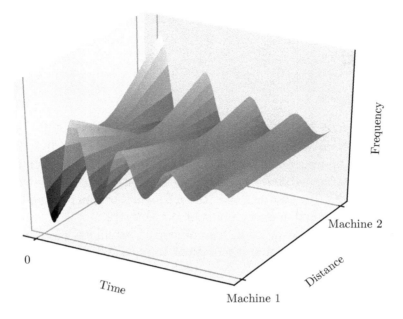

Space-time representation of the frequency in a two-machine system.

in power flow and fault analysis of power systems. Such a linear expression has been really under the eyes of everybody, but only in recent years, with the development of reliable phase-measurement units, its applications for modeling, state estimation, and control have become relevant. We have been lucky, thus, to have lived in these years and to have been the first to recognize the implications of the frequency divider formula.

Organization

The matter of the book is organized in three parts: background, theory, and applications; and ten chapters, as follows.

Part I: Background

Chapter 1 defines frequency as commonly understood and intended in AC circuit analysis and its fundamental role in power systems. In the first part, the chapter discusses the well-known steady-state representation, the Fourier transform, and the dynamic phasors representation. In the second part, the chapter provides an overview of relevant transforms utilized in power system analysis, namely, symmetrical-component theory, the forward-backward transform, the $\alpha\beta$o-transform by Clarke, and the dqo-transform by Park. The latter transform is the most relevant for the theoretical developments of the book.

Chapter 2 introduces the most relevant modeling approaches for power system analysis. The quasi-steady-state model based on a set of nonlinear semi-implicit differential algebraic equations is formally defined and chosen as the reference formulation for all examples considered in the book. The chapter also presents the most relevant device models that are utilized in the examples and case studies discussed in the book, namely, synchronous machines, loads, transmission lines, and transformers.

Chapter 3 outlines the basic concepts of conventional state estimation and dynamic state estimation, and Kalman filtering. Next, the chapter provides a comprehensive discussion on how to estimate bus frequency variations in simulations through the numerical derivative of the phase angle of voltage phasors and in practice by measuring AC voltages and using phase-locked loop devices or the discrete Fourier transform.

Chapter 4 justifies the need for conventional frequency control in power systems. Two kinds of regulators that utilize the rotor speeds of synchronous machines as input signals are described: primary and secondary frequency controllers that are aimed at compensating active power unbalances; and power system stabilizers that are aimed at damping the electromechanical oscillations of synchronous machines. The last section of the chapter introduces a variety of frequency controllers of nonsynchronous devices that require the estimation of bus frequency variations.

Part II: Theory

Chapter 5 constitutes the first part of the theoretical foundation of the book. It develops a simple, yet effective general analytical expression to estimate the frequency variations at every location of the network based on synchronous machine rotor speeds. This expression is called *frequency divider formula*. After discussing the theoretical background and the rationale behind the proposed formula, the chapter provides its formal derivation and discusses, as relevant byproducts of the proposed formula, the utilization of network equivalents and PMU measurements. A discussion on the participation factors of the rotor speeds of synchronous machines to bus frequencies completes the chapter.

Chapter 6 constitutes the second part of the theoretical foundation of the book. It presents an alternative derivation of the frequency divider formula, which leads us to determine, in transient conditions, whether a device provides inertial response and/or frequency control (*frequency maker*) or not (*frequency taker*). Such a formulation is shown to be formally equivalent to the DC power flow equations and leads to the definition of the concept of "rate of change of power," which refers to the "regulated" power injected at a bus. The chapter then presents a taxonomy of frequency maker and frequency taker devices for a variety of technologies, including synchronous machines, loads, renewable generation with and without frequency control, and energy storage systems.

The proposal of a criterion to quantitatively discriminate between frequency makers and frequency takers is also discussed.

Part III: Applications

Chapter 7 revisits the frequency control of nonsynchronous devices in view of the concepts discussed in Chapters 5 and 6. The first part of the chapter focuses on the impact of the quality of the frequency estimation used by the controllers of fast and slow distributed energy resources on the dynamic performance of the system. The discussion takes into account realistic measurements as obtained using PLLs and compares their performance using the "ideal" signals obtained using the frequency divider formula. The second part of the chapter compares, through a case study based on a distribution network, different strategies to estimate the frequency, namely centralized, decentralized, and averaged.

Chapter 8 revisits the dynamic state estimation through applications of the frequency divider formula and rate-of-change-of-power concept. First, a linear rotor-speed estimation problem based on the frequency divider is presented. Such a state estimation is *model-agnostic*, i.e. generator and load dynamic models and parameters need not to be known except for the capacity and technology of the synchronous machine. This approach only requires the bus frequencies measured by, e.g. phasor measurement units, instead of voltage and current phasors. Based on the same model-agnostic approach, a technique to estimate the frequency of the center of inertia based on bus frequency measurements, rather than on machine rotor angular speeds, is presented. Finally, the chapter discusses the ability of nonsynchronous devices to modify the local frequency. An expression to estimate the equivalent inertia of nonsynchronous devices is also provided.

Chapter 9 revisits the quasi-steady-state model of power systems for angle and voltage stability analysis and discusses how the frequency divider formula can be utilized to define a *frequency dependent model*, whose accuracy and computational burden are in between the conventional quasi-steady-state and electromagnetic transient models.

Chapter 10 proposes two new definitions of frequency based on the concepts of frequency divider and rate of change of power.

Software Tools

For the reader interested in software technicalities, all simulations included in the book are obtained using the Python-based software tool Dome [128]. The Dome version utilized for this book is based on Fedora Linux 25, Python 3.6.2, CVXOPT 1.1.9, KLU 1.3.8, and MAGMA 2.2.0. The hardware consists of two 20-core 2.2 GHz Intel Xeon CPUs and one NVidia Tesla K80 GPU.

Lessons Learned

The germ of the idea of this book was born about eight years ago, when the first author was an associate professor in Spain. The beginnings were not simple. When he joined the University College Dublin, the first author assigned the topic to a research master student. After one year, the master student left for irreconcilable differences with his supervisor. The student, in fact, thought that the whole concept of the frequency divider formula was wrong. The second author, by then PhD student, was more than happy to take over and give it a try. This collaboration happened to be a continuous brainstorming and led to a number of, we believe, remarkable findings. This short story introduces the main lesson learned by writing this book. Research is a very social matter. To bloom, it needs dialogue and trust. Writing this book would not have been possible without the many interminable discussions between the authors.

We hope that the readers will find the book as exciting as it was for us to develop the ideas contained in the following pages.

Federico Milano and Álvaro Ortega Manjavacas

Dublin, Madrid, Sori, Chiusa di Pesio

August 2019

Acknowledgments

We wish to acknowledge and thank the following individuals for their assistance and support toward the preparation of the book.

- Professor Mario Paolone, EPFL, Switzerland.

- Dr. Giacomo Severini, Dr. Ioannis Dassios and Dr. Muyang Liu, UCD, Ireland.

Funding

This work is part of a project that has received funding from the European Union's Horizon 2020 research and innovation program under Grant No. 727481.

Disclaimer

The opinions, findings, conclusions, and recommendations expressed in this work are those of the authors and do not necessarily reflect the views of the European Commission. The European Commission is not responsible for any use that may be made of the information that this work contains.

Acronyms and Abbreviations

AC	alternating current
AEMO	Australian Energy Market Operator
AGC	automatic generation control
AIITS	all-island Irish transmission system
AVR	automatic voltage regulator
CIG	converter-interfaced generation
CoI	center of inertia
CSWT	constant-speed wind turbine
DAE	differential algebraic equation
DC	direct current
DER	distributed energy resource
DFIG	doubly-fed induction generator
DFT	discrete Fourier transform
DP	dynamic phasor
DSE	dynamic state estimation
DST	discrete sine transform
EKF	extended Kalman filter
ELF	extremely low frequency
EMF	electromotive force
EMT	electromagnetic transients
ENTSO-E	European Network of Transmission System Operators for Electricity
ERCOT	Electric Reliability Council of Texas
ESS	energy storage system

FACTS   flexible AC transmission system

FBT   forward-backward transform

FD   frequency dependent

FDF   frequency divider formula

FDMS   frequency data management system

FFR   fast frequency response

FPF   frequency participation factor

FRRS   fast responding regulation service

GPS   global positioning system

HVDC   high-voltage direct current

IEEE   Institute of Electrical and Electronics Engineers

KKT   Karush-Kuhn-Tucker

LF   loop filter

LPF   low-pass filter

MPC   model predictive control

MPPT   maximum power point tracking

NNZ   number of nonzero (elements of a sparse matrix)

OPT   optimization problem

PCC   point of common coupling

PD   phase detector

PDC   phasor data concentrator

PDF   probability density function

PFC   primary frequency control

PI   proportional integral

PLL   phase-locked loop

PMU   phasor measurement unit
POD   power oscillator damper
PPF   power participation factor
PSS   power system stabilizer
PV   photovoltaic
PWM   pulse-width modulation

QSS   quasi steady state

REI   radial, equivalent, and independent
RMS   root mean square
RoCoF   rate of change of frequency
RoCoP   rate of change of power
RTDS   real-time digital simulator

SDE   stochastic differential equation
SFC   secondary frequency control
SPVG   solar photovoltaic generation
SRF   synchronous reference frame
STFT   short-time Fourier transform

TCL   thermostatically controlled load
TDS   time domain simulation
TG   turbine governor
TS   transient stability
TSO   transmission system operator

ULTC   under-load tap changer
UTC   coordinated universal time

VCO   voltage controlled oscillator
VDL   voltage dependent load
VPF   voltage participation factor

VSC   voltage-sourced converter

WASC wide-area stabilizing control

WECS wind energy conversion system

WF   washout filter

WLS   weighted least square

WSCC Western Systems Coordinating Council

Notation

Scalars, Vectors, and Matrices

$v, V, \mathcal{V}$	scalar
$\boldsymbol{v}, \boldsymbol{V}$	vector
$\mathbf{V}$	matrix
$\bar{v}$	phasor or complex quantity, i.e. $\bar{v} = v\angle\theta = v\big(\cos(\theta) + j\sin(\theta)\big)$
$\bar{v}^*$	conjugate of $\bar{v}$, i.e. $\bar{v}^* = v\angle-\theta = v\big(\cos(\theta) - j\sin(\theta)\big)$
$\langle v \rangle$	average quantity
$\tilde{v}$	measurement
$\check{v}$	optimal value; value estimated using measurements
$\boldsymbol{v}^>$	vector with elements sorted in descending order
$\mathbf{V}^+$	Moore-Penrose pseudo-inverse of $\mathbf{V}$
$\mathbf{V}^\dagger$	transpose of $\mathbf{V}$
$\mathbf{V}^{-1}$	inverse of $\mathbf{V}$
$\mathbf{V}_{\boldsymbol{x}}$	Jacobian matrix of $\boldsymbol{v}$ with respect to $\boldsymbol{x}$, i.e., $\mathbf{V}_{\boldsymbol{x}} = \dfrac{\partial \boldsymbol{v}}{\partial \boldsymbol{x}}$

Reference Frames and Domains

$v(t)$	continuous time-domain quantity (v is used if the context is unambiguous and in schemes for simplicity)
$v(s)$	frequency-domain quantity
$v[n\Delta t]$	sampled quantity
$v\{k\omega_{\mathrm{s}}\}$	harmonic quantity
$\boldsymbol{v}_{\mathrm{abc}}(t)$	vector of three-phase time domain quantities
$\bar{v}_{\mathrm{dq}}$	Park vector in dq-axis reference frame, i.e. $\bar{v}_{\mathrm{dq}} = v_{\mathrm{d}} + jv_{\mathrm{q}}$

Time Derivatives

$\dfrac{d}{dt}$ time derivative in time domain

s time derivative in frequency domain (Laplace transform)

$j\omega_o$ time derivative in phasor domain

$\dfrac{d}{dt} + j\omega_o$ time derivative in dq-axis reference frame (Park transform)

$\dot{v}$ rate of change with respect to time

Common Quantities

$\mathbf{0}_{n,m}$ null matrix with n rows and m columns

$\mathbf{1}_n$ vector of 1s of order n

a transformer tap ratio, drift term of stochastic process

b diffusion term of stochastic process

B susceptance

C capacitance

D damping coefficient due to viscous friction

e electromotive force

$\mathcal{E}$ expectation

f electrical frequency

$\boldsymbol{f}$ vector of differential equations

$\boldsymbol{g}$ vector of algebraic equations

G conductance

$\boldsymbol{h}$ vector of measurement equations

$\mathcal{H}$ Heaviside function

H transfer function

i, I current

$\mathbf{I}_n$	identity matrix of order n
J	objective function
K	controller gain
L	inductance
$\mathcal{L}$	Lagrangian function
m	modulation amplitude of AC/DC converters, normalized inertia constant
M	starting time of synchronous machine (loosely called *inertia*)
p	active power
$\mathbf{P}$	Park tensor
$\mathcal{P}$	hard limit function (windup or antiwindup)
q	reactive power
r	time within integrals
R	resistance
$\mathcal{R}$	droop of primary frequency control
$\bar{s}$	complex power
s	variable of the Laplace transform
$\mathbf{S}$	covariance matrix of residues
t	time
T	time constant
$\boldsymbol{u}$	vector of input variables
$\mathbf{U}$	covariance matrix of measurements
v, V	voltage
w	windowing function
$\boldsymbol{w}$	weighting coefficient vector
$\boldsymbol{x}$	vector of state variables
X	reactance

y vector of algebraic variables

$\bar{Y}$ admittance

z boolean variable

z states of the state estimation problem

$\bar{Z}$ impedance

α rate of change of frequency

β exponent of the frequency in load models

γ exponent of the voltage in load models

δ angular position

Δ variation of quantity

ε measurement error

ϵ control error

ζ damping coefficient of an oscillation mode

η measurement

ϑ phase angle

θ phase angle relative to a reference angular speed

μ observation dual variable

ν process dual variable

ξ white noise

ρ density

σ standard deviation

τ torque, time delay

φ magnetic flux

ϕ phase shift

ψ total magnetic flux

ω angular frequency, angular speed

Common Superscripts and Subscripts

ac	AC quantity
b	base quantity
B	bus
CoI	center of inertia
d	direct axis of the dqo transform
D	load (demand)
dc	DC quantity
dq	dq-frame (Park) vector
e	electrical
FD	frequency divider
G	generator
L	transmission line
m	mechanical
max	maximum value
min	minimum value
n	nominal or rated quantity
o	reference, initial or base-case condition
o	zero axis quantity of the dqo transform
q	quadrature-axis quantity of the dqo transform
r	rotor
ref	reference
s	stator
T	transformer
tot	total
w	washout filter

Constants

$\bar{a} = \exp(j2\pi/3)$

$\exp(x) = (2.71828182846)^x$

$j = \sqrt{-1}$

$\pi = 3.14159265359$

Sets

$\mathbb{B}$ set of buses

$\mathbb{G}$ set of generators

$\mathbb{L}$ set of transmission lines

$\mathbb{T}$ set of transformers

Numbers

The order of vector and matrices is indicated with n and a suffix to indicate the variable to which such a number refers. For example, n_x indicates the order of the vector of state variables $x(t)$.

Units

The units of absolute quantities undergo the International System of Units (SI). Unless explicitly indicated, however, the equations that describe AC circuits are in per unit values, as usual in power system analysis. The bases are the three-phase apparent power s_b, the phase-to-phase voltage v_b, and the frequency f_b. All other bases are derived from these three quantities. For example, the bases of the impedance and the line current are, respectively:

$$Z_b = \frac{v_b^2}{s_b}, \qquad i_b = \frac{s_b}{\sqrt{3}\,v_b}.$$

In per-unit, the *frequency*, f, has the same numerical value of the *angular frequency* or *angular speed*, ω. Hence, for simplicity and if the context is unambiguous, the term *frequency* is used to indicate both quantities. For the same reason and because they are often plotted together, f and ω are both indicated in pu(Hz).

Part I
Background

Chapter 1
Frequency in Power Systems

1.1. Conventional Definitions

Frequency is a fundamental quantity used in several fields of science and engineering and is utilized to characterize a huge variety of oscillatory and periodic phenomena. These include mechanical vibrations, sound, radio waves, and light.

Frequency is usually denoted as f [Hz] and is defined in the dictionaries as *the number of occurrences of a repeating event per unit of time*. This definition implies that, whatever is the event to be counted, the frequency cannot be measured until such an event has repeated at least twice. This definition is thus not adequate for transient phenomena and, apparently, indicates that the frequency cannot be defined for "fractions" of an event. The definition above is generally coupled with that of *period*, T [s], which is *the duration of time of one cycle in a repeating event*, thus $T = 1/f$.

The definition of frequency above is only adequate in stationary conditions, where events (or signals) repeat forever. This is sometimes called *stationary frequency* or *Fourier frequency*.

In electrical engineering, it is more useful to define an *instantaneous angular frequency*, ω [rad/s], as *the rate of change of the phase*, ϑ [rad], *of a sinusoidal wave* [12, 20]. If one has a signal:

$$u(t) = \sin\left(\vartheta(t)\right) , \tag{1.1}$$

the relationship between ϑ, ω, and f is:

$$\boxed{\omega(t) = 2\pi f(t) = \frac{d}{dt}\vartheta(t)} , \tag{1.2}$$

and, consequently, also the period becomes time-varying:

$$T(t) = \frac{2\pi}{\omega(t)} = \frac{1}{f(t)} . \tag{1.3}$$

Frequency Variations in Power Systems: Modeling, State Estimation, and Control, First Edition.
Federico Milano and Álvaro Ortega Manjavacas.

Finally, the rate of change of frequency (RoCoF) [Hz/s] is defined as:

$$2\pi \, \text{RoCoF} = \alpha(t) = \frac{d}{dt}\omega(t) = \frac{d^2}{dt^2}\vartheta(t) \, . \tag{1.4}$$

To properly interpret (1.2) is not always trivial. If the angular frequency is constant, the definition is also consistent with the intuition. In fact, if ϑ is:

$$\vartheta(t) = \omega_o \, t + \theta_o \, , \tag{1.5}$$

where ω_o and θ_o are constant, then, as expected, the angular frequency is ω_o. However, if ϑ is:

$$\vartheta(t) = \hat{\omega}(t) \, t + \theta_o \, , \tag{1.6}$$

where $\hat{\omega}$ is a smooth, i.e. derivable at least once, function of time, then the time derivative of (1.6) gives:

$$\frac{d}{dt}\vartheta(t) = \omega(t) = \hat{\omega}(t) + t \frac{d}{dt}\hat{\omega}(t) \, . \tag{1.7}$$

In (1.7), the instantaneous angular frequency is not $\hat{\omega}$, as one would expect.[1] The definition (1.2) is consistent. The notation is misleading as one associates $\hat{\omega}$ with the frequency because it is the term that multiplies the time in (1.6). However, $\hat{\omega}$ is the frequency only if it is constant.

While commonly accepted in the power system community [217], (1.2) does not really solve the problem but simply "shifts" it to the definition and measurement of the phase angle. This is certainly possible, e.g. [172], but conceptually, it is as challenging as estimating the frequency.

In a recent paper [105], the authors approach the problem of the definition of the frequency from the metrological point of view and indicate two approaches: *representationalist* and *operationalist*. The former states that the measurement of the frequency is a unique number that can be deduced with some function. The latter states that the process with which the frequency is measured determines its value.

---

[1] In an attempt to conciliate common intuition and math, the IEEE Std. C37.118.1-2011 [94], assumes $\frac{d}{dt}\hat{\omega} \approx 0$ in (1.7), which implies a quasi-steady-state approximation. However, $\frac{d}{dt}\hat{\omega} \approx 0$ does not hold in general. For example, if $\hat{\omega}$ has the following expression:

$$\hat{\omega}(t) = \omega_o + A\sin(\varpi t) \, ,$$

then the term $\frac{d}{dt}\hat{\omega}$ is negligible only if $\varpi \ll A$. Also because of this clear inconsistency, the IEEE Std. C37.118.1-2011 has been recently superseded by the standard IEC/IEEE 60255-118-1:2018, where the instantaneous frequency is defined as in (1.2).

(a)

(b)

——— 200 km

— 50 km

Figure 1.1 The length of the coast of Britain depends on the measuring method. (a) With a rod of 200 km, the British coast measures about 2,400 km; (b) with a rod of 50 km, the British coast measures about 3,400 km.

To better understand these concepts, it is worth recalling the famous article on fractal geometry "How long is the coast of Britain?" by Benoît Mandelbrot [121]. The first reaction to this question is generally to think of a well-defined number in km or miles. Unconsciously, one tends to use a representational-ist approach. This is also the typical approach of engineers, who like simple answers to complex questions. In his paper, however, Mandelbrot suggests that the length of the coast of Britain really depends on the method that one uses to measure it (operationalist approach). If the minimum resolution is 200 km, the coast measures about 2,400 km, whereas, if the minimum resolution is 50 km, one finds a length about 1,000 km higher. The process is illustrated in Figure 1.1.[2] Extending this reasoning to the limit of an infinitesimal measur-ing unit, one has to conclude that the length of the coast of Britain must be infinite. Another possible answer is that the coast line of Britain has actually all possible lengths until ones tries to measure it.[3] Mandelbrot concluded that the measurement approach *per se* made no sense and invented the concept of *fractal dimension* of geometrical objects.

Measuring the frequency of a real-world signal with an operationalist ap-proach is similar to measuring the length of a coast line. Depending on the num-ber of samples, on the filters utilized and, last but not least, on the model

---

[2]Figure 1.1 has been redrawn based on: http://en.wikipedia.org/wiki/Image:Britain-fractal-coastline-combined.jpg.

[3]This is the well-known *Copenhagen interpretation* of quantum mechanics [83].

that is imposed for the signal,[4] the estimation of the frequency changes. As a consequence, about 250 instantaneous signal models and, consequently, instantaneous frequency definitions, can be counted in the review papers [20] and [38]. Such an operationalist approach, apart from leading to endless discussions, is conceptually similar to the myth of Procrustes, a bandit from Attica who stretched or cut off the legs (sampling and filtering) of his victims (signals) to make them fit the size of an iron bed (signal model). Moreover, if the signal is not periodic, Fourier theory dictates that it should be sampled for an infinite time, leading to a spectrum of infinite frequencies. The analogy between nonperiodic signals and the infinite coast line is apparent.

Should the conclusion be that the measurement of a time-varying frequency has no meaning as in Mandelbrot's paper? In [105], the authors go further in the paradox, and state that *frequency is a parameter that has no physical existence.* If this assertion is taken literally, then it makes no sense to attempt measuring it. This conclusion is consistent only with a pure "measurist" approach, which is concerned more on *what* it is measured rather than *how* it is measured. Engineers, however, must be pragmatic.

The working assumption of the book is that the instantaneous frequency *exists* and is a variable of the system. In this context, *variable* means both the argument of a function and a quantity that can be measured either directly or indirectly. In Chapter 5, the frequency at any point of a power system is shown to be a linear combination of the rotor speeds of synchronous machines. Then, in Chapter 6, a second, more general, relationship between frequency variations and rate of change of power is deduced. These relationships lead to representationalist frequency definitions, which are given in Chapter 10.

1.2. Alternating Current

Electric power systems are almost exclusively based on alternating current (AC) at a unique frequency, e.g. 50 Hz in Europe and 60 Hz in North America. The "superiority" of AC systems over direct current (DC) ones to generate, transmit, and distribute electric energy has been under no discussion for about a century.

---

[4]For example, a signal model utilized in power systems is:

$$u(t) = A_1(t) \cos \left(0.5 \, \alpha(t) \, t^2 + \omega(t) \, t + \theta_1 \right) + \sum_{h \in \{2,3\}} A_h(t) \cos \left(h \, \omega(t) \, t + \theta_h \right) .$$

Another generalized signal model that also include an exponentially decaying component and white noise (ξ) is:

$$u(t) = \sum_{h=1}^{N} A_h(t) \cos \left(\omega_h(t) \, t + \theta_h \right) + A_{\mathrm{dc}}(t) \exp(-t/T_{\mathrm{dc}}) + \xi(t) .$$

Alternating current technology was developed by the Hungarian company Ganz Works in the 1870s, and in the 1880s by engineers and researchers such as Sebastian Ziani de Ferranti, Lucien Gaulard, and Galileo Ferraris. In 1876, the Russian engineer Pavel Yablochkov utilized an "open core" transformer to feed, through an AC generator, a lighting system. This technology was further developed by Ganz Works that, by 1885, patented the "closed core" transformer, from which modern transformers have been inspired since then. The first industrial applications of synchronous machines date from 1891, when the first long-distance transmission of high-voltage, three-phase power was demonstrated at the International Exposition at Frankfurt, Germany, using a machine with a power output of 210 kW [149]. The first AC commutator-free two- and three-phase induction motors were independently invented by Galileo Ferraris and Nikola Tesla, in 1885 and 1887, respectively. However, it was the induction motor design patented by Tesla in 1888 the one to be industrially exploited by Westinghouse and that eventually imposed the development of AC power systems all around the world.

Periodic and, in particular, sinusoidal currents and voltages are what make possible the electromagnetic energy conversion through the well-known Faraday's law of induction:

$$e(t) = -\frac{d}{dt}\varphi(t) \ , \tag{1.8}$$

where e is the electromotive force (EMF) and φ the magnetic flux.

All electrical machines exploit (1.8). However, to really be of practical use, it was realized that the flux has to be a periodic function of time and/or space, as it would be impractical to make it increase or decrease indefinitely.

The Faraday's law coupled with the understanding of the importance of alternating current for practical applications led to the invention of the three fundamental machines that constitute the bulk of any modern power system: (i) transformers, where the fluxes vary periodically in time due to the alternating currents flowing in fixed windings; (ii) synchronous machines, where the rotor flux varies periodically in space in the rotor due to a direct current flowing in a winding that rotates with the rotor shaft; and (iii) induction machines, where rotor fluxes rotate in both space and time in the air gap.

In recent years, DC transmission and generation have gained interest due to power electronic converters and it is expected that, in the future, power systems will be more and more a hybrid AC and DC circuit with a mix of conventional electrical machines and AC/DC and DC/DC converters. However, while power electronics-based devices are, slowly but steadily, taking over the conventional ones, AC power systems have been around for about a century and a half. To build the AC power systems that exist today all around the world has cost the equivalent of several trillions of dollars. For this reason, AC systems and conventional AC electrical devices, comprising synchronous generators, overhead lines, transformers, and induction motors, still have a long life ahead of them.

1.3. Reference Frequency

Most technologies that exploit periodic waves are generally based on a range (or *band*) of frequencies. The radio spectrum, for example, is divided into 12 bands, each spanning a decade of wavelengths, ranging from 3–30 Hz (extremely low frequency, ELF) to 300–3,000 GHz (tremendously high frequency, THF). Electromagnetic radiations that lead to visible light comprise wavelengths between 380 nm (violet) and 760 nm (red), which correspond to 790 and 400 THz, respectively. Audible sounds are waves of pressure in the range between 20 Hz and 20 kHz.

Power systems, instead, must be operated at a single nominal reference frequency, and the more constant and closer to its nominal value such a frequency is, the better. Any voltage or current with any other frequency than the nominal one leads to increase losses and decrease quality and, in turn, has to be removed or, at least, minimized.

Even small deviations with respect to the nominal frequency have to be corrected. For example, in Europe, the nominal frequency of electric energy systems is 50 Hz. The standard frequency ranges accepted in continental Europe (ENTSO-E transmission system) and UK (National Grid) are ±50 and ±200 mHz, respectively, whereas the maximum steady-state frequency deviations for these two systems are ±200 mHz and ±500 mHz, respectively [33].

In ideal operating conditions, the frequency of an AC power system is unique everywhere, constant and equal to the nominal value. Due to the variations of demand or generation, network operations, e.g. line disconnections, stochastic noise, harmonics, nonlinearity, e.g. saturation of the iron core of transformers, and, sporadically, faults, the frequency in a power system is actually never constant nor exactly the same everywhere.

The rationale why it is so important that the frequency in a power system must be unique everywhere and constant is twofold.

The first reason is that, due to the nature of synchronous machines, the system can be in steady-state only if the angular frequency is also constant and, thus, the same everywhere. In fact, in its roughest approximation, the dynamic of an interconnected AC power system can be represented as:

$$M \frac{d}{dt}\omega(t) \approx p_{\text{gen}}(t) - p_{\text{load}}(t) - p_{\text{loss}}(t) \ , \tag{1.9}$$

which expresses the balance of the power generation, p_{gen}, and the power consumption, p_{load}, and losses, p_{loss}. Any power unbalance on the right-hand side of (1.9) imposes a nonnull rate of change of the angular frequency $\frac{d}{dt}\omega$ of the system. Such a variation of the frequency depends not only on the power unbalance itself but also on the total inertia M of the synchronous machines.

The other reason why the frequency has to be unique is due to a property of sinusoidal functions. Such a property is exploited in all AC electrical machines and can be stated as follows: *the product of two sinusoids has nonzero average only if they are isofrequential.* This fact is crucial to be actually able to generate, transmit, and deliver electric power. This property is also at the base of the definition of active (or real) power.

Given the following voltage and current:

$$v(t) = \sqrt{2}V \sin\left(\vartheta_v(t)\right) = \sqrt{2}V \sin(\omega_v t) \, ,$$
$$i(t) = \sqrt{2}I \sin\left(\vartheta_i(t)\right) = \sqrt{2}I \sin(\omega_i t - \phi) \, , \qquad (1.10)$$

the instantaneous power is:

$$p(t) = v(t)\, i(t)$$
$$= VI \cos(\omega_v t - \omega_i t + \phi) + VI \cos(\omega_v t + \omega_i t - \phi) \, , \qquad (1.11)$$

which has zero average over the least common multiple of the periods $2\pi/\omega_v$ and $2\pi/\omega_i$ of the voltage and current respectively.

The only case for which the average value of the instantaneous power p is nonnull is for $\omega_v = \omega_i = \omega$, for which, in fact, one obtains the well-known expression:

$$p(t) = v(t)i(t) = VI \cos(\phi) + VI \cos(2\omega_o t - \phi) \, , \qquad (1.12)$$

that has average $\langle p \rangle = VI \cos(\phi)$ over a period $T = 2\pi/\omega$ and is nonnull for $\phi \neq \pm\pi/2$. It is important to note that the nonzero average of the instantaneous power is achieved as long as $\omega_v = \omega_i$. It does not matter whether these frequencies are also equal to the "nominal" angular speed of the system, ω_o.[5] The condition $\omega = \omega_o$ is required for regulation purposes and power quality requirements.

The focus of the book is on the dynamics that lead to "small" variations of the frequency around the synchronous reference value. Such variations of the frequency are to be expected in large national power systems following contingencies, e.g. three-phase faults and line outages. In particular, the book investigates what happens to the frequency in transient conditions, i.e. when the frequency varies in time and is not exactly the same everywhere in the grid.

---

[5] As a matter of fact, the same expression of the average power holds for every harmonic, which, thus, carries an energy. Ultimately, this is why the frequency of an AC system has to be unique.

1.3.1. **Example**

Figures 1.2 and 1.3 show the instantaneous power p and average power $\langle p \rangle$ obtained with a voltage and a current that are nonisofrequential and isofrequential, respectively. As expected only the average power shown in Figure 1.3 is nonnull.

Since the focus is on "small" frequency variations with respect to a nominal value, it is worth further elaborating on (1.11). With this aim, assume an "almost" isofrequential voltage and current, as follows:

$$v(t) = \sqrt{2}V \sin\big((\omega_o + \Delta\omega_o)t\big)$$
$$i(t) = \sqrt{2}I \sin(\omega_o t - \phi) \; , \tag{1.13}$$

where $\Delta\omega_o \ll \omega_o$. The instantaneous power obtained for the voltage and current in (1.13) is:

$$p(t) = \quad VI\cos(\Delta\omega_o t + \phi) - VI\cos\big((2\omega_o + \Delta\omega_o)t - \phi\big) \; . \tag{1.14}$$

This example is "unusual" as, in conventional electrical machines, the current and the voltage in a same winding have to be isofrequential, at least at their fundamental frequency component. This is because the EMF that appears in (1.8) is obtained from a magnetic flux originated with an alternating current.

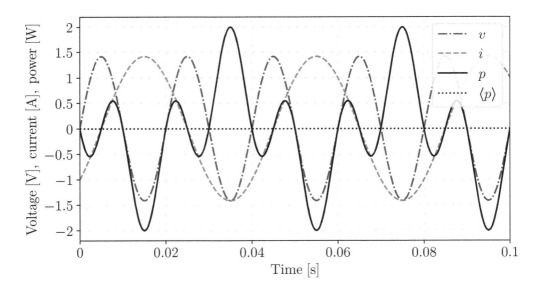

Figure 1.2 Instantaneous power obtained with nonisofrequential voltage ($V = 1$ V, $\omega_v = 2\pi\,50$ rad/s) and current ($I = 1$ A, $\phi = \pi/4$ rad, $\omega_i = 2\pi\,25$ rad/s).

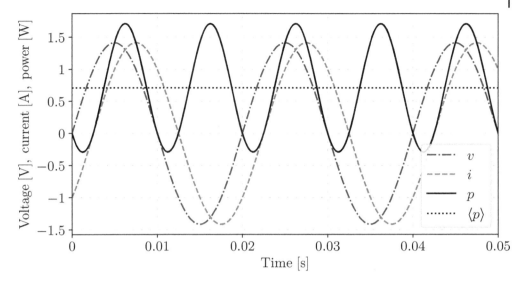

Figure 1.3 Instantaneous power obtained with isofrequential voltage ($V = 1$ V, $\omega_v = 2\pi\,50$ rad/s) and current ($I = 1$ A, $\phi = \pi/4$ rad, $\omega_i = 2\pi\,50$ rad/s).

One can always assume, however, that these quantities represent different harmonics, due, for example, to saturation of the iron core.

Since the voltage and the current are not isofrequential, all terms of the instantaneous power have zero average. In particular, the first term has zero average in the period $T_\Delta = 2\pi/\Delta\omega_o$, which, if $\Delta\omega_o \ll \omega_o$, satisfies the condition $T_\Delta \gg T$. For example, if $\omega_o = 2\pi 50$ rad/s, and $\Delta\omega_o = 2\pi\,0.1$ rad/s (i.e. 100 mHz), $T = 0.02$ ms and $T_\Delta = 10$ s, hence $T_\Delta = 500\,T$. Figure 1.4 confirms that, for $\Delta\omega_o \neq 0$, the average value $\langle p \rangle$ of the active power is null, as expected as the voltage and the current are not exactly isofrequential.

Another scenario, whose relevance is clarified in Chapter 2, is where the frequency difference between voltage and current is also a function of time. Assume that the phase angle variation is characterized by an exponentially decaying oscillation:

$$\Delta\theta(t) = \Delta\theta_o\big(1 - \cos(\omega_\theta t)\big)\exp(-\lambda_\theta t)\ . \tag{1.15}$$

This function resembles the damped oscillations that occur in a power system following a large disturbance and is thus of relevance for the transient phenomena discussed in this book. Figure 1.5 shows an example that considers:

$$\begin{aligned} v(t) &= \sqrt{2}V\sin\big(\omega_o t + \Delta\theta(t)\big)\\ i(t) &= \sqrt{2}I\sin(\omega_o t - \phi)\ . \end{aligned} \tag{1.16}$$

The average value of the active power is not null and eventually converges to the expected value $VI\cos(\phi)$.

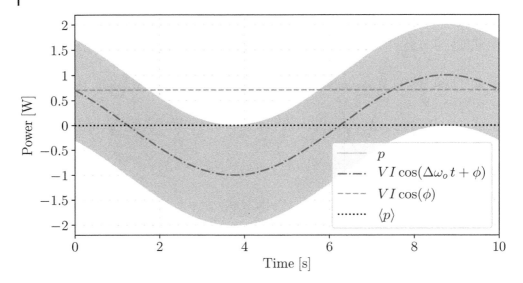

Figure 1.4 Instantaneous power obtained with almost isofrequential voltage ($V = 1$ V, $\omega_o = 2\pi\,50$ rad/s) and current ($I = 1$ A, $\phi = \pi/4$ rad), with $\Delta\omega_o = 2\pi\,0.1$ rad/s.

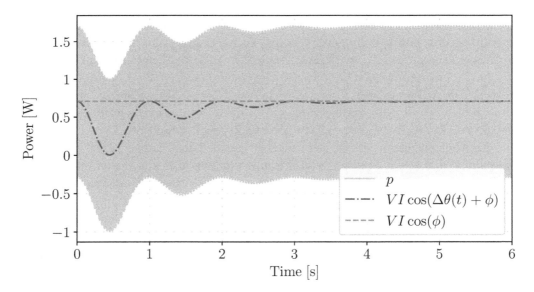

Figure 1.5 Instantaneous power obtained with almost isofrequential voltage ($V = 1$ V, $\omega_o = 2\pi\,50$ rad/s, with $\Delta\theta_o = 0.2\pi$ rad, $\omega_\theta = 2\pi$ rad/s and $\lambda_\theta = 1$ Hz) and current ($I = 1$ A, $\phi = \pi/4$ rad).

This example serves to introduce, using a quantity that has a clear physical meaning, the behavior of the product of "slightly" nonisofrequential sinusoids. Understanding well this behavior is relevant for the applications of the Park transform, which is discussed at the end of this chapter.

1.4. Transforms

The fully fledged electromagnetic model of an electric circuit provides the highest fidelity but also, in most circumstances, a very large and often unnecessary amount of information. If one is interested in the dynamic response of synchronous machines following a typical electromechanical transient lasting, say, 10 s, knowing in detail the trajectories of the voltages and currents in each phase of each line of a large interconnected network with thousands of lines is generally not required. Moreover, in 10 s an alternating quantity at the frequency of 50 Hz shows 500 oscillations. Even just visualizing such a trajectory is a challenge. It is thus common practice to use some "transform" to both simplify the equation (and thus reduce the computational burden of the simulation) and facilitate the interpretation of the results.

Several transforms have been defined in power system analysis. It is important to note that a transform does not bring new information. It just changes a set of variables into another set. Often, in the process, the physical meaning of the original variables is lost, and hence a transform requires some sort of abstraction. The other consequence is that, while computer programs can easily run simulations using transformed quantities (*de facto*, this computer-based implementation of power systems is also often the main motivation for the utilization of some transform), reality is tied to physical quantities. This means that transformed quantities cannot be measured directly but, at most, estimated based on physical ones.[6] The remainder of this section briefly discusses some relevant transforms that have wide applications.

1.4.1. Phasor Representation

The first transform that has been utilized in power system is the *phasor* representation.[7] The word "phasor" merges the words "phase" and "vector"

---

[6]For this reason, the estimation of the frequency based on the measurements of AC quantities poses both theoretical and practical challenges.

[7]The introduction of phasors in circuit theory is commonly attributed to C.P. Steinmetz in 1897 in the famous book [197]. A recent article indicates that the idea was presented by Steinmetz four years before, in 1893 at the International Electrical Congress of the American Institute of Electrical Engineers but not published in that year because of lack of funds [9]. Interestingly, while inventing the phasors, Steinmetz did not like "phasor diagrams" such as the one shown in Figure 1.6, which he called "crank diagrams" (see, for example, the fifth edition of [197]).

and refers to the representation of a complex number in terms of a complex exponential.[8] By definition, this representation works exclusively if stationary conditions are satisfied and the grid has same frequency everywhere.

The idea is simple and powerful at the same time: if all currents and voltages are oscillating at the very same angular frequency, say ω_o, then carrying on the information on the frequency is not really necessary. If one uses a reference frame that is rotating at ω_o and, in fact, all currents and voltages are described by a constant magnitude and a constant phase angle.

Consider the following sinusoidal voltage v:

$$v(t) = \sqrt{2}\,V\cos(\omega_o t + \theta_o)\ . \tag{1.17}$$

In phasor domain, v can be represented with a much simpler expression:

$$\bar{v} = V\,\angle\theta_o\ , \tag{1.18}$$

where the fact that the angular frequency is ω_o is implicit (the reference frame is in fact rotating at angular speed ω_o) and the phasor retains the information on the phase angle position θ_o, which is relative to a given reference angle and the Root Mean Square (RMS) value of the sinusoid, defined as:[9]

$$V = \sqrt{\frac{1}{T_o}\int_{-T_o/2}^{T_o/2} v^2(r)dr}\ , \tag{1.19}$$

where $T_o = 2\pi/\omega_o$.

One can certainly use other definitions for the magnitude of the phasors, such as the peak value $\sqrt{2}V$, but the RMS value has relevant properties when it comes to define the power. And since in power systems everything is about generation, transmission and consumption of power, the RMS is the most natural and probably the best choice after all.

The conceptual "leap" that enables the transformation of (1.17) into (1.18) is the transformation from a real time-varying quantity $v \in \mathbb{R}$ to a constant complex quantity $\bar{v} \in \mathbb{C}$. The phasor domain, in practice, implements a

---

[8]The term "phasor" was introduced during WWII to avoid using "vector," which has a meaning in electromagnetic field theory [198].

[9]The origin of the term "RMS" is uncertain. One of the first mentions to this term is found in the paper "The Measurement of Re-dressed or Adjusted Currents," by Édouard Hospitalier, The Electrician And Electrical Engineer, pp. 87–88, March 1886. The RMS value was already a well-accepted term by the mid-1890s (see, for example, [68]). Other terms also in use in the last decade of the nineteenth century were "effective" or "virtual." In a discussion published in 1893 on a paper by John Ambrose Fleming on transformers, William Ayrton proposed to drop both terms in favor of RMS value.

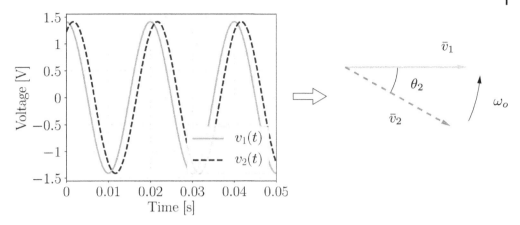

Figure 1.6 Time dimensionality reduction in the phasor domain. $v_1(t) = \sqrt{2}V_1 \cos(\omega_o t)$ and $v_2(t) = \sqrt{2}V_2 \cos(\omega_o t + \theta_2)$.

reduction of the temporal dimensionality of the original time domain representation. This is actually a smart mathematical "trick" that enables the representation of time-varying sinousoids as constant vectors (see Figure 1.6).

A property of phasors, which will be utilized and thoroughly discussed in the remainder of this book, is their representation of the time derivative. In time domain, the derivative of the voltage in (1.17) is readily given by:

$$
\begin{aligned}
\frac{d}{dt}v(t) &= -\omega_o \sqrt{2}\,V \sin(\omega_o t + \theta_o) \\
&= \omega_o \sqrt{2}\,V \cos(\omega_o t + \theta_o + \pi/2) \\
&= \omega_o \sqrt{2}\,V \cos(\omega_o \cdot (t + T_o/4) + \theta_o) \\
&= \omega_o\, v(t + T_o/4)\ ,
\end{aligned} \tag{1.20}
$$

where $\omega_o = 2\pi/T_o$. Equation (1.20) indicates that the time derivative shifts by a quarter of a period the original waveform. From (1.20), it descends that in phasor domain the time derivative is given by a $\pi/2$ rad rotation in the complex plain and a deformation proportional to the angular frequency ω_o. The time derivative is thus expressed in phasor domain by an algebraic linear operator $j\omega_o$, where j is the imaginary unit, as follows:

$$
j\omega_o\,\bar{v} = \omega_o\,V\angle(\theta_o + \pi/2)\ . \tag{1.21}
$$

The phasor representation has two major drawbacks: (i) it assumes a constant unique angular frequency in the whole circuit; and (ii) the power, in such a representation, also becomes a complex quantity.

The first issue can be resolved using the Laplace transform, which works for any transient condition, and where the operator $j\omega_o$ is substituted with the Laplace operator s. The Laplace transform is one of the theoretical pillars of circuit analysis and automatic control. The time-domain formulation of differential algebraic equations, however, is preferred in the remainder of this book except for control schemes.

The second major issue of the phasor domain, as it is said above, is that the power becomes a complex quantity. The following product gives the power in phasor domain:

$$
\begin{aligned}
\bar{s} &= P + jQ \\
&= \bar{v}\bar{i}^* = VI\cos(\phi) + jVI\sin(\phi) ,
\end{aligned} \tag{1.22}
$$

where $*$ indicates the conjugate operation; $\cos(\phi)$ is the *power factor*, term coined, as it is well-known, by Fleming in 1892; and ϕ is the phase angle difference between the voltage and current phasors, with $\bar{v} = v\angle 0$ and $\bar{i} = i\angle - \phi$. $\bar{s}$ is also called "apparent" power, which is an unfortunate expression as electric power is all but apparent.

Consistently, $P = \text{Re}\{\bar{s}\} = \langle p \rangle$, i.e. the real part of the complex power is the average power of the product of the isofrequential voltage and current given in (1.12). The problem is the imaginary term, $Q = \text{Im}\{\bar{s}\}$, known as *reactive power*. This power is required by inductors and capacitors to work, goes back and forth these components but cannot be really "used." The reactive power, however, has to be transmitted and thus contributes to system losses.

While the reactive power can be somehow explained in steady state, during a transient it is hard to find a meaningful physical interpretation or even a convincing mathematical definition, basically because this is another example of operationalist approach. The reader can find a very well-written introduction to the challenges implicit in the definition of reactive power in Appendix B of [205].

1.4.2. Harmonic Analysis

One of the first approaches that likely comes to mind when dealing with frequency variations and AC systems in not perfectly synchronous operating conditions is the harmonic or Fourier analysis, that is, *the analysis of a complex waveform expressed as a series of sinusoidal functions, the frequencies of which form a harmonic series.*

The Fourier series and its generalization for nonperiodic functions, the Fourier transform, both in their continuous and discrete forms, are well-known, ubiquitous transforms utilized in many fields of physics and engineering. The impact of Fourier cannot be overestimated. Just to cite a few applications:

efficient algorithms to multiply large numbers, compression of image, audio and video files, signal processing for telecommunication, and the solution of the Schrödinger equation for quantum physics problems.

The story of the birth of the Fourier series is an excellent example of pragmatism. The exact solution of the heat equation, i.e. the parabolic partial differential equation that describes how temperature varies in time and space in a given region of matter, is known only for a reduced set of boundary conditions, e.g. if the heat source is a sinusoidal wave. Instead of solving the heat function for a general case, Fourier looked for an analytical approach to express a general function as a series of sines and cosines.[10] Of course, the underlying assumption made by Fourier is that the superposition principle holds for the heat equation. But the idea of transforming a generic function into a series of other functions is a powerful tool whose applications, as said above, go well beyond the initial problem solved by Fourier.

Since AC systems involve periodic functions, the Fourier series and, more in general, the Fourier transform are broadly utilized, especially for harmonic analysis. Some reference books on the topic are [3, 10, 44].

There are several works that attempt to describe time-varying and/or "instantaneous" spectra using special time-frequency distributions (see, for example, the reviews given in [21] and [38]). A relevant approach is the *spectrogram*, which, in turn, is a short-time Fourier spectrum of a signal around a specific time. The concept of *sliding-window* Fourier series/transform is a similar approach. The main idea of all these approaches, however, is the same as the one invented by Fourier, i.e. a signal is assumed to be a superposition of (infinite) sinusoids, whose spectrum can vary in time but, at any given time, is "stationary." This section describes exclusively the conventional Fourier approach.

Consider a periodic function h, with period T_o and fundamental frequency $\omega_o = 2\pi/T_o$. This function can be expressed as a series with complex coefficients:[11]

$$h(t) = \sum_{n=-\infty}^{\infty} \bar{c}_n \exp(j\, n\, \omega_o t) \,, \qquad (1.23)$$

where the complex coefficients $\bar{c}_n$ are given by:

$$\bar{c}_n = \frac{1}{T_o} \int_{-T_o/2}^{T_o/2} h(r) \exp(-j\, n\, \omega_o r)\, dr \,. \qquad (1.24)$$

---

[10] Jean-Baptiste Joseph Fourier first introduced the series that takes his name in 1807, in the work "Mémoire sur la propagation de la chaleur dans les corps solides," and later on in 1822 in the monograph "Théorie analytique de la chaleur."

[11] The Hartley series, which is based on the function $\mathrm{cas}(x) = \sin(x) + \cos(x)$, is arguably more elegant than the Fourier series as it enables the expression of the same function using real coefficients [3].

Similarly to the phasor-domain transform, the Fourier series implements a time dimensionality reduction but, for nonsinusoidal periodic functions, there are infinite discrete complex components for each of the angular frequencies $n\omega_o$.

Arguably, the main reason why the Fourier series and transform have been so successful and have found so many practical applications is the fact that they can be applied to time series of measurement data with constant sampling rate.

The discrete Fourier transform (DFT), in complex value, takes the form:

$$\hat{h}\{k\,\omega_s\} = \frac{1}{N}\sum_{n=0}^{N-1} h[n\Delta t]\exp(-j\,k\,\omega_s n\Delta t), \quad k = 0, 1, 2, \ldots, N-1, \quad (1.25)$$

where Δt and N are the sampling time and the number of samples of the time series with sampling frequency $\omega_s = 2\pi/(N\Delta t)$. Knowing $\hat{h}\{k\,\omega_s\}$, the original time series can be reconstructed as follows:

$$h[n\Delta t] = N\sum_{k=0}^{N-1} \hat{h}\{k\,\omega_s\}\exp(-j\,k\,\omega_s n\Delta t), \quad n = 0, 1, 2, \ldots, N-1. \quad (1.26)$$

From (1.25) and (1.26), it can be seen why the DFT has found several applications, especially in data compression. If one accepts a certain percentage of loss of information, in fact, only the most relevant components of $\hat{h}\{k\,\omega_s\}$ can be stored and used to reconstruct the original signal through the inverse DFT.

The following two examples are relevant. First, note that for the purely sinusoidal voltage in (1.17), the application of (1.24) gives:

$$\bar{c}_0 = 0, \quad \bar{c}_1 = \sqrt{2}\,V\angle\theta_o, \quad \bar{c}_n = 0, \quad \text{for } n = 2, 3, \ldots, \infty. \quad (1.27)$$

The coefficient $\bar{c}_0$ is null because the average of the cosine along its period is null, and all other coefficients for $n \geq 2$ are zero because, as discussed in Section 1.3, the product of nonisofrequential sinusoids have zero average value.

A more interesting case consists in a periodical variation of the frequency of the voltage, as follows:

$$v_\theta(t) = \sqrt{2}V \sin\left(\omega_o t + \Delta\theta(t)\right), \quad (1.28)$$

where $\Delta\theta$ is the test function (1.15) with $\lambda_\theta = 0$. v_θ is a periodic function, an example of which is depicted in Figure 1.7. The small amplitude of $\Delta\theta$ prevents appreciating the variations of the period of the overall voltage v_θ. The signal v_θ is certainly periodic but its frequency oscillates sinusoidally between 49.9 and 50.1 Hz. Thus, determining an analytical expression of the coefficients of the Fourier series of (1.28) is not trivial.

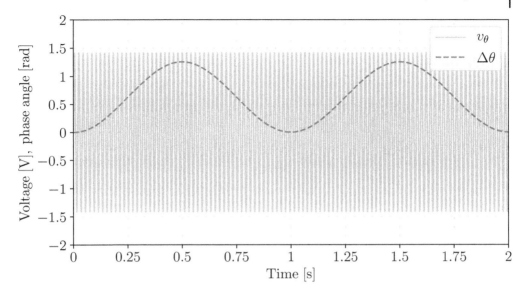

Figure 1.7 Periodic voltage as in (1.28), with $V = 1$ V, $\omega_o = 2\pi 50$ rad/s, and $\Delta\theta$ as in (1.15), with $\Delta\theta_o = 0.2\pi$ rad, $\omega_\theta = 2\pi$ rad/s and $\lambda_\theta = 0$.

For simplicity and since v_θ is an odd function, a discrete sine transform (DST), which takes real values, can be used instead of the complex DFT, as follows:

$$\hat{v}_\theta\{k\,\omega_s\} = \frac{2}{N} \sum_{n=0}^{N-1} v_\theta[n\Delta t] \sin(k\,\omega_s n\Delta t), \qquad k = 0, 1, 2, \ldots, N-1 . \quad (1.29)$$

Figure 1.8 shows the DST obtained with the open-source library FFTW[12] for the voltage expression in (1.28).

The results of the DST show the major drawback of the Fourier series approach, at least for the objectives of this book. The number of harmonics with nonnull magnitude is large. At the end of the day, we had just added a small sinusoidal phase variation on top of the fundamental frequency of the voltage signal. Moreover (1.28) represents a stationary periodic signal. During an electromechanical transient, the actual evolution of $\Delta\theta$ is expected to decay in time, similarly to the trajectory shown in Figure 1.5, which is obtained for $\lambda_\theta \neq 0$. On the other hand, the Fourier series assumes that the periodic signal is stationary in the observed window.

It is interesting to note that "frequency" is often associated with the hypothesis of stationarity or, in other words, with the idea that, to be defined,

---

[12]FFTW is available at www.fftw.org. See also the paper [72].

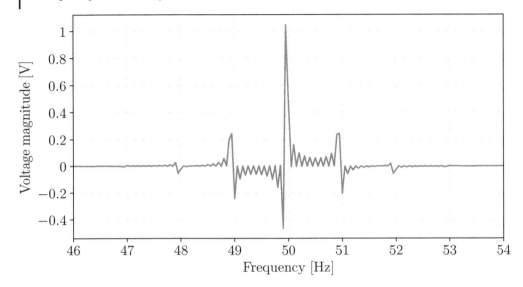

Figure 1.8 Results of the DST of $v_\theta[n\Delta t]$, with $\Delta t = 0.0001$ s and $N = 100,000$, $V = 1$ V, $\omega_o = 2\pi 50$ rad/s, and with $\Delta\theta$ defined in (1.15), with $\Delta\theta_o = 0.2\pi$ rad/s, $\omega_\theta = 2\pi$ rad/s and $\lambda_\theta = 0$.

one needs to wait at least one full period. This idea is reinforced by the Fourier analysis that requires a certain number of samples of a signal to be able to identify its harmonic content.[13] It appears, however, that a signal with a fast time-varying frequency is incompatible with such an approach as, while sampling the signal, its frequency and, thus, its harmonic content, has already changed. While the Fourier interpretation of the frequency is also common in the power system community (see the discussion in [105]), the frequency should be treated just as any other continuous variable of a system of differential equations and not just as a property of another signal.

Despite the caveat above, the Fourier approach is unquestionably a fundamental tool for power system analysis, measurement, and monitoring. A common solution to determine the frequency variations of a signal through the Fourier analysis is to use a sliding window of measurements. For each window, the DFT still defines a snapshot of the harmonic content of the signal. The perception of time-varying signal, however, is given by the fact that the window is updated with the sampling rate of the measurements. An in-depth description of the techniques to determine frequency variations using the Discrete Fourier Transform (DFT) is beyond the scope of this book but we briefly revisit this issue in Section 3.3.3.

---

[13]This is in turn the main results of the well-known Nyquist-Shannon theorem and the aliasing phenomenon.

1.4.3. Dynamic Phasors

In recent years, the concept of *dynamic phasors* (DPs) has been introduced in the study of asymmetrical faults [196] and in the modeling of power systems, especially those with a high number of power electronic converters [116]. The main idea is, in turn, an application of the Fourier series to a signal $h(r)$ along a sliding time window $r \in (t - T_o, t]$:

$$h(r) = \sum_{n=-\infty}^{\infty} \bar{c}_n(t) \exp(jn\,\omega_o t) , \tag{1.30}$$

where the complex coefficients $\bar{c}_n(t)$ are called *phasors*, which depend on time and are given by:

$$\bar{c}_n(t) = \frac{1}{T_o} \int_{t-T_o}^{t} h(r) \exp(-jn\,\omega_o r)\, dr = \langle \bar{c} \rangle_n(t) , \tag{1.31}$$

where $\langle \bar{c} \rangle_n$ is the n-th time-dependent average phasor in the interval $(t - T_o, t]$. The two main properties that are at the base of the DP approach are:

- The time derivative of the coefficients (1.31) is given by:

$$\frac{d}{dt}\bar{c}_n(t) = \left\langle \frac{d}{dt}\bar{c} \right\rangle_n (t) - jn\,\omega_o \bar{c}_n(t) . \tag{1.32}$$

- The product of two time-domain variables can be obtained as the convolution of the phasors (Fourier coefficients) of the two variables.

The interest in DPs is mainly due to the observation that, in many practical cases, only very few coefficients are needed to approximate a signal with a reasonable accuracy. The idea is thus to use DPs to have a better approximation than using conventional phasors while avoiding the computational burden of fully fledged electromagnetic models. So far, however, the expectation raised by DPs, especially the promise to reduce the computational burden with respect to standard electromagnetic transients models, has not been fulfilled.

DPs suffer of the same methodological issue of the Fourier series, i.e. the need to define a fundamental frequency and multiples of it. The Park transform that is described later in this chapter enables the definition of a sort of dynamic vector (called Park vector) with a derivative similar to (1.32) but without its limitations and drawbacks.

1.4.4. Transforms for Three-Phase Systems

The phasor representation is particularly suitable for poly-phase (most commonly, three-phase) systems as it enables a great simplification of the notation with respect to the time-domain form. Transmission and distribution networks are mostly balanced three-phase systems, as follows:

$$\boldsymbol{v}_{\mathrm{abc}}(t) = \begin{bmatrix} v_{\mathrm{a}}(t) \\ v_{\mathrm{b}}(t) \\ v_{\mathrm{c}}(t) \end{bmatrix} = \begin{bmatrix} \sqrt{2}V\cos(\omega_o t + \theta_o) \\ \sqrt{2}V\cos(\omega_o t + \theta_o - 2\pi/3) \\ \sqrt{2}V\cos(\omega_o t + \theta_o + 2\pi/3) \end{bmatrix} . \tag{1.33}$$

In phasor representation, (1.33) becomes:

$$\bar{\boldsymbol{v}}_{\mathrm{abc}} = \begin{bmatrix} \bar{v}_{\mathrm{a}} \\ \bar{v}_{\mathrm{b}} \\ \bar{v}_{\mathrm{c}} \end{bmatrix} = \begin{bmatrix} V\angle\theta_o \\ V\angle\theta_o - 2\pi/3 \\ V\angle\theta_o + 2\pi/3 \end{bmatrix} . \tag{1.34}$$

As it is well-known, under the hypothesis of balanced and symmetrical operating conditions, each phase contains all necessary information and one can thus study an equivalent single-phase system, e.g. phase "a."

In unbalanced conditions, all three phases have to be retained. Since the early years of the twentieth century and the infancy of three-phase electric power systems, however, it was recognized the convenience of transforming the original equations into simpler representations. Actually, the main original purpose of all transforms is to simplify the solution of some relevant problem, e.g. short-circuit analysis. With modern computers, the need to "simplify" calculations is not compeling anymore. However, some transforms have a second feature, i.e. the ability to retain and emphasize a specific feature or behavior of the original three-phase system. It is worth noticing that it was actually a fortune that modern computers were not available at the beginning of the twentieth century: the need to "simplify" calculations while retaining the fundamental behavior of AC systems led to the development of the theoretical foundations and mathematical tools that are still utilized today.[14]

The common feature among all transforms is the definition of some sort of "coordinate change" (or *homomorphism*) through a second order tensor and a set of linearly independent components that form a *base* for the three-dimensional[15] vector space. A consequence of being homomorphisms is that it is always possible to change from one base to another, which in turn means that all transforms are equivalent, i.e. carry same information. A byproduct of this property is that the matrices that describe the transform tensors are invertible, i.e. have full rank. The possible independent bases in $\mathbb{C}^3$ are also

---

[14]With this regard, the interested reader can refer to, for example, the inspiring books [104] and [215], which are classical examples of power system stability analysis made with "paper and pencil".

[15]The dimension can be higher than three if poly-phase systems are considered.

Table 1.1 Taxonomy of transforms utilized for the analysis of three-phase power systems.

Transform	Notation	Tensor		Application
		Domain	Time dependency	
Symmetrical component theory [69]	+−o	complex	no	steady state
Forward-backward transform [111]	+−o	complex	no	transient
Clarke transform [37]	$\alpha\beta$o	real	no	transient
Park transform [168]	dqo	real	yes	transient

infinite. However, there are not many homomorphisms that have an apparent physical meaning and/or make calculations simpler.

Table 1.1 shows the taxonomy of three well-known and also one not-so-well-known transforms based on the domain and the time-dependency of the tensor. Note that the Park transform is the only one with a time-dependent tensor. The remainder of this section describes the rationale and the features of each transform included in Table 1.1.

1.4.4.1. Symmetrical Components

The basic idea behind Fortescue's theory is that any set of vectors can be described by the linear combination of symmetrical components.[16] In a three-phase system, these components are three: a positive sequence "+," a negative sequence "−," and a zero sequence "o." The notation "012" is also often used for the symmetrical component sequence, where "1" and "2" are the positive and the negative sequence, respectively.

The symmetrical component theory is commonly applied to the phasor representation of three-phase AC circuits, as follows:

$$\bar{v}_{+-o} = \bar{A}\,\bar{v}_{\mathrm{abc}} , \qquad (1.35)$$

where

$$\bar{v}_{+-o} = \begin{bmatrix} \bar{v}_+ \\ \bar{v}_- \\ \bar{v}_o \end{bmatrix}, \quad \bar{A} = \frac{1}{3}\begin{bmatrix} 1 & \bar{a} & \bar{a}^2 \\ 1 & \bar{a}^2 & \bar{a} \\ 1 & 1 & 1 \end{bmatrix},$$

---

[16] While symmetrical components theory is attributed to Fortescue for its paper [69], this was already proposed by Stovkis in a series of papers, the first of which is in German and dates 1912, as it is indicated in Chapter 2 of [37] and, as a matter of fact, the symmetrical components are called "Stovkis-Fortescue scheme" in [111] but the attribution to Stovkis has been eventually dropped.

and $\bar{a} = \exp(j2\pi/3)$. The inverse of the symmetrical component transform is given by:

$$\bar{\boldsymbol{v}}_{\mathrm{abc}} = \bar{\mathbf{A}}^{-1}\,\bar{\boldsymbol{v}}_{+-\mathrm{o}}\ , \tag{1.36}$$

where

$$\bar{\mathbf{A}}^{-1} = \begin{bmatrix} 1 & 1 & 1 \\ \bar{a}^2 & \bar{a} & 1 \\ \bar{a} & \bar{a}^2 & 1 \end{bmatrix}.$$

Figure 1.9 shows an example of symmetrical components for a set of unbalanced voltages $\bar{\boldsymbol{v}}_{\mathrm{abc}}$. Note that the zero sequence is null, namely $\bar{v}_{\mathrm{o}} = 0$, if:

$$\bar{v}_{\mathrm{a}} + \bar{v}_{\mathrm{b}} + \bar{v}_{\mathrm{c}} = 0\ , \tag{1.37}$$

which is satisfied for triangle connections and for star connections with no neutral or ground currents. The condition (1.37) is always assumed in this book. The zero sequence is considered in the remainder of this section only for completeness.

The symmetrical component theory is applied to phasors and, hence, assumes the frequency to be unique and constant for the three sequences. This

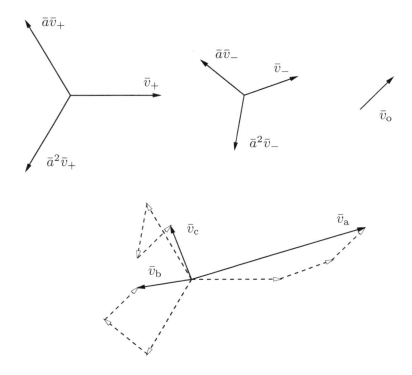

Figure 1.9 Graphical representation of symmetrical components.

makes the symmetrical components not adequate for the analysis of electrome-chanical transients, i.e. when the frequency is time variant.

1.4.4.2. Forward-Backward Transform

Among the four transforms described in this section, the forward-backward transform (FBT) is certainly the less known and used. The main reason is likely that, while it is a mix between the Fortescue theory of symmetrical components and the Park two-reaction theory, i.e. the one that leads to the well-known Park-Concordia model of synchronous machines, it does not have the clear physical interpretations of the two.

The FBT uses "positive" and "negative" dynamic components of the voltage and the current in the synchronous machines [111]. In turn, the FBT is the time domain version of the symmetrical component approach, as follows:

$$\bar{v}_{+-\mathrm{o}}(t) = \bar{\mathbf{A}}\, \boldsymbol{v}_{\mathrm{abc}}(t) \,, \tag{1.38}$$

where $\bar{\mathbf{A}}$ is defined as in (1.35). For example, assuming $v_{\mathrm{a}} = \sqrt{2}V\sin(\omega_o t)$, one has:

$$\bar{v}_{+}(t) = -j\frac{V}{\sqrt{2}}\exp(j\omega_o t) \,,$$

$$\bar{v}_{-}(t) = j\frac{V}{\sqrt{2}}\exp(-j\omega_o t) \,, \tag{1.39}$$

$$\bar{v}_{\mathrm{o}}(t) = 0 \,.$$

The unusual mix of complex and time-dependent quantities can be seen as a drawback of this approach, although a similar approach is also utilized in the definition of the Park vector (see Section 1.4.4.4). More likely, it is the lack of a clear physical meaning of a dynamic positive and negative sequence, in particular, in the synchronous machine (see [111]), what makes this approach not particularly interesting.

1.4.4.3. Clarke Transform

The transform discussed in this section is often referred to as Clarke transform, although in [37], Clarke indicates that the idea of the $\alpha\beta$o transform,[17] although not called in that way, dated back to 1917 and was proposed in a paper by Lewis [114] and later on by a few others. However, Clarke was the first to actually use this transform in transient conditions, which is the most relevant practical application of this transform today.

---

[17]Although in more recent works it is often called $\alpha\beta\gamma$ transform, the original notation $\alpha\beta$o better delivers the physical meaning of the "o" component.

The transform works both in time and phasor domains. In time domain and using the notation of [37], one has:

$$\boldsymbol{v}_{\alpha\beta o}(t) = \mathbf{C}\,\boldsymbol{v}_{abc}(t) \, , \tag{1.40}$$

where

$$\boldsymbol{v}_{\alpha\beta o}(t) = \begin{bmatrix} v_\alpha(t) \\ v_\beta(t) \\ v_o(t) \end{bmatrix}, \quad \mathbf{C} = \frac{2}{3}\begin{bmatrix} 1 & -1/2 & -1/2 \\ 0 & \sqrt{3}/2 & -\sqrt{3}/2 \\ 1/2 & 1/2 & 1/2 \end{bmatrix} .$$

The inverse Clarke transform is:

$$\boldsymbol{v}_{abc}(t) = \mathbf{C}^{-1}\,\boldsymbol{v}_{\alpha\beta o}(t) \, , \tag{1.41}$$

where

$$\mathbf{C}^{-1} = \begin{bmatrix} 1 & 0 & 1 \\ -1/2 & \sqrt{3}/2 & 1 \\ -1/2 & -\sqrt{3}/2 & 1 \end{bmatrix} .$$

For example, for the symmetrical voltages in (1.33), the components of the Clarke transform are:

$$\begin{aligned} v_\alpha(t) &= \sqrt{2}V\cos(\omega_o t + \theta_o) \, , \\ v_\beta(t) &= -\sqrt{2}V\sin(\omega_o t + \theta_o) \, , \\ v_o(t) &= 0 \, . \end{aligned} \tag{1.42}$$

Figures 1.10.b and 1.10.c show the Clarke transform for the set of symmetrical voltages $\boldsymbol{v}_{abc}$ as in (1.33) with constant angular frequency ω_o shown in Figure 1.10.a. Figure 1.10.d, on the other hand, shows the α and β components of the Clarke transform for a set $\boldsymbol{v}_{abc}$ rotating at $\omega_o + \frac{d}{dt}\Delta\theta$, with $\Delta\theta$ defined in (1.15). The fact that Figures 1.10.c and 1.10.d do not show any noticeable difference is an issue for the analysis carried out in this book.

1.4.4.4. Park Transform

The Park transform is conceptually different from the transforms seen so far because its tensor is time variant and projects the reference frame of a three-phase system onto a direct ("d") and quadrature ("q") axes rotating at a given, possibly time-dependent, angular speed and on a zero ("o") axis [168].[18]

---

[18]The Park transform is also often called dqo transform. Strictly speaking, the Park transform differs from the dqo one for a factor that affects the power invariance of the transformation. In particular, the Park transform is not power invariant, while the dqo is power invariant. This difference, however, disappears if per-unit quantities with proper bases are utilized.

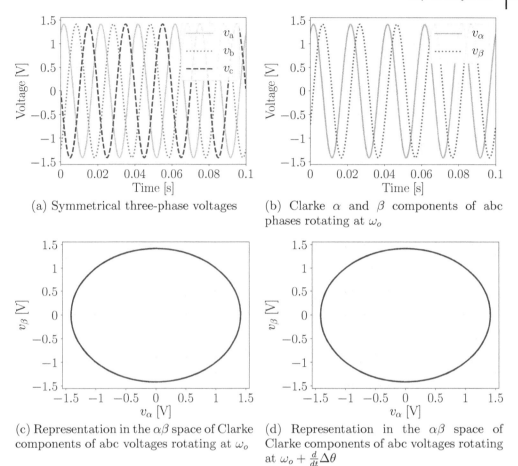

(a) Symmetrical three-phase voltages

(b) Clarke α and β components of abc phases rotating at ω_o

(c) Representation in the $\alpha\beta$ space of Clarke components of abc voltages rotating at ω_o

(d) Representation in the $\alpha\beta$ space of Clarke components of abc voltages rotating at $\omega_o + \frac{d}{dt}\Delta\theta$

Figure 1.10 Panel (a): Three-phase voltages $\boldsymbol{v}_{\mathrm{abc}}$ rotating with constant angular speed $\omega_o = 2\pi 50$ rad/s and phase $\theta_o = -\pi/6$ rad. Panel (b): α and β components of the Clarke transform as a function of time. Panel (c): State-space representation of the α and β components of the Clarke transform. Panel (d): State-space representation of the α and β components of the Clarke transform for a set $\boldsymbol{v}_{\mathrm{abc}}$ rotating at $\omega_o + \frac{d}{dt}\Delta\theta$, with $\Delta\theta$ defined in (1.15) with $\Delta\theta_o = 0.2\pi$ rad, $\omega_\theta = 2\pi$ rad/s and $\lambda_\theta = 0$.

The main idea, i.e. using a *rotating reference frame*, is very similar to the principle of the phasor representation and, in fact, the Park transform can be actually thought as a generalization of the phasors. The generalization consists in that the Park transform is valid instantaneously, in any transient condition of the original three-phase system. No assumption on stationarity, symmetry or balance is made.

Due the arbitrariness of the definition of the direct and quadrature axes, in the literature, there are several variants of Park transforms, with different axis coordinates. Of course, while the choice of the axes changes the signs of the resulting equations and the elements of the tensor that defines the Park transform, the concept is the same.

There are two main groups: "d-lags-q" and "q-lags-d." The configuration "d-lags-q" is preferred by practitioners [80, 113, 123] and power electronics books, e.g. [32, 228], but also monographs on power system stability, e.g. [213], and Park himself [168]. The configuration "d-lags-q" is also the most typical choice in power system software tools. On the other hand, the configuration "q-lags-d" is recommended by IEEE [95] and can be found in several classic monographs on power systems, e.g. [8, 120, 127, 186] as well as electrical machines, e.g. [110].

Within each group, there are two possible choices of the reference angle of the Park transform and phase "a" of the original three-phase system, i.e. "d-aligned-with-a," e.g. [8, 32, 113, 120, 123, 127, 213, 228] and "q-aligned-with-a," e.g. [80, 95, 110, 186]. Depending on this choice the signs of the dq-axis coordinates change.

This arbitrariness leads to a great variety of equations for the synchronous machine and, possibly, to some confusion. Nevertheless, one has to choose one set of coordinates and a reference angle. The convention shown in Figure 1.11 is utilized in this book, namely "d-lags-q" and "q-aligned-with-a."

Adopting the notation of Figure 1.11, the Park transform is given by:

$$\boldsymbol{v}_{\mathrm{dqo}}(t) = \mathbf{P}(t)\,\boldsymbol{v}_{\mathrm{abc}}(t) \;, \tag{1.43}$$

where:

$$\mathbf{P}(t) = \frac{2}{3}
\begin{bmatrix}
\sin \delta_{\mathrm{dq}}(t) & \sin \delta'_{\mathrm{dq}}(t) & \sin \delta''_{\mathrm{dq}}(t) \\
\cos \delta_{\mathrm{dq}}(t) & \cos \delta'_{\mathrm{dq}}(t) & \cos \delta''_{\mathrm{dq}}(t) \\
1/2 & 1/2 & 1/2
\end{bmatrix}, \tag{1.44}$$

and δ_{dq} is the phase angle of the dq-reference frame; and $\delta'_{\mathrm{dq}} = \delta_{\mathrm{dq}} - 2\pi/3$ and $\delta''_{\mathrm{dq}} = \delta_{\mathrm{dq}} + 2\pi/3$.

The phase angle of the dq-reference frame is defined as:

$$\delta_{\mathrm{dq}}(t) = \int_0^t \omega_{\mathrm{dq}}(r)\,dr + \delta_{\mathrm{dq},o} \;, \tag{1.45}$$

where ω_{dq} is the Park reference angular frequency and $\delta_{\mathrm{dq},o}$ is an offset that depends on the position of phase "a" at $t = 0$ (see Figure 1.11).

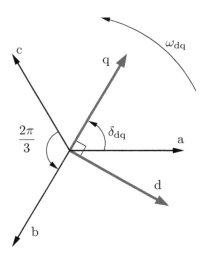

Figure 1.11 Graphical representation of the Park transform. The q-axis leads the d-axis. The q-axis is initially aligned with phase "a."

The inverse transformation from the dqo to the abc representation is:

$$v_{abc}(t) = [\mathbf{P}(t)]^{-1} v_{dqo}(t) , \qquad (1.46)$$

where:

$$[\mathbf{P}(t)]^{-1} = \begin{bmatrix} \sin \delta_{dq}(t) & \cos \delta_{dq}(t) & 1 \\ \sin \delta'_{dq}(t) & \cos \delta'_{dq}(t) & 1 \\ \sin \delta''_{dq}(t) & \cos \delta''_{dq}(t) & 1 \end{bmatrix} . \qquad (1.47)$$

The angular frequency ω_{dq} of the dq-reference frame can be any arbitrary function of time. Of course, certain choices make more sense than others, as a proper ω_{dq} can lead to simplified equations. For example, if ω_{dq} is the rotor speed of a synchronous generator, the dynamic equations of the machine are highly simplified and this is, in fact, the motivation of the two-reaction theory of the synchronous machine proposed by Park in [168] (see also Section 2.4.1).

When studying large interconnected systems, the most convenient choice, however, is $\omega_{dq} = \omega_o$, where ω_o is the fundamental (constant) nominal frequency of the system. This is the choice adopted in this book to model loads and the transmission system in Sections 2.4.2 and 2.4.3, respectively.

Moreover, from the observation of (1.43) and (1.40) as well as of (1.46) and (1.41), it is immediate to observe that the Clarke transform is just a special case of the Park transform, where $\delta_{dq} = \omega_{dq} = 0$.

In turn, the Park transform merges together two different concepts, as follows [122]:

- *Space-dimensionality reduction*: From the original three phases abc, the transform returns two coordinates in the dq-reference frame.[19]

- *Time-dimensionality reduction*: Under certain conditions, the rotation of the dq-axis frame enables the removal of the time dependency from the Park d- and q-axis components.

The space-dimensionality reduction is conceptually similar to the Clarke "α" and "β" components. As matter of fact, the Clarke transform can be considered as an intermediate step of the Park transform, as follows:

$$\boldsymbol{v}_{\mathrm{dqo}}(t) = \mathbf{P}_{\mathrm{C}}(t)\,\boldsymbol{v}_{\alpha\beta\mathrm{o}}(t) = \mathbf{P}_{\mathrm{C}}(t)\,\mathbf{C}\,\boldsymbol{v}_{\mathrm{abc}}(t) = \mathbf{P}(t)\,\boldsymbol{v}_{\mathrm{abc}}(t)\,, \tag{1.48}$$

where:

$$\mathbf{P}_{\mathrm{C}}(t) = \begin{bmatrix} \sin\delta_{\mathrm{dq}}(t) & -\cos\delta_{\mathrm{dq}}(t) & 0 \\ \cos\delta_{\mathrm{dq}}(t) & \sin\delta_{\mathrm{dq}}(t) & 0 \\ 0 & 0 & 1 \end{bmatrix}\,. \tag{1.49}$$

On the other hand, $\mathbf{C}$ can be obtained from $\mathbf{P}$ by imposing $\delta_{\mathrm{dq}} = 0$. This is also the reason why, in several practical implementations, i.e. the control of power electronics converters [32], the Clarke transform is used as an intermediate step to obtain the d- and q-axis components of the Park transform of the original abc signal.

The matrix in equation (1.49) can be generalized to define the transformation between any two reference frames, say x and y, of the Park transform, as follows:

$$\boldsymbol{v}_{\mathrm{abc}}(t) = [\mathbf{P}^x(t)]^{-1}\,\boldsymbol{v}_{\mathrm{dqo}}^x(t)\,,$$

$$\boldsymbol{v}_{\mathrm{dqo}}^y(t) = \mathbf{P}^y(t)\,\boldsymbol{v}_{\mathrm{abc}}(t)\,, \tag{1.50}$$

$$\boldsymbol{v}_{\mathrm{dqo}}^y(t) = \mathbf{P}^y(t)\,[\mathbf{P}^x(t)]^{-1}\,\boldsymbol{v}_{\mathrm{dqo}}^x(t) = \mathbf{P}^{yx}(t)\,\boldsymbol{v}_{\mathrm{dqo}}^x(t)\,,$$

where:

$$\mathbf{P}^{yx}(t) = \mathbf{P}^y(t)[\mathbf{P}^x(t)]^{-1} = \begin{bmatrix} \cos\delta_{\mathrm{dq}}^{yx}(t) & -\sin\delta_{\mathrm{dq}}^{yx}(t) & 0 \\ \sin\delta_{\mathrm{dq}}^{yx}(t) & \cos\delta_{\mathrm{dq}}^{yx}(t) & 0 \\ 0 & 0 & 1 \end{bmatrix}\,, \tag{1.51}$$

with $\delta_{\mathrm{dq}}^{yx} = \delta_{\mathrm{dq}}^y - \delta_{\mathrm{dq}}^x$. Note that the sine and cosine functions are swapped in (1.49) and (1.51) because of the different convention of the $\alpha\beta$ axes with respect to the dq axes defined in Figure 1.11.

---

[19]The component of the zero sequence is hereinafter assumed to be null and, thus, dropped.

The space-dimensionality reduction is certainly useful for studying symmetrical systems. However, the true added value of the Park transform with respect to the other approaches discussed in this section is the time-dimensionality reduction. This is, in turn, the same idea that motivates the phasor representation of AC quantities but with an important difference: the Park transform is valid in any transient condition.

To better understand the importance of the time reduction, it is convenient to define the *Park vector*, as follows:

$$\bar{v}_{dq}(t) = v_d(t) + j v_q(t) \ . \tag{1.52}$$

The Park vector, similarly to the components of the FBT, is a time-dependent complex quantity. It enables the reformulation of the equations of the electrical machines and other components of the system in an elegant and compact way (see Chapter 2 for some examples). More importantly, the Park vector also constitutes the link between steady-state and dynamic analysis.

If the three-phase system is symmetrical, balanced and in stationary conditions, and one chooses $\omega_{dq} = \omega_o$, then the Park vector *is* a phasor. For example, applying the Park transform to the stationary three-phase system in (1.33) and using the convention of Figure 1.11, one has:

$$\bar{v}_{dq} = \sqrt{2}V\left(-\sin(\theta_o) + j\cos(\theta_o)\right) = j\sqrt{2}\,\bar{v}_a = \sqrt{2}\,\bar{v}_a \exp(j\pi/2) \ , \tag{1.53}$$

where $\bar{v}_a = V\left(\cos(\theta_o) + j\sin(\theta_o)\right)$. It is relevant to note that the magnitude of the Park vector is equal to the peak value of the original three-phase voltages and its phase leads by $\pi/2$ rad the reference given by phase "a." The phase angle of the Park vector depends on the convention utilized for the dq-axis coordinates but so does the phase angle of any phasor. The Park vector, however, is much more than a phasor, as it is defined for any transient condition. This is illustrated with the following example.

For comparison, the same data utilized in the examples shown in Figure 1.10 that illustrate the Clarke transform are assumed. Results are shown in Figure 1.12. For all plots, $\omega_{dq} = \omega_o$ holds.

If the three-phase system is rotating at the same angular frequency as the dq-axis reference frame, then the d- and q-axis components of the Park transform are constant, as expected (stationary conditions, see Figure 1.12.b).

If there is a linear slip, namely $\Delta\omega_o\, t$, between the phase angle of the three-phase system and the dq-axis reference frame, then the d- and q-axis components of the Park transform oscillate with angular frequency $\Delta\omega_o$ (see Figure 1.12.c). This is indeed the same behavior that can be observed in Figure 1.4.

Finally, if there is a time-decaying slip, namely $\Delta\theta$ as defined in (1.15), between the phase angle of the three-phase system and the dq-axis reference

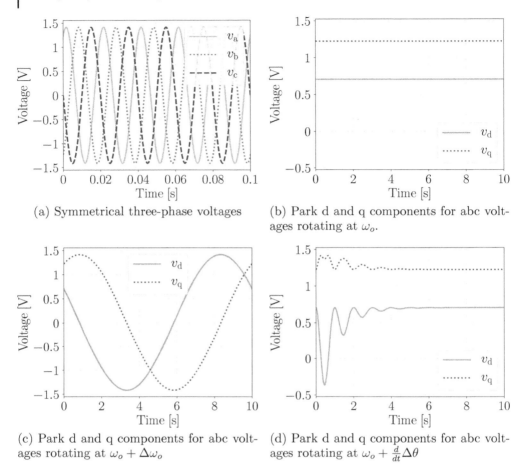

(a) Symmetrical three-phase voltages

(b) Park d and q components for abc voltages rotating at ω_o.

(c) Park d and q components for abc voltages rotating at $\omega_o + \Delta\omega_o$

(d) Park d and q components for abc voltages rotating at $\omega_o + \frac{d}{dt}\Delta\theta$

Figure 1.12 Panel (a): Three-phase voltages $\boldsymbol{v}_{\text{abc}}$ rotating with constant angular speed $\omega_o = 2\pi 50$ rad/s and phase $\theta_o = -\pi/6$ rad. Panel (b): d- and q-axis components of the Park transform as a function of time for a set $\boldsymbol{v}_{\text{abc}}$ rotating at ω_o. Panel (c): d- and q-axis components of the Park transform as a function of time for a set $\boldsymbol{v}_{\text{abc}}$ rotating at $\omega_o + \Delta\omega_o$ with $\Delta\omega_o = 2\pi 0.1$ rad/s; Panel (d): d- and q-axis components of the Park transform as a function of time for a set $\boldsymbol{v}_{\text{abc}}$ rotating at $\omega_o + \frac{d}{dt}\Delta\theta$ with $\Delta\theta_o = 0.2\pi$ rad, $\omega_\theta = 2\pi$ rad/s and $\lambda_\theta = 1$ Hz. The angular speed of the Park transform is $\omega_{\text{dq}} = \omega_o$ in all cases.

frame, then the d- and q-axis components of the Park transform show a damped oscillation, with frequency $\frac{d}{dt}\Delta\theta$, around their steady-state value (see Figure 1.12.d). Note that this is the same transient behavior observed in Figure 1.5 discussed in Section 1.3.1.

1.4.5. Time Derivative of Park Components

The expression of the time derivative of the dqo quantities, can be obtained starting from the definition of the inverse Park transform (1.46) and multiplying both right- and left-hand sides by the Park tensor, as follows:

$$
\mathbf{P}(t)\frac{d}{dt}\boldsymbol{v}_{abc}(t) = \mathbf{P}(t)\frac{d}{dt}\left\{[\mathbf{P}(t)]^{-1}\,\boldsymbol{v}_{dqo}(t)\right\}
$$
$$
= \left\{\mathbf{P}(t)\frac{d}{dt}[\mathbf{P}(t)]^{-1}\right\}\boldsymbol{v}_{dqo}(t) + \mathbf{P}(t)[\mathbf{P}(t)]^{-1}\frac{d}{dt}\boldsymbol{v}_{dqo}(t)\,,
$$

$$(1.54)$$

where $\mathbf{P}\,\mathbf{P}^{-1} = \mathbf{I}_3$ is the identity matrix of order 3 and

$$
\mathbf{P}(t)\frac{d}{dt}[\mathbf{P}(t)]^{-1} = \omega_{dq}(t)\begin{bmatrix} 0 & -1 & 0 \\ 1 & 0 & 0 \\ 0 & 0 & 0 \end{bmatrix} = \omega_{dq}(t)\,\mathbf{P}_\omega\,, \qquad (1.55)
$$

where ω_{dq} is the angular frequency of the dqo-reference frame as defined in (1.45).

Merging (1.54) and (1.55), the time derivative of the dqo variables can be written in the compact form:

$$
\mathbf{P}(t)\frac{d}{dt}\boldsymbol{v}_{abc}(t) = \frac{d}{dt}\boldsymbol{v}_{dqo}(t) + \omega_{dq}(t)\,\mathbf{P}_\omega\,\boldsymbol{v}_{dqo}(t)\,. \qquad (1.56)
$$

Then, dropping the "o" component, which it is assumed to be null, and using the definition of the Park vector in (1.52), the time derivative can be formally defined as a complex time-dependent operator, as follows:

$$
\bar{p}(t) = \frac{d}{dt} + j\omega_{dq}(t)\,. \qquad (1.57)
$$

In stationary conditions and using $\omega_{dq} = \omega_o$, i.e. a Park reference frame rotating at the constant angular frequency ω_o, one has:

$$
\bar{p} = j\omega_o\,, \qquad (1.58)
$$

which is the derivative of phasors, as discussed in Section 1.4.1 (see equation (1.20)).

When studying transmission systems with many transformers, it is convenient to use per unit quantities for the network and device parameters. In per unit, the time derivative of the Park vector becomes:

$$
\bar{p}_1(t) = \omega_b^{-1}\left[\frac{d}{dt} + j\omega_{dq}(t)\right]\,, \qquad (1.59)
$$

where ω_b is the base angular speed in rad/s. Assuming $\omega_{dq} = \omega_b = \omega_o$, (1.59) becomes:

$$\bar{p}_1(t) = \omega_o^{-1}\frac{d}{dt} + j \ , \tag{1.60}$$

which, in stationary conditions, further simplifies as:

$$\bar{p}_1 = j \ . \tag{1.61}$$

Expressions (1.60) and (1.61) are utilized in Chapter 2 for the description of power system models.

1.4.5.1. Park Transform with $\omega_{dq} = \omega_o$

The formalism of the Park vector (1.52) and the complex time derivative operator (1.57) allow elaborating on the example shown in Figure 1.12. The scenario for which the angular speed of the Park transform is $\omega_{dq} = \omega_o$ is illustrated below through three relevant cases.

Case (a): $\boldsymbol{v}_{abc}$ rotating at constant angular frequency ω_o.

The Park vector has constant components (see Figure 1.12.b), however, its time derivative is not null due to the rotation of the dq-reference frame. From (1.52) and (1.58), one obtains:

$$\begin{aligned} \bar{p}\,\bar{v}_{dq} &= j\omega_o\,(v_d + jv_q) \\ &= -\omega_o v_q + j\omega_o v_d \ , \end{aligned} \tag{1.62}$$

where $v_d = -\sqrt{2}V\sin(\theta_o)$ and $v_q = \sqrt{2}V\cos(\theta_o)$.

Equation (1.62) states that, in stationary conditions and with an isofrequential dq-reference frame, the result of the time derivative of a Park vector is a Park vector shifted $\pi/2$ rad leading with respect to the original vector.

Case (b): $\boldsymbol{v}_{abc}$ rotating at constant angular frequency $\omega_o + \Delta\omega_o$.

The components of the Park vector vary sinusoidally with period $2\pi/\Delta\omega_o$ (see Figure 1.12.c):

$$\begin{aligned} v_d(t) &= -\sqrt{2}V\sin(\Delta\omega_o t + \theta_o) \ , \\ v_q(t) &= \sqrt{2}V\cos(\Delta\omega_o t + \theta_o) \ . \end{aligned} \tag{1.63}$$

The time derivatives of the components above are:

$$\begin{aligned} \frac{d}{dt}v_d(t) &= -\Delta\omega_o\sqrt{2}V\cos(\Delta\omega_o t + \theta_o) = -\Delta\omega_o v_q(t) \ , \\ \frac{d}{dt}v_q(t) &= -\Delta\omega_o\sqrt{2}V\sin(\Delta\omega_o t + \theta_o) = \Delta\omega_o v_d(t) \ , \end{aligned} \tag{1.64}$$

which lead to write the time derivative of the Park vector in the dq-reference frame as:

$$\bar{p}(t)\,\bar{v}_{dq}(t) = \left(\frac{d}{dt} + j\omega_o\right)\bar{v}_{dq}(t)$$
$$= j(\Delta\omega_o + \omega_o)\,\bar{v}_{dq}(t) \tag{1.65}$$
$$= \bar{p}_{\Delta,o}\,\bar{v}_{dq}(t) \ ,$$

where $\bar{p}_{\Delta,o}$ is the time derivative operator for the stationary condition considered in this case, namely:

$$\bar{p}_{\Delta,o} = j(\omega_o + \Delta\omega_o) \ . \tag{1.66}$$

Case (c): $\boldsymbol{v}_{abc}$ rotating at angular frequency $\omega_o + \frac{d}{dt}\Delta\theta$.

If $\Delta\theta$ is defined by the exponentially decaying sinusoidal function (1.15), the components of the Park vector vary in time as (see Figure 1.12.d):

$$v_d(t) = -\sqrt{2}V\sin\left(\Delta\theta(t) + \theta_o\right) ,$$
$$v_q(t) = \sqrt{2}V\cos\left(\Delta\theta(t) + \theta_o\right) . \tag{1.67}$$

The time derivatives of the two components above are:

$$\frac{d}{dt}v_d(t) = -\Delta\omega(t)\,\sqrt{2}V\cos\left(\Delta\theta(t) + \theta_o\right) = -\Delta\omega(t)\,v_q(t) ,$$
$$\frac{d}{dt}v_q(t) = -\Delta\omega(t)\,\sqrt{2}V\sin\left(\Delta\theta(t) + \theta_o\right) = \Delta\omega(t)\,v_d(t) , \tag{1.68}$$

where:

$$\Delta\omega(t) = \frac{d}{dt}\Delta\theta(t) = \Delta\theta_o\big[\omega_\theta\sin(\omega_\theta t) + \lambda_\theta\big(\cos(\omega_\theta t) - 1\big)\big]\exp(-\lambda_\theta t) . \tag{1.69}$$

Equations (1.67) and (1.68) lead to write the time derivative of the Park vector in the dq-reference frame as follows:

$$\bar{p}(t)\,\bar{v}_{dq}(t) = \left(\frac{d}{dt} + j\omega_o\right)\bar{v}_{dq}(t)$$
$$= j\big(\Delta\omega(t) + \omega_o\big)\,\bar{v}_{dq}(t) \tag{1.70}$$
$$= \bar{p}_\Delta(t)\,\bar{v}_{dq}(t) \ ,$$

where $\bar{p}_\Delta$ is the time derivative operator for the transient conditions considered in this case, namely:

$$\bar{p}_\Delta(t) = j\big(\omega_o + \Delta\omega(t)\big) . \tag{1.71}$$

Case (a) is an ideal scenario, i.e. the perfect balance of generation and demand and perfect stationary conditions at exactly the reference angular speed ω_o. This scenario is rarely encountered in practice.

Case (b), a stationary condition with a frequency different from the reference one, is not uncommon in power systems. The primary frequency regulation is not perfect tracking and, after an event, e.g. contingency or load variation, the frequency of the system does not recover exactly to the reference ω_o. It is the duty of the secondary frequency control to adjust the power set point of the synchronous machines to impose the nominal frequency. Since the secondary frequency control is slower than the primary one, the frequency can steadily deviate from ω_o for relatively a long time with respect to the time scale of transient stability analysis, i.e. several seconds to a few minutes.

Case (c) qualitatively illustrates the behavior of the frequency variations for an electromechanical transient, i.e. in the first seconds after a contingency. If no persistent oscillation or loss of synchronism occurs after the contingency, the oscillations of the frequency eventually damp and the system recovers a steady-state operating condition.

1.4.5.2. Park Transform with $\omega_{dq} = \frac{d}{dt}\vartheta_a$

Another relevant scenario assumes that the reference speed of the Park transform is the phase angle position of the voltage of phase "a" shown in Figure 1.11, i.e. $\omega_{dq} = \frac{d}{dt}\vartheta_a$. In this scenario, all three cases discussed in Section 1.4.5.1 are characterized by constant dq-axis components:

$$
\begin{aligned}
v_d &= -\sqrt{2}V\sin(\theta_o)\ , \\
v_q &= \sqrt{2}V\cos(\theta_o)\ ,
\end{aligned}
\tag{1.72}
$$

where θ_o is an angular offset with respect to the reference and the time derivative of the Park vector is:

$$
\bar{p}_\vartheta(t) = j\omega_{dq}(t) = j\frac{d}{dt}\vartheta_a(t)\ .
\tag{1.73}
$$

Particularizing ϑ and $\bar{p}_\omega$ for each case, the following expressions are obtained:

Case (a):

$$
\begin{aligned}
\vartheta_a(t) &= \omega_o t + \theta_o\ , \\
\bar{p}_\vartheta &= j\omega_o\ .
\end{aligned}
\tag{1.74}
$$

Case (b):

$$
\begin{aligned}
\vartheta_a(t) &= (\omega_o + \Delta\omega_o)t + \theta_o\ , \\
\bar{p}_\vartheta &= j(\omega_o + \Delta\omega_o) = \bar{p}_{\Delta,o}\ .
\end{aligned}
\tag{1.75}
$$

Case (c):

$$\vartheta_a(t) = \omega_o t + \Delta\theta(t) + \theta_o \,,$$
$$\bar{p}_\vartheta(t) = j(\omega_o + \Delta\omega(t)) = \bar{p}_\Delta(t) \,. \tag{1.76}$$

The choice of the reference frame does not affect the time derivative, as expected.[20] It does affect, however, the values of the dq-axis components. This has a crucial role in the modeling of the synchronous machine, which is key for the developments that are carried out in Part II. This is discussed in the next chapter.

---

[20]The choice of the reference frame is also invariant with respect to the power.

Chapter 2
Power System Model

2.1. Time Scales

The formulation of a dynamic power system model, as any physical model, depends on the time scale of interest. Figure 2.1 shows relevant power system models classified based on the time scale. These models are briefly outlined below.

The most detailed model considers electromagnetic transients (EMT) and takes into account fast phenomena, such as transmission line dynamics and the switching logics of electronic converters. The EMT model is utilized to study specific devices in the time scale from a few hundreds of milliseconds to a few seconds. Its computational burden prevents the utilization of the EMT model for the simulation of large networks, but thanks to parallelism, diakoptics and the increasing computing capabilities of CPUs, this situation is rapidly changing. The fundamental work on the EMT model is the book by Hermann W. Dommel [51], whereas an up-to-date monograph that describes industry-grade implementations can be found in [5].

The stability analysis of large interconnected networks, however, is generally carried out with quasi-steady-state (QSS) models.[1] These can be divided into short-term or transient stability (TS) and long-term models. All QSS models are based on a hybrid formulation, where some dynamics are retained and described by differential equations and others are neglected and assumed to be fast enough to be *always* in steady state. The latter dynamics are thus described with algebraic equations.

In the conventional TS model, the grid and loads are modeled using QSS phasors, whereas synchronous machines are modeled using the Park transform, where fast rotor flux dynamics are neglected. The purpose of the TS model is to determine the dynamic interaction between synchronous machines and primary and secondary controllers of a transmission grid. The time scale of interest ranges from a few seconds to a few minutes. These are the models considered in several books on power system stability and control, e.g. [8, 113, 127, 186] and is also the model considered in the remainder of this book.

---

[1]In some works, the expression "RMS models" is also used.

Frequency Variations in Power Systems: Modeling, State Estimation, and Control, First Edition.
Federico Milano and Álvaro Ortega Manjavacas.

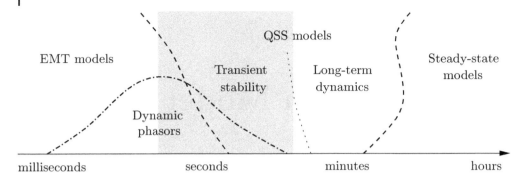

Figure 2.1 Classification of power system models based on the time scale. The gray region indicates the time scale of interest in this book.

In [213] and other references by the same authors, a long-term power system model is proposed. The time scale spans several minutes. This model is also based on QSS phasors and, in fact, in [213], it is called QSS model and is considered different from the TS model. The approach, however, is the same. Retained dynamics, which, in this case, are those of "slow" controllers, such as thermostatic load recovery, secondary frequency control, under-load tap changers and overexcitation limiters are described through differential equations. Everything else, including synchronous machines and primary regulators, are described by algebraic equations.

A variety of dynamic phasor (DP) models proposed in recent years, e.g. [116] and [196], are an attempt to define a trade-off between the EMT and the TS models. There exist several kinds of DP models, depending on the number of harmonics considered and the representation of the network (single-phase equivalent or three phase). While, in principle, these models take the best of the two worlds, namely the accuracy of the EMT model and the computational efficiency of the TS one, in practice, they have not been able so far to establish themselves because they are not as efficient nor as accurate as expected.

All models above consider the network topology and are described by sets of differential algebraic equations (DAEs). In the literature, there are also a variety of models that either neglect the network (single node) or neglect the dynamics and are thus purely steady state. The latter models include continuation power flow and optimal power flow analysis. There are also steady-state single-node models, e.g. unit commitment and adequacy studies. These models, however, are not discussed here as they are inadequate for the analysis that is carried out in this book.

The remainder of the chapter is dedicated to the description of the conventional TS model that is discussed in the literature and implemented in most software tools for angle and voltage stability analyses.

2.2. Quasi-Steady-State Model

Among the models that have been briefly introduced in the previous section, only the EMT and the steady-state ones are fully consistent. The former because it takes into account all dynamics, the latter because it neglects them all. Since the focus of the book is on dynamic phenomena, there is no need to explain why the steady-state model cannot be taken into consideration. It is, however, less evident why the EMT model is not used.

The main reason for not considering the EMT model is that the extra information that it provides with respect to QSS models is not relevant to capture the variations of the fundamental frequency that occur in the time scale of a few seconds. Supporters of the EMT model claim that it provides *all* information. However, such an information has then to be properly measured, filtered and extracted. When it comes to the determination of frequency variations, this procedure ultimately removes the effect of electromagnetic transients and high-order harmonics, which are precisely the phenomena that are approximated with the TS model.

To illustrate this concept, consider the circuit shown in Figure 2.2. The equations that describe the behavior of the circuit are:

$$\bar{v}_{1,\mathrm{dq}}(t) = (R_1 + \bar{\mathrm{p}}(t)L)\,\bar{i}_{L,\mathrm{dq}}(t) + \bar{v}_{C,\mathrm{dq}}(t) + \bar{v}_{2,\mathrm{dq}}(t)\;, \tag{2.1}$$

$$\bar{i}_{L,\mathrm{dq}}(t) = \bar{\mathrm{p}}(t)\,C\,\bar{v}_{C,\mathrm{dq}}(t)\;, \tag{2.2}$$

$$\bar{v}_{2,\mathrm{dq}}(t) = \begin{cases} \dfrac{R_2 R_3}{R_2 + R_3}\,\bar{i}_{L,\mathrm{dq}}(t), & \text{with switch closed,} \\[2mm] R_3\,\bar{i}_{L,\mathrm{dq}}(t), & \text{with switch open,} \end{cases} \tag{2.3}$$

where, $\bar{\mathrm{p}}$ is the time derivative of the Park vector as defined in (1.57).

If the time scale of the RLC transients is fast with respect to the dynamics of interest, as it is the case of typical AC transmission lines with respect to

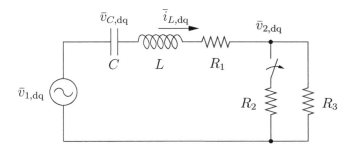

Figure 2.2 Equivalent single-phase diagram of a balanced three-phase circuit.

the electromechanical transients of synchronous machines and their primary controllers, a common approximation for the time derivative of the Park vector is to neglect the term $\frac{d}{dt}$, thus leading to:

$$\bar{p}(t) = \frac{d}{dt} + j\omega_o \approx j\omega_o \,, \tag{2.4}$$

which is the QSS approximation of the Park vector as defined in (1.58). This is equivalent to the phasor representation except for the fact that voltages and currents are time dependent and their magnitude and phase angle (but not their frequency) can vary during a transient.[2]

Using QSS phasors, equations (2.1) and (2.2) become:

$$
\begin{aligned}
\bar{v}_{1,\text{dq}}(t) &= \left(R_1 + j\omega_o L - j\frac{1}{\omega_o C} \right) \bar{i}_{L,\text{dq}}(t) + \bar{v}_{2,\text{dq}}(t) \\
&= \left(R_1 + j\left(X_L - \frac{1}{B_C} \right) \right) \bar{i}_{L,\text{dq}}(t) + \bar{v}_{2,\text{dq}}(t) \,,
\end{aligned} \tag{2.5}
$$

where the terms $X_L = \omega_o L$ and $B_C = \omega_o C$ are a reactance and a susceptance, respectively, calculated at the reference frequency ω_o. These values are constant if ω_o is constant.

For the sake of example, the magnitude of the voltage source $\bar{v}_{1,\text{dq}}$ is assumed to vary "slowly" as the output of the controller shown in Figure 2.3.

The differential equations that describe the controller in Figure 2.3 are:

$$
\begin{aligned}
\frac{d}{dt}v_1(t) - T_2\frac{d}{dt}x_1(t) &= x_1(t) \,, \\
T_1\frac{d}{dt}x_1(t) &= K_1[v_1^{\text{ref}}(t) - v_1(t)] - x_1(t) \,,
\end{aligned} \tag{2.6}
$$

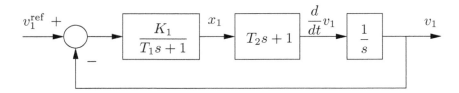

Figure 2.3 Controller driving the magnitude of the voltage source of the circuit shown in Figure 2.2.

---

[2]The Park vector and the phasor may also differ for a constant phase shift, depending on the choice of the dq-reference frame.

or, in frequency domain:

$$v_1(s) = \frac{K_1(1 + T_2 s)}{T_1 s^2 + (1 + K_1 T_2)s + K_1} v_1^{\text{ref}}(s)$$

$$= \frac{1 + T_2 s}{s^2/\hat{\omega}_v^2 + 2\zeta_v/\hat{\omega}_v + 1} v_1^{\text{ref}}(s) , \tag{2.7}$$

with

$$\frac{1}{\hat{\omega}_v^2} = \frac{T_1}{K_1}, \qquad \frac{2\zeta_v}{\hat{\omega}_v} = \frac{1 + K_1 T_2}{K_1} ,$$

where $\hat{\omega}_v$ and ζ_v are the natural angular frequency and the damping coefficient, respectively, of the closed-loop transfer function of the controller.

The link between $\bar{v}_{1,\text{dq}}$ and v_1 is:

$$|\bar{v}_{1,\text{dq}}(t)| = v_1(t) , \tag{2.8}$$

or, equivalently:

$$v_1^2(t) = v_{1,\text{d}}^2(t) + v_{1,\text{q}}^2(t) . \tag{2.9}$$

As any AC system, a phase reference has to be defined. For example, assuming that node 1 is chosen as the reference, and its phase angle is $\theta_{1,o}^{\text{ref}}$:

$$0 = \cos(\theta_{1,o}^{\text{ref}}) v_{1,\text{d}} - \sin(\theta_{1,o}^{\text{ref}}) v_{1,\text{q}} . \tag{2.10}$$

If the reference voltage magnitude v_1^{ref} has a step variation:

$$v_1^{\text{ref}}(t) = v_{1,o}^{\text{ref}} + \Delta v_1 \mathcal{H}(t) , \tag{2.11}$$

where $\mathcal{H}$ is the Heaviside step function:

$$\mathcal{H}(t) = \int_{-\infty}^{t} \partial(r) dr = \begin{cases} 0, & t < 0, \\ 1, & t \geq 0 , \end{cases} \tag{2.12}$$

where ∂ is the Dirac delta function, then the following expression solves the set of differential equations (2.6):

$$v_1(t) = v_{1,o}^{\text{ref}} + \Delta v_1 \left[1 - \left(\cos(\omega_v t) + \frac{\zeta_v - T_2 \hat{\omega}_v}{1 + \zeta_v^2} \sin(\omega_v t) \right) \exp(-\lambda_v t) \right] , \tag{2.13}$$

where $\omega_v = \sqrt{1 - \zeta_v^2} \, \hat{\omega}_v$ and $\lambda_v = \zeta_v \hat{\omega}_v$ are the pseudo angular frequency and the damping, respectively, of the closed-loop control transfer function.

Assuming $T_2 = \zeta_v/\hat{\omega}_v$, one finally obtains:

$$v_1(t) = v_{1,o}^{\text{ref}} + \Delta v_1 \left[1 - \cos(\omega_v t) \exp(-\lambda_v t) \right] , \tag{2.14}$$

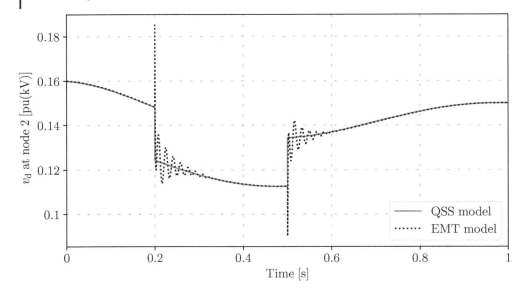

Figure 2.4 Comparison of the dynamic behavior of Park vectors and QSS phasors for the circuit of Figure 2.2. The switch opens at $t = 0.2$ s and recloses at $t = 0.5$ s.

which reproduces, on purpose, the test function (1.15) utilized in the examples of Chapter 1.

Figure 2.4 compares the transient behavior of a component of the Park vector at node 2 for the two models (2.1)–(2.3), and (2.5) and (2.3). In the example, the following parameters have been assumed:

$$v_{1,o}^{\text{ref}} = 1 \text{ pu(kV)}, \qquad \theta_{1,o}^{\text{ref}} = 0 \text{ rad}, \qquad \Delta v_1 = 0.1 \text{ pu(kV)},$$

$$R_1 = 0.1 \text{ pu}(\Omega), \qquad R_2 = 5 \text{ pu}(\Omega), \qquad R_3 = 1.25 \text{ pu}(\Omega),$$

$$X_L = 0.3 \text{ pu}(\Omega), \qquad B_C = 10 \text{ pu}(\Omega),$$

$$\omega_o = 2\pi 60 \text{ rad/s}, \qquad \omega_v = 2\pi \text{ rad/s}, \qquad \lambda_v = 1 \text{ s}^{-1},$$

where per unit values are referred to the bases $s_b = 100$ MVA, $v_b = 220$ kV and $\omega_b = 2\pi 60$ rad/s.

The values of ω_v and λ_v lead to the following parameters of the controller of v_1 shown in Figure 2.3:

$$K_1 = \hat{\omega}_v/\zeta_v = 38.44, \quad T_1 = \lambda_v^{-1} = 1 \text{ s}, \quad T_2 = \hat{\omega}_v^{-2} = 0.026 \text{ s},$$

which can be easily deduced from the expressions of ω_v and λ_v and by observing that $\hat{\omega}_v = \sqrt{\omega_v^2 - 1}$ if $\zeta_v \hat{\omega}_v = 1$.

The oscillations due to the coupling of the inductance and the capacitance last a few tens of milliseconds and are fast with respect to the typical electromechanical oscillations of power systems, whose time scale is a few seconds and that, in this example, are represented by the time-varying magnitude of the voltage source $\bar{v}_{1,\mathrm{dq}}$. Note that the effect of the switch is to create an "instantaneous" variation of the QSS phasors, while the original Park vector quantities are continuous. In other words, QSS phasors are algebraic variables, whereas Park vectors are states.

This example shows that a QSS model is not an odd mix of phasors and differential equations but, rather, an approximation of the original dynamic system, transformed with Park, where fast dynamics, mostly electromagnetic ones, are neglected.

2.3. Differential Algebraic Equations

Dynamic systems can be conveniently formulated as a set of DAEs:

$$
\begin{bmatrix} \mathbf{T} & \mathbf{0}_{n_x,n_y} \\ \mathbf{R} & \mathbf{0}_{n_y,n_y} \end{bmatrix} \begin{bmatrix} \dfrac{d}{dt}\boldsymbol{x}(t) \\ \mathbf{0}_{n_y,1} \end{bmatrix} = \begin{bmatrix} \boldsymbol{f}(\boldsymbol{x}(t), \boldsymbol{y}(t), \boldsymbol{u}(t)) \\ \boldsymbol{g}(\boldsymbol{x}(t), \boldsymbol{y}(t), \boldsymbol{u}(t)) \end{bmatrix}, \tag{2.15}
$$

where $\boldsymbol{x}$ ($\boldsymbol{x} \in \mathbb{R}^{n_x}$) are the state variables; $\boldsymbol{y}$ ($\boldsymbol{y} \in \mathbb{R}^{n_y}$) are the algebraic variables; $\boldsymbol{u}$ ($\boldsymbol{u} \in \mathbb{R}^{n_u}$) are input variables; $\boldsymbol{f}$ ($\boldsymbol{f} : \mathbb{R}^{n_x+n_y+n_u} \mapsto \mathbb{R}^{n_x}$) are the differential equations; $\boldsymbol{g}$ ($\boldsymbol{g} : \mathbb{R}^{n_x+n_y+n_u} \mapsto \mathbb{R}^{n_y}$) are the algebraic equations; and $\mathbf{T}$ and $\mathbf{R}$ are $n_x \times n_x$ and $n_y \times n_x$ matrices, respectively.

Equations (2.15) are written in a semi-implicit form, where matrices $\mathbf{T}$ and $\mathbf{R}$ comprise the coefficients of the first time derivatives. In general, $\mathbf{T}$ and $\mathbf{R}$ are time variant, nondiagonal, and nonfull rank. In the models considered in this chapter, however, $\mathbf{T}$ and $\mathbf{R}$ have constant elements.

The semi-implicit formulation has several advantages with respect to the conventional explicit one. Most importantly, it enables the "transformation" of any state variable into an algebraic one by just setting to zero the corresponding column of the $\mathbf{T}$ and $\mathbf{R}$ matrices. Another advantage is the ability to reduce the number of operations and increase the sparsity of the Jacobian matrices of the DAE.

From a mere mathematical point of view, there is no need to distinguish between state and algebraic variables, as the latter are simply state variables with null time constants. From the numerical point of view, however, it is useful to maintain this distinction as one can impose $\frac{d}{dt}\boldsymbol{y} = \mathbf{0}_{n_y,1}$ rather than obtaining this equality as a byproduct of the integration scheme [129]. The interested reader can find further details on the implementation of numerical integration schemes to solve (2.15) in [23], which is probably the best reference on the numerical integration of DAEs to date.

The circuit of Figure 2.2 is a good example to illustrate the DAE formulation. The EMT model, which includes all dynamics, consists of equations (2.1)–(2.3), (2.6), (2.9), and (2.10). Using the notation of (2.15) and per unit quantities, and omitting for simplicity the dependency of the variables on time, one obtains:

$$\boldsymbol{x}^\dagger = [v_1, \, x_1, \, i_{L,\mathrm{d}}, \, i_{L,\mathrm{q}}, \, v_{C,\mathrm{d}}, \, v_{C,\mathrm{q}}] \, ,$$

$$\boldsymbol{y}^\dagger = [v_{1,\mathrm{d}}, \, v_{1,\mathrm{q}}, \, v_{2,\mathrm{d}}, \, v_{2,\mathrm{q}}] \, ,$$

$$\boldsymbol{u}^\dagger = [v_1^{\mathrm{ref}}, \, \theta_1^{\mathrm{ref}}, \, z_{\mathrm{s}}] \, ,$$

$$\boldsymbol{f} = \begin{bmatrix} x_1 \\ K_1(v_1^{\mathrm{ref}} - v_1) - x_1 \\ v_{1,\mathrm{d}} - R_1 i_{L,\mathrm{d}} - v_{C,\mathrm{d}} - v_{2,\mathrm{d}} + X_L \, i_{L,\mathrm{q}} \\ v_{1,\mathrm{q}} - R_1 i_{L,\mathrm{q}} - v_{C,\mathrm{q}} - v_{2,\mathrm{q}} - X_L \, i_{L,\mathrm{d}} \\ i_{L,\mathrm{d}} + B_C \, v_{C,\mathrm{q}} \\ i_{L,\mathrm{q}} - B_C \, v_{C,\mathrm{d}} \end{bmatrix} \, ,$$

$$\boldsymbol{g} = \begin{bmatrix} v_{1,\mathrm{d}}^2 + v_{1,\mathrm{q}}^2 - v_1^2 \\ \cos(\theta_{1,o}^{\mathrm{ref}}) \, v_{1,\mathrm{d}} - \sin(\theta_{1,o}^{\mathrm{ref}}) \, v_{1,\mathrm{q}} \\ [z_{\mathrm{s}} R_{23} + (1 - z_{\mathrm{s}}) R_3] \, i_{L,\mathrm{d}} - v_{2,\mathrm{d}} \\ [z_{\mathrm{s}} R_{23} + (1 - z_{\mathrm{s}}) R_3] \, i_{L,\mathrm{q}} - v_{2,\mathrm{q}} \end{bmatrix} \, ,$$

$$\boldsymbol{T} = \begin{bmatrix} 1 & -T_2 & 0 & 0 & 0 & 0 \\ 0 & T_1 & 0 & 0 & 0 & 0 \\ 0 & 0 & \omega_o^{-1} X_L & 0 & 0 & 0 \\ 0 & 0 & 0 & \omega_o^{-1} X_L & 0 & 0 \\ 0 & 0 & 0 & 0 & \omega_o^{-1} B_C & 0 \\ 0 & 0 & 0 & 0 & 0 & \omega_o^{-1} B_C \end{bmatrix} \, ,$$

$$\boldsymbol{R} = \boldsymbol{0}_{4,6} \, ,$$

where $R_{23} = R_2 R_3 / (R_2 + R_3) = 1 \, \mathrm{pu}(\Omega)$ and z_{s} is defined by:[3]

$$z_{\mathrm{s}}(t) = 1 - \mathcal{H}(t - 0.2) + \mathcal{H}(t - 0.5) \, . \tag{2.16}$$

---

[3]Discrete events such as those defined by (2.16) can be modeled with a more general approach using hybrid automata, Petri nets, or Filippov theory, which enable the handling of both time- and state-driven events with a formal and systematic approach. The interested reader can find some applications of the methods above to power system models in [86, 87, 146, 147].

In matrix $\mathbf{T}$, the expressions of the diagonal elements of the third to sixth rows are obtained from the expression of the per-unit time derivative of the Park vector given in (1.60) and using the per-unit values of the reactance X_L and susceptance B_C. For example, considering the time derivative of the current $\bar{i}_{L,\text{dq}}$, in absolute values one has:

$$\bar{p}(t)\, L\, \bar{i}_{L,\text{dq}}(t) = \left(\frac{d}{dt} + j\omega_o \right) L \left(i_{L,\text{d}}(t) + j i_{L,\text{q}}(t) \right) . \tag{2.17}$$

The expression (2.17) has the units of a voltage or, which is the same, the product of an impedance and a current. Hence, the term $\bar{p}L$ has the units of an impedance. Assuming that the base of the impedance is Z_b, one has:

$$\begin{aligned} \frac{\bar{p}(t)\, L}{Z_b} &= \frac{L}{Z_b} \frac{d}{dt} + j\frac{\omega_o L}{Z_b} \\ &= \frac{\omega_o}{\omega_o} \frac{L}{Z_b} \frac{d}{dt} + j X_L = \omega_o^{-1} X_L \frac{d}{dt} + j X_L , \end{aligned} \tag{2.18}$$

where X_L is the reactance in per unit. A similar expression can be obtained for the time derivative of $\bar{v}_{C,\text{dq}}$.

The EMT model consists of $n_x = 6$ state variables, $n_y = 4$ algebraic variables and $n_u = 3$ inputs. The QSS model can be straightforwardly obtained from the EMT one by simply setting to zero X_L/ω_o and B_C/ω_o in the diagonal elements of $\mathbf{T}$. In this way, the QSS model consists of $n_x = 2$ state variables, $n_y = 8$ algebraic variables and $n_u = 3$ inputs. Of the total 8 algebraic variables, 4 are "demoted" states, namely, $i_{L,\text{d}}$, $i_{L,\text{q}}$, $v_{C,\text{d}}$, and $v_{C,\text{q}}$. Finally, the steady-state model is given by setting $\mathbf{T} = \mathbf{0}_{6,6}$, which leads to a system with $n_y = 10$ algebraic variables and $n_u = 3$ inputs.

While the values of the elements of matrix $\mathbf{T}$ cannot be compared directly, it is always possible to solve an eigenvalue analysis of the system at an equilibrium point and compare the eigenvalues of the state matrix of the system. With the switch closed, the eigenvalue analysis at the equilibrium point of the system, namely, a point $(\boldsymbol{x}_o, \boldsymbol{y}_o, \boldsymbol{u}_o)$ that satisfies the conditions:

$$\begin{aligned} \mathbf{0}_{n_x,1} &= \boldsymbol{f}(\boldsymbol{x}_o, \boldsymbol{y}_o, \boldsymbol{u}_o) , \\ \mathbf{0}_{n_y,1} &= \boldsymbol{g}(\boldsymbol{x}_o, \boldsymbol{y}_o, \boldsymbol{u}_o) , \end{aligned} \tag{2.19}$$

leads to the results shown in Table 2.1. The state variables associated with the eigenvalues were determined through the calculation of the eigenvectors and, consequently, of the participation factors of the eigenvalues in the state variables, as thoroughly discussed, for example, in [127] or [186].

Table 2.1 shows that the dynamics of $\bar{i}_{L,\text{dq}}$ and $\bar{v}_{C,\text{dq}}$ are, respectively, about three and one orders of magnitude faster than the dynamic of v_1, thus confirming the qualitative information and conclusions drawn from Figure 2.4.

Table 2.1 Eigenvalues of the system composed of the circuit of Figure 2.2 and the controller of the voltage magnitude v_1 of Figure 2.3.

Eigenvalue	Damping [%]	Frequency [Hz]	Assoc. variables
$-1.00 \pm j6.12$	16.3	0.987	v_1, x_1
$-35.17 \pm j376.99$	9.29	60.26	$v_{C,\mathrm{d}}, v_{C,\mathrm{q}}$
$-1347.12 \pm j376.99$	96.3	222.6	$i_{L,\mathrm{d}}, i_{L,\mathrm{q}}$

2.4. Conventional Devices

Synchronous machines, loads, transmission lines, and transformers constitute the bulk of conventional power systems. This section presents the models of these devices and discusses the assumptions and simplifications adopted in transient stability analysis. All equations are written assuming per-unit values. The angular frequency base, expressed in rad/s, is also the reference angular frequency of the system, namely $\omega_b = \omega_o$.

2.4.1. Synchronous Machine

The well-known Park-Concordia model of the synchronous machine is based on the Park transform. This consists in projecting abc quantities onto a dq-axis reference frame. As discussed Section 1.4.4.4, the common choice for the angular frequency of the dq-axis is the constant reference angular frequency ω_o. Synchronous machines, however, are a relevant exception to this rule. The main goal of the Park transform is to obtain a set of electromagnetic equations of the machine with constant parameters. This can be achieved only if, for each machine, the dq-axis reference frame rotates at the machine rotor angular speed ω_G. This also means, however, that each machine has its own reference and hence they cannot be connected directly to the system. A dq-axis transformation is needed to allow each machine stator voltage and currents to be referred to the common dq-axis reference frame of the system, which is rotating at ω_o.

According to the notation given in Figure 1.11 and the transformation matrix between two reference frames given in (1.51), the following link between the stator voltage $\bar{v}_{\mathrm{s,dq}}$ of the machine on the machine dq-axis frame rotating at ω_G and the same voltage on the dq-axis frame of the system rotating at ω_o is given by:

$$\bar{v}_{\mathrm{s,dq}}(t) = \bar{v}_{h,\mathrm{dq}}(t) \exp\left(-j(\delta_G(t) - \delta_o)\right), \tag{2.20}$$

or, equivalently:

$$v_{s,d}(t) = v_{h,d}(t) \cos(\delta_G(t) - \delta_o) + v_{h,q}(t) \sin(\delta_G(t) - \delta_o) \,,$$
$$v_{s,q}(t) = v_{h,q}(t) \cos(\delta_G(t) - \delta_o) - v_{h,d}(t) \sin(\delta_G(t) - \delta_o) \,, \tag{2.21}$$

where the subscript h indicates the index of the bus to which the machine is connected and δ_G is the angle position of the rotor of the synchronous machine, which is referred to the reference angle δ_o of the system. In the following, $\delta_o = 0$ is always assumed.

Similarly, the link of the stator current $\bar{\imath}_{s,dq}$ of the machine on the machine dq-axis frame rotating at ω_G and the same current, assumed to be injected into bus h, on the dq-axis frame of the system rotating at ω_o is given by:

$$\bar{\imath}_{s,dq}(t) = \bar{\imath}_{h,dq}(t) \exp(-j\delta_G(t)) \,, \tag{2.22}$$

which has a similar expansion as (2.21), namely:

$$i_{h,d}(t) = i_{s,d}(t) \cos(\delta_G(t)) - i_{s,q}(t) \sin(\delta_G(t)) \,,$$
$$i_{h,q}(t) = i_{s,q}(t) \cos(\delta_G(t)) + i_{s,d}(t) \sin(\delta_G(t)) \,. \tag{2.23}$$

As to be expected, the angular rotation by δ_G is invariant with respect to the complex power injected by the synchronous machine into bus h of the grid. In fact, from (2.20) and (2.22):

$$\begin{aligned} \bar{s}_h(t) &= \bar{v}_{h,dq}(t) \, \bar{\imath}^*_{h,dq}(t) \\ &= \bar{v}_{s,dq}(t) \exp(j\delta_G(t)) \, \bar{\imath}^*_{s,dq}(t) \exp(-j\delta_G(t)) \\ &= \bar{v}_{s,dq}(t) \, \bar{\imath}^*_{s,dq}(t) \,, \end{aligned} \tag{2.24}$$

or, equivalently:

$$\begin{aligned} p_h(t) &= v_{h,d}(t) \, i_{h,d}(t) + v_{h,q}(t) \, i_{h,q}(t) \\ &= v_{s,d}(t) \, i_{s,d}(t) + v_{s,q}(t) \, i_{s,q}(t) \,, \\ q_h(t) &= v_{h,q}(t) \, i_{h,d}(t) - v_{h,d}(t) \, i_{h,q}(t) \\ &= v_{s,q}(t) \, i_{s,d}(t) - v_{s,d}(t) \, i_{s,q}(t) \,. \end{aligned} \tag{2.25}$$

The synchronous machine constitutes the fundamental link between active power balance of a power system and frequency variations (see equation (1.9)). This link is given by the electromechanical equations of the machine:

$$M \frac{d}{dt} \omega_G(t) = \tau_m(t) - \tau_e(t) - D \left(\omega_G(t) - 1 \right) \,, \tag{2.26}$$

where M is the inertia constant; D is the damping; 1 is the reference angular frequency in per unit; τ_m is the mechanical torque as imposed by the turbine connected to the machine shaft; and τ_e is the electromagnetic torque, which is a function of the dq-axis components of stator flux, $\bar{\psi}_{s,dq}$, and current, $\bar{i}_{s,dq}$, as follows:

$$\tau_e(t) = \psi_{s,d}(t)\, i_{s,q}(t) - \psi_{s,q}(t)\, i_{s,d}(t) \ . \tag{2.27}$$

The rotor angle of the machine, as any angle, is a relative quantity with respect to the phase reference. Such an angle has to be computed to be able to properly connect the machine to the rest of the system. This is possible by calculating the "slip" between the machine rotor speed and the reference, as follows:

$$\omega_o^{-1}\frac{d}{dt}\delta_G(t) = \omega_G(t) - 1 \ , \tag{2.28}$$

where 1 is again the per-unit value of the reference angular frequency. Note that, in (2.28), ω_o is expressed in rad/s whereas ω_G is in per unit.

The electrical equations that link stator fluxes, voltages and currents are, using per unit quantities, the generation convention and the Park vector:

$$\bar{v}_{s,dq}(t) = -R_a\, \bar{i}_{s,dq}(t) - \left(\omega_o^{-1}\frac{d}{dt} + j\,\omega_G(t)\right)\bar{\psi}_{s,dq}(t) \ , \tag{2.29}$$

where $\omega_o^{-1}\frac{d}{dt} + j\omega_G(t)$ is the time derivative of the Park vector in the dq-axis frame of the synchronous machine; and R_a is the armature resistance. Separating real and imaginary components, (2.29) becomes:

$$\begin{aligned}
\omega_o^{-1}\frac{d}{dt}\psi_{s,d}(t) &= R_a\, i_{s,d}(t) + \omega_G(t)\,\psi_{s,q}(t) + v_{s,d}(t) \ , \\
\omega_o^{-1}\frac{d}{dt}\psi_{s,q}(t) &= R_a\, i_{s,q}(t) - \omega_G(t)\,\psi_{s,d}(t) + v_{s,q}(t) \ .
\end{aligned} \tag{2.30}$$

As discussed in the example of Section 2.3, electromagnetic (flux) dynamics are fast with respect to the time scale of electromechanical modes and of the TS model. A common simplification is thus to assume $\omega_o^{-1} \approx 0$ in (2.30), which leads to:

$$\begin{aligned}
0 &= R_a\, i_{s,d}(t) + \omega_G(t)\,\psi_{s,q}(t) + v_{s,d}(t) \ , \\
0 &= R_a\, i_{s,q}(t) - \omega_G(t)\,\psi_{s,d}(t) + v_{s,q}(t) \ .
\end{aligned} \tag{2.31}$$

Finally, in the conventional TS model, the variations of the rotor speed are considered "small" with respect to the reference, i.e. $\omega_G \approx 1$ pu, which enables a further simplification of (2.31), as follows:

$$\begin{aligned}
0 &= R_a\, i_{s,d}(t) + \psi_{s,q}(t) + v_{s,d}(t) \ , \\
0 &= R_a\, i_{s,q}(t) - \psi_{s,d}(t) + v_{s,q}(t) \ ,
\end{aligned} \tag{2.32}$$

which is clearly consistent only in per unit. However, the approximation $\omega_G \approx 1$ pu(rad/s) in (2.32) *cannot* be extended to the mechanical equations (2.26) and

(2.28), otherwise the main dynamic behavior of the synchronous machine would be lost. This observation is key to the TS model, which, in turn, assumes that all frequency variations are negligible *except for* rotor speed variations and even those are considered not null *only* in the electromechanical equations of the synchronous machine.

Using (2.32) rather than (2.31) is an approximation that has mainly an historical reason. Equation (2.32) is linear and time-invariant, which enables the determination of the explicit expression of the flux components in terms of the components of stator voltages and currents and, thus, it also enables the elimination of the fluxes from the machines equations. A few decades ago, when the TS models were implemented for the first times in digital mainframes, any simplification that led to reduce the computational burden of the overall power system model was relevant. With the computing capability of modern CPUs adopting or not such an approximation does not lead to any significant difference in the performance of the simulation, even for large networks.

The model of the machine is completed with the equations that link the rotor fluxes with the rotor field voltage and stator currents and voltages.

A common and relatively detailed model that includes four rotor dynamics is presented in [186]. This model considers a dynamic of the DC field winding, excited by the excitation field voltage v_{ef}, and three fictitious damper windings that represent rotor-core induced currents, one on the direct and two on the quadrature axis. Using the semi-implicit form of DAEs discussed in Section 2.3, the equations that describe the transient ($'$) and subtransient ($''$) flux dynamics are, respectively:

$$T'_{do}\frac{d}{dt}e'_{r,q}(t) + \widehat{T}''_{do}\frac{d}{dt}e''_{r,q}(t) = -e'_{r,q}(t) - (X_d - X'_d)\,i_{s,d}(t) + v_{ef}(t)\ ,$$

$$T'_{qo}\frac{d}{dt}e'_{r,d}(t) - \widehat{T}''_{qo}\frac{d}{dt}e''_{r,d}(t) = -e'_{r,d}(t) + (X_q - X'_q)\,i_{s,q}(t)\ ,$$

(2.33)

and:

$$T''_{do}\frac{d}{dt}e''_{r,q}(t) = -e''_{r,q}(t) + e'_{r,q}(t) - (X'_d - X_\ell)\,i_{s,d}(t)\ ,$$

$$T''_{qo}\frac{d}{dt}e''_{r,d}(t) = -e''_{r,d}(t) - e'_{r,d}(t) - (X'_q - X_\ell)\,i_{s,q}(t)\ ,$$

(2.34)

where

$$\widehat{T}''_{do} = (X_d - X'_d)\,\gamma_{d2}\,T''_{do}\ , \qquad\qquad \widehat{T}''_{qo} = (X_q - X'_q)\,\gamma_{q2}\,T''_{qo}\ ,$$

$$\gamma_{d2} = \frac{X'_d - X''_d}{(X'_d - X_\ell)^2} = \frac{1 - \gamma_{d1}}{X'_d - X_\ell}\ , \qquad \gamma_{q2} = \frac{X'_q - X''_q}{(X'_q - X_\ell)^2} = \frac{1 - \gamma_{q1}}{X'_q - X_\ell}\ , \quad (2.35)$$

$$\gamma_{d1} = \frac{X''_d - X_\ell}{X'_d - X_\ell}\ , \qquad\qquad \gamma_{q1} = \frac{X''_q - X_\ell}{X'_q - X_\ell}\ ,$$

and variables and parameters are defined in Tables 2.2 and 2.3, respectively.

Table 2.2 Variables of the synchronous machine model.

Variable	Description	Unit
$\bar{e}'_{\mathrm{r,dq}}$	Rotor dq-axis transient voltage	pu(kV)
$\bar{e}''_{\mathrm{r,dq}}$	Rotor dq-axis subtransient voltage	pu(kV)
$\bar{i}_{\mathrm{s,dq}}$	Stator terminal-bus dq-axis current	pu(kA)
v_{ef}	Excitation field voltage	pu(kV)
$\bar{v}_{\mathrm{s,dq}}$	Stator terminal-bus dq-axis voltage	pu(kV)
δ_{G}	Rotor angular position	rad
τ_{e}	Electromagnetic torque	pu(MNm)
τ_{m}	Mechanical torque	pu(MNm)
$\bar{\psi}_{\mathrm{s,dq}}$	Stator dq-axis flux	pu(kWb)
ω_{G}	Rotor angular speed	pu(rad/s)

Table 2.3 Parameters of the synchronous machine model.

Parameter	Description	Unit
D	Damping coefficient	pu(MW)
M	Inertia constant	s pu(MW)
R_{a}	Armature resistance	pu(Ω)
T'_{do}	d-axis transient time constant	s
T''_{do}	d-axis subtransient time constant	s
T'_{qo}	q-axis transient time constant	s
T''_{qo}	q-axis subtransient time constant	s
X_{d}	d-axis synchronous reactance	pu(Ω)
X'_{d}	d-axis transient reactance	pu(Ω)
X''_{d}	d-axis subtransient reactance	pu(Ω)
X_{ℓ}	Leakage reactance	pu(Ω)
X_{q}	q-axis synchronous reactance	pu(Ω)
X'_{q}	q-axis transient reactance	pu(Ω)
X''_{q}	q-axis subtransient reactance	pu(Ω)

The following algebraic equations that link stator and rotor magnetic fluxes and stator currents complete the model of the machine:

$$
\begin{aligned}
0 &= \psi_{\mathrm{s,d}}(t) + X''_{\mathrm{d}}\, i_{\mathrm{s,d}}(t) - \gamma_{\mathrm{d1}}\, e'_{\mathrm{r,q}}(t) - (1 - \gamma_{\mathrm{d1}})\, e''_{\mathrm{r,q}}(t) \;, \\
0 &= \psi_{\mathrm{s,q}}(t) + X''_{\mathrm{q}}\, i_{\mathrm{s,q}}(t) + \gamma_{\mathrm{q1}}\, e'_{\mathrm{r,d}}(t) - (1 - \gamma_{\mathrm{q1}})\, e''_{\mathrm{r,d}}(t) \;.
\end{aligned}
\tag{2.36}
$$

The resulting balanced EMT model of the synchronous machine is a two direct- and two quadrature-axis machine model that consists of the following equations: (2.26), (2.27), (2.21), (2.23), (2.25), (2.28), (2.30), (2.33), (2.34), and (2.36).

With the notation of (2.15) and omitting for simplicity the dependency on time, one has a DAE system with $n_x = 8$, $n_y = 9$ and $n_u = 4$, where the vectors of variables are:

$$
\begin{aligned}
\boldsymbol{x}^\dagger &= [\delta_G, \omega_G, e'_{r,q}, e'_{r,d}, e''_{r,q}, e''_{r,d}, \psi_{s,d}, \psi_{s,q}] , \\
\boldsymbol{y}^\dagger &= [\tau_e, v_{s,d}, v_{s,q}, i_{s,d}, i_{s,q}, v_{h,d}, v_{h,q}, p_h, q_h] , \\
\boldsymbol{u}^\dagger &= [\tau_m, v_{ef}, i_{h,d}, i_{h,q}] ,
\end{aligned}
\tag{2.37}
$$

with equations:

$$
\boldsymbol{f} =
\begin{bmatrix}
\omega_G - 1 \\
\tau_m - \tau_e - D\left(\omega_G - 1\right) \\
-e'_{r,q} - (X_d - X'_d)\, i_{s,d} + v_{ef} \\
-e'_{r,d} + (X_q - X'_q)\, i_{s,q} \\
-e''_{r,q} + e'_{r,q} - (X'_d - X_\ell)\, i_{s,d} \\
-e''_{r,d} - e'_{r,d} - (X'_q - X_\ell)\, i_{s,q} \\
R_a\, i_{s,d} + \hat{\omega}\, \psi_{s,q} + v_{s,d} \\
R_a\, i_{s,q} - \hat{\omega}\, \psi_{s,d} + v_{s,q}
\end{bmatrix}
,
$$

$$
\boldsymbol{g} =
\begin{bmatrix}
\psi_{s,d}\, i_{s,q} - \psi_{s,q}\, i_{s,d} \\
\psi_{s,d} + X''_d\, i_{s,d} - \gamma_{d1}\, e'_{r,q} - (1 - \gamma_{d1})\, e''_{r,q} \\
\psi_{s,q} + X''_q\, i_{s,q} + \gamma_{q1}\, e'_{r,d} - (1 - \gamma_{q1})\, e''_{r,d} \\
v_{h,d}\, \cos(\delta_G) + v_{h,q}\, \sin(\delta_G) - v_{s,d} \\
v_{h,q}\, \cos(\delta_G) - v_{h,d}\, \sin(\delta_G) - v_{s,q} \\
i_{s,d}\, \cos(\delta_G) - i_{s,q}\, \sin(\delta_G) - i_{h,d} \\
i_{s,q}\, \cos(\delta_G) + i_{s,d}\, \sin(\delta_G) - i_{h,q} \\
v_{s,d}\, i_{s,d} + v_{s,q}\, i_{s,q} - p_h \\
v_{s,q}\, i_{s,d} - v_{s,d}\, i_{s,q} - q_h
\end{bmatrix}
,
\tag{2.38}
$$

where $\hat{\omega} = \omega_G$, and left-hand-side matrices:

$$
\boldsymbol{T} =
\begin{bmatrix}
\omega_o^{-1} & 0 & 0 & 0 & 0 & 0 & 0 & 0 \\
0 & M & 0 & 0 & 0 & 0 & 0 & 0 \\
0 & 0 & T'_{do} & 0 & \widehat{T}''_{do} & 0 & 0 & 0 \\
0 & 0 & 0 & T'_{qo} & 0 & -\widehat{T}''_{qo} & 0 & 0 \\
0 & 0 & 0 & 0 & T''_{do} & 0 & 0 & 0 \\
0 & 0 & 0 & 0 & 0 & T''_{qo} & 0 & 0 \\
0 & 0 & 0 & 0 & 0 & 0 & T_\psi & 0 \\
0 & 0 & 0 & 0 & 0 & 0 & 0 & T_\psi
\end{bmatrix}
,
\tag{2.39}
$$

$$
\boldsymbol{R} = \boldsymbol{0}_{9,8} ,
$$

where $T_\psi = \omega_o^{-1}$.

The TS model is given by the following equations: (2.26), (2.27), (2.21), (2.23), (2.25), (2.28), (2.32)–(2.34), and (2.36). Hence, the TS model can be described by the same set of variables (2.37), equations (2.38), and matrices (2.39) as the EMT but for $\hat{\omega} \approx 1$ pu(rad/s) and $T_\psi \approx 0$.

2.4.1.1. Simplified Models

Other simplified models of the synchronous machine have been defined in the literature. For example, a commonly used one direct- and one quadrature-axis model is obtained by assuming:

$$X_d'' \approx X_d' , \qquad X_q'' \approx X_q' , \tag{2.40}$$

which leads to $\gamma_{d1} \approx \gamma_{q1} \approx 1$ and $\gamma_{d2} \approx \gamma_{q2} \approx 0$, which also implies $\widehat{T}_{do}'' \approx \widehat{T}_{qo}'' \approx 0$. Consequently, the dynamics of the transient and subtransient fluxes are decoupled. Subtransient dynamics can be thus neglected.

Imposing the simplifications above in (2.33) and (2.36), one obtains the following dq-axis equations:

$$T_{do}' \frac{d}{dt} e_{r,q}'(t) = -e_{r,q}'(t) - (X_d - X_d') \, i_{s,d}(t) + v_{ef}(t) , \tag{2.41}$$
$$T_{qo}' \frac{d}{dt} e_{r,d}'(t) = -e_{r,d}'(t) + (X_q - X_q') \, i_{s,q}(t) ,$$

and:

$$0 = \psi_{s,d}(t) + X_d' \, i_{s,d}(t) - e_{r,q}'(t) , \tag{2.42}$$
$$0 = \psi_{s,q}(t) + X_q' \, i_{s,q}(t) + e_{r,d}'(t) .$$

The one direct- and one quadrature-axis machine model consists of the following equations: (2.26), (2.27), (2.21), (2.23), (2.25), (2.28), (2.32), (2.41), and (2.42). The very same model can be also straightforwardly obtained by imposing (2.40) as well as $T_{do}'' \approx T_{qo}'' \approx 0$ in (2.38) and (2.39), along with the approximations of the TS two direct- and two quadrature-axis machine model.

Finally the classical synchronous machine model, which only retains the dynamics of the mechanical variables δ_G and ω_G, is obtained with the approximations of the one direct- and one quadrature-axis model plus the conditions:

$$X_d \approx X_d' , \qquad X_q \approx X_q' , \tag{2.43}$$
$$T_{do}' \approx 0 , \qquad T_{qo}' \approx 0 .$$

In conclusion, the semi-implicit formulation of the set of DAEs (2.38)–(2.39) enables the seamless switching from the fully-fledged EMT to the TS model, and thus the implementation of all synchronous machine models with a unique set of DAEs, namely the one presented above for the two direct- and two quadrature-axis machine model.

2.4.1.2. Dynamic Circuit Representation

For the developments discussed in Part II, it is convenient to provide a circuit representation of the synchronous machine. This is possible only with the QSS models of the machine, i.e. if it is assumed that $\frac{d}{dt}\psi_{s,d} \approx 0$ and $\frac{d}{dt}\psi_{s,q} \approx 0$. Substituting (2.36) into (2.32) leads to:

$$
\begin{aligned}
0 &= R_a\, i_{s,d}(t) - X_q''\, i_{s,q}(t) - \gamma_{q1}\, e_{r,d}'(t) + (1-\gamma_{q1})e_{r,d}''(t) + v_{s,d}(t) \ , \\
0 &= R_a\, i_{s,q}(t) + X_d''\, i_{s,d}(t) - \gamma_{d1}\, e_{r,q}'(t) - (1-\gamma_{d1})e_{r,q}''(t) + v_{s,q}(t) \ ,
\end{aligned}
\tag{2.44}
$$

which can be conveniently rewritten in complex form by taking the first equation plus j times the second equation and multiplying by $\exp(j\delta_G)$:

$$
\begin{aligned}
\Big[[\gamma_{q1}e_{r,d}'(t) &- (1-\gamma_{q1})e_{r,d}''(t) + (X_q'' - X_d'')\, i_{s,q}(t)] \\
&+ j[\gamma_{d1}e_{r,q}'(t) + (1-\gamma_{d1})e_{r,q}''(t)]\Big] \exp(j\delta_G(t)) \\
= (R_a + jX_d'')\,\bar{\imath}_{s,dq}(t) &\exp(j\delta_G(t)) + \bar{v}_{s,dq}(t)\exp(j\delta_G(t)) \ ,
\end{aligned}
\tag{2.45}
$$

and, substituting the expressions of (2.21) and (2.23):

$$
\bar{e}_{dq}''(t)\exp(j\delta_G(t)) = (R_a + jX_d'')\,\bar{\imath}_{h,dq}(t) + \bar{v}_{h,dq}(t) \ ,
\tag{2.46}
$$

where $\bar{e}_{dq}''$ is the equivalent internal EMF of the machine which is a function of both transient and subtransient quantities:

$$
\begin{aligned}
\bar{e}_{dq}''(t) &= [\gamma_{q1}e_{r,d}'(t) - (1-\gamma_{q1})e_{r,d}''(t) + (X_q'' - X_d'')\, i_{s,q}(t)] \\
&\quad + j[\gamma_{d1}e_{r,q}'(t) + (1-\gamma_{d1})e_{r,q}''(t)] \ .
\end{aligned}
\tag{2.47}
$$

Equations (2.46) and (2.47) take into account transient and subtransient dynamics of the machine and can be represented through the equivalent circuit shown in Figure 2.5.a.

Simplified models of the machine can also be represented through a circuit similar to that of Figure 2.5.a by imposing the approximations discussed in the previous section. For example, substituting (2.40) into (2.47) leads to the one d- and one q-axis model with an internal EMF, $\bar{e}_{dq}'$, that is a function of

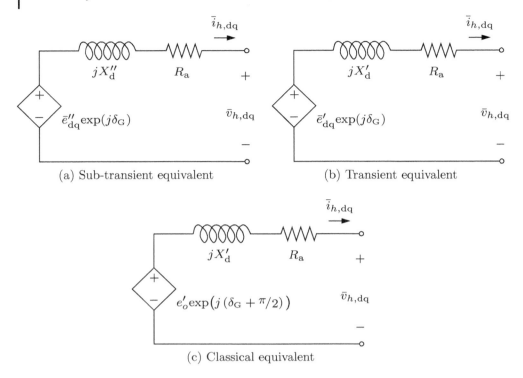

(a) Sub-transient equivalent

(b) Transient equivalent

(c) Classical equivalent

Figure 2.5 Equivalent circuit representations of the synchronous machine.

transient quantities only:

$$\bar{e}'_{\mathrm{dq}}(t) = [e'_{\mathrm{r,d}}(t) + (X'_{\mathrm{q}} - X'_{\mathrm{d}})i_{\mathrm{s,q}}(t)] + je'_{\mathrm{r,q}}(t) , \tag{2.48}$$

which is graphically represented by the equivalent circuit shown in Figure 2.5.b.

The well known classical model of the synchronous machine is obtained from (2.48) by imposing both (2.40) and (2.43), and assuming a constant excitation voltage. The resulting internal EMF is:

$$\bar{e}'_{o} = jv_{\mathrm{ef,o}} = v_{\mathrm{ef,o}} \exp\left(j\frac{\pi}{2}\right) , \tag{2.49}$$

which leads to the circuit shown in Figure 2.5.c.

Figure 2.6 shows the approximated circuit of the synchronous machine utilized for the determination of the extended admittance matrix of the system (see Section 2.4.3.3). In this equivalent circuit, the EMF is approximated as $je_{\mathrm{G}} \approx je'_{\mathrm{q}}$ and the internal reactance is taken as the average of the d- and q-axis reactances, as:

$$\langle X \rangle_{\mathrm{dq}} = \begin{cases} \dfrac{1}{2}\left(X''_{\mathrm{d}} + X''_{\mathrm{q}}\right) & \text{for subtransient dynamics,} \\[2mm] \dfrac{1}{2}\left(X'_{\mathrm{d}} + X'_{\mathrm{q}}\right) & \text{otherwise.} \end{cases} \tag{2.50}$$

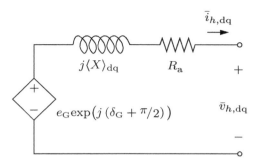

Figure 2.6 Approximated equivalent circuit representation of the synchronous machine.

2.4.1.3. Approximated Power and Swing Equation

The dynamic circuit representation of the classical model of the synchronous machine given in Figure 2.5.c can be utilized to deduce approximated expressions that will be useful in Chapter 6.

First, according to the notations used for the Park transform, the following equality holds:

$$\bar{v}_{h,\mathrm{dq}}(t) = v_h(t)\angle\left(\theta_h(t) + \frac{\pi}{2}\right) , \tag{2.51}$$

or, equivalently:

$$\begin{aligned} v_{h,\mathrm{d}}(t) &= -v_h(t)\sin\left(\theta_h(t)\right) , \\ v_{h,\mathrm{q}}(t) &= v_h(t)\cos\left(\theta_h(t)\right) . \end{aligned} \tag{2.52}$$

Then, the generator current can be written as:

$$\bar{i}_h(t) = \frac{\bar{e}'_o - \bar{v}_h(t)}{R_\mathrm{a} + jX'_\mathrm{d}} , \tag{2.53}$$

and the complex power injected by the generator into the grid:

$$\bar{s}_h(t) = \bar{v}_h(t)\,\bar{i}^*_h(t) = \frac{(e'_o)^2 - e'_o v_h(t)\exp\left(j\delta_\mathrm{G}(t) - j\theta_h(t)\right)}{R_\mathrm{a} - jX'_\mathrm{d}} , \tag{2.54}$$

from where, neglecting losses, one can obtain the active power as:

$$p_h(t) = \frac{e'_o v_h(t)}{X'_\mathrm{d}}\sin\left(\delta_\mathrm{G}(t) - \theta_h(t)\right) . \tag{2.55}$$

The latter expression is often utilized in stability studies as an approximation of the electromagnetic power generated by the machine and is included in the swing equation of the machine. Recalling (2.26), and neglecting for simplicity the damping, one has:

$$M\frac{d}{dt}\omega_\mathrm{G}(t) = \tau_\mathrm{m}(t) - \tau_\mathrm{e}(t) . \tag{2.56}$$

In per unit, since the variations of the rotor speed have to be small and, hence $\omega_G \approx 1$ pu, one can assume:

$$\tau_m(t) = \frac{p_m(t)}{\omega_G(t)} \approx p_m(t) \ ,$$

$$\tau_e(t) = \frac{p_h(t)}{\omega_G(t)} \approx p_h(t) \ ,$$

(2.57)

from where one obtains the well-known expression of the swing equation that is utilized in several transient stability studies:

$$M \frac{d}{dt} \omega_G(t) = p_m(t) - p_h(t) \ ,$$

(2.58)

where p_h is given in (2.55).

2.4.1.4. Center of Inertia

The need to switch from the local dq-axis frame of the synchronous machine to the dq-axis reference frame of the system is probably one of the most confusing points of the dynamic analysis of power systems. As discussed in the previous section, the system dq-axis frame appears in three equations of the machine: in (2.28), where the rotor angular speed of the machine is compared with the reference angular speed ω_o; and in (2.21) and (2.23), where the dq-axis components of the voltage and currents have to be rotated to refer to the reference phase angle δ_o of the system. δ_o never really appears in any equation and its value is irrelevant, but it is important to note that it is constant, as ω_o (see discussion of equation (2.21)).

While ω_o and δ_o certainly appear as reasonable references, they are not the only possible choice. One can choose any other angular frequency and phase angle, even time varying, to which refer synchronous machine rotor speeds and angles.

The very first alternative that one may think of is to choose as references $\omega_{G,h}$ and $\delta_{G,h}$ of the h-th machine that is connected to the system. This setup enables the removal of equation (2.28), which defines $\delta_{G,h}$. If $\delta_{G,h}$ is the phase reference, in fact, its variations cannot be evaluated. Then, equations (2.26) and (2.28) of all machines have to be referred to $\omega_{G,h}$.

Using one machine as reference, however, is not common as it poses implementation issues that are not easy to solve. For example, the model of the reference machine is different from that of the other machines and there is the need to change reference if the reference machine undergoes an outage.

A classical solution is the utilization of the center of inertia (CoI), which is an arithmetic mean of rotor speeds of synchronous machines weighted through their inertia constants, as follows [203]:

$$
\delta_{\text{CoI}}(t) = \frac{\sum_{h \in \mathbb{G}} M_h \, \delta_{\text{G},h}(t)}{\sum_{h \in \mathbb{G}} M_h} \ ,
$$

$$
\omega_{\text{CoI}}(t) = \frac{\sum_{h \in \mathbb{G}} M_h \, \omega_{\text{G},h}(t)}{\sum_{h \in \mathbb{G}} M_h} = \frac{d}{dt} \delta_{\text{CoI}}(t) \ ,
\tag{2.59}
$$

where $\mathbb{G}$ is the set of synchronous generators and M_h are the (normalized) inertia constants. Using the CoI, the mechanical equations of the machine become:

$$
\omega_o^{-1} \frac{d}{dt} \delta_{\text{G}}(t) = \omega_{\text{G}}(t) - \omega_{\text{CoI}}(t) \ ,
$$

$$
M \frac{d}{dt} \omega_{\text{G}}(t) = \tau_{\text{m}}(t) - \tau_{\text{e}}(t) - D \left(\omega_{\text{G}}(t) - \omega_{\text{CoI}}(t) \right) \ .
\tag{2.60}
$$

The definition of the CoI, while still posing some implementation issues in case of generator outages, is utilized in several works on primary and secondary frequency control, e.g. [169, 181, 182, 224], and is also further elaborated throughout the remainder of this book.

2.4.2. Loads

Load models are as crucial as generator ones for the proper stability analysis of power systems. However, while there is substantial agreement on the model of the synchronous machine, how to model loads has been and still is an open field of research. This is due to the fact that loads are rarely just a single device connected to the grid, at least at the transmission and distribution system levels. The power consumption at a node of the network is thus the combined effect of several devices, each of them with a potentially different dynamic behavior. Another difficulty is given by the stochastic nature of loads, which makes their power consumption and dynamic behavior volatile. To further complicate the problem, the trend of recent years to install distributed energy resources (DERs) at the medium and low voltage levels introduces new dynamics and increases the randomness of the power consumption as seen at the load buses at the transmission system level.

The intrinsic difficulty of modeling loads has led to the proliferation of load model proposals, for both steady state and dynamic analyses. In the best scenario, gray models and statistical analysis based on measurements are used. However, in practice, only very generic steady-state models, mostly based on hourly time series of the power consumption, are available.

Fortunately, the theory developed in Part II and the applications of such theory discussed in Part III do not really rely on detailed load models, at least no more than any other dynamic analysis of power systems. Thus, only a simple but versatile voltage and frequency dependent load model is presented in this section, as follows:

$$p_{\mathrm{D}}(t) = p_{\mathrm{D},o}\, \omega_h^{\beta_p}(t) \left(\frac{v_h(t)}{v_{h,o}}\right)^{\gamma_p},$$

$$q_{\mathrm{D}}(t) = q_{\mathrm{D},o}\, \omega_h^{\beta_q}(t) \left(\frac{v_h(t)}{v_{h,o}}\right)^{\gamma_q},$$

(2.61)

where v_h and ω_h are the voltage magnitude and frequency, respectively, at the bus h where the load is connected. The voltage magnitude is defined as

$$v_h^2(t) = v_{h,\mathrm{d}}^2(t) + v_{h,\mathrm{q}}^2(t)\ ;$$

(2.62)

$p_{\mathrm{D},o}$ and $q_{\mathrm{D},o}$ are the load active and reactive powers at the reference voltage magnitude $v_{h,o}$; and γ_p, γ_q, β_p, and β_q are parameters that depend on the load behavior. The determination of ω_h is more elaborated. The reader is referred to Section 3.3 for a description of conventional methods and to Chapter 5 for the unconventional approach discussed in this book.

The current injections from the load into the bus h are:

$$\begin{bmatrix} i_{h,\mathrm{d}}(t) \\ i_{h,\mathrm{q}}(t) \end{bmatrix} = -\frac{1}{v_h^2(t)} \begin{bmatrix} v_{h,\mathrm{d}}(t) & v_{h,\mathrm{q}}(t) \\ v_{h,\mathrm{q}}(t) & -v_{h,\mathrm{d}}(t) \end{bmatrix} \begin{bmatrix} p_{\mathrm{D}}(t) \\ q_{\mathrm{D}}(t) \end{bmatrix}.$$

(2.63)

For example, if $\beta_p = \beta_q = 0$, then $\gamma_p = \gamma_q = 0$ gives constant power consumption, whereas $\gamma_p = \gamma_q = 2$ leads to a constant admittance. Empirical values of γ_p, γ_q, β_p, and β_q for some industrial loads can be found, for example, in [14]. However, since loads are typically represented through aggregated models, the determination of precise values for the exponents of (2.61) is not an easy task.

As a general rule, $\beta_p = \beta_q = 0$ is assumed,[4] also because of the difficulty to define ω_h in conventional QSS models. Then $\gamma_p = \gamma_q = 2$ is appropriate for short-term simulations, e.g. transient stability analysis, whereas $\gamma_p = \gamma_q = 0$ is the common choice for medium to long-term simulations, e.g. frequency stability analysis and control, as well as steady-state analysis, e.g. power flow analysis.

The rationale behind the values above is that, at the distribution and transmission levels, loads are connected to the grid through an under-load tap changer (ULTC) that regulates the voltage on the low-voltage winding. The time scale of the response of the regulator of the ULTCs is of the order of

---

[4]Unless indicated otherwise, $\beta_p = \beta_q = 0$ is assumed by default for all loads in the examples included in this book.

minutes. In the short term, the tap ratio can be assumed to be constant and, generally, the voltage dependency of the power consumption on the voltage becomes quite evident. The exact values of the exponents γ_p and γ_q is still unknown, but since a large percentage of the load is, *de facto*, an impedance, in the short term loads can thus be approximated as constant admittances. On the other hand, in the long term, the ULTC control is able to regulate the secondary voltage to an almost constant value and, hence, the load can be adequately modeled as a constant power consumption. An in-depth discussion on the dynamic response of ULTCs and their effect on load modeling and behavior can be found in [127].

It is often required to define the equivalent impedance of a load model from its complex power consumption. This is typical, for example, to carry out a transient stability analysis after solving the power flow problem. Given the active and reactive power consumption, the equivalent per-unit impedance of a load is given by:

$$\bar{Y}_{\mathrm{D},o} = \frac{\bar{s}^*_{\mathrm{D},o}}{v^2_{h,o}} , \qquad (2.64)$$

where $\bar{s}_{\mathrm{D},o} = p_{\mathrm{D},o} + jq_{\mathrm{D},o}$.

The EMT models of the load are also hard to define. Even when assuming an equivalent admittance model as in (2.64), there is an ambiguity on whether the susceptance that composes such an admittance is purely inductive or a blend of an inductive load and a capacitive compensation. The topology of the load, i.e. series or parallel connection of the elements that composes it, also makes a difference in its EMT dynamic behavior.

For example, assume the circuit shown in Figure 2.7, which represents an inductive load composed of the series of an inductance and a resistance, and

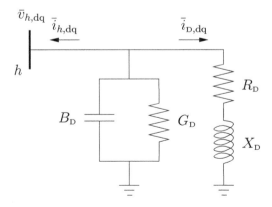

Figure 2.7 Single-phase equivalent circuit of a load for EMT simulations.

a shunt capacitor that provides reactive power support and whose losses are modeled as a conductance.

The EMT per-unit equations of the circuit are:

$$
\begin{aligned}
\bar{i}_{h,\mathrm{dq}}(t) &= -\bar{i}_{\mathrm{D,dq}}(t) - (G_{\mathrm{D}} + \bar{\mathrm{p}}_1(t)B_{\mathrm{D}})\,\bar{v}_{h,\mathrm{dq}}(t) \;, \\
\bar{v}_{h,\mathrm{dq}}(t) &= (R_{\mathrm{D}} + \bar{\mathrm{p}}_1(t)X_{\mathrm{D}})\,\bar{i}_{\mathrm{D,dq}}(t) \;,
\end{aligned}
\tag{2.65}
$$

where $\bar{\mathrm{p}}_1$ is defined in (1.60).

Using the notation of (2.15) and dropping for simplicity the dependency on time, (2.65) becomes:

$$
\boldsymbol{x}^\dagger = [v_{h,\mathrm{d}},\; v_{h,\mathrm{q}},\; i_{\mathrm{D,d}},\; i_{\mathrm{D,q}}] \;,
$$

$$
\boldsymbol{y} = \varnothing \;,
$$

$$
\boldsymbol{u}^\dagger = [i_{h,\mathrm{d}},\; i_{h,\mathrm{q}}] \;,
$$

$$
\boldsymbol{f} =
\begin{bmatrix}
-i_{h,\mathrm{d}} - i_{\mathrm{D,d}} - G_{\mathrm{D}}v_{h,\mathrm{d}} + B_{\mathrm{D}}v_{h,\mathrm{q}} \\
-i_{h,\mathrm{q}} - i_{\mathrm{D,q}} - G_{\mathrm{D}}v_{h,\mathrm{q}} - B_{\mathrm{D}}v_{h,\mathrm{d}} \\
v_{h,\mathrm{d}} - R_{\mathrm{D}}i_{\mathrm{D,d}} + X_{\mathrm{D}}i_{\mathrm{D,q}} \\
v_{h,\mathrm{q}} - R_{\mathrm{D}}i_{\mathrm{D,q}} - X_{\mathrm{D}}i_{\mathrm{D,d}}
\end{bmatrix} ,
$$

$$
\boldsymbol{g} = \varnothing \;,
\tag{2.66}
$$

$$
\mathbf{T} = \omega_o^{-1}
\begin{bmatrix}
B_{\mathrm{D}} & 0 & 0 & 0 \\
0 & B_{\mathrm{D}} & 0 & 0 \\
0 & 0 & X_{\mathrm{D}} & 0 \\
0 & 0 & 0 & X_{\mathrm{D}}
\end{bmatrix} ,
$$

$$
\mathbf{R} = \varnothing \;.
$$

The QSS model of the load is obtained by setting $\mathbf{T} = \mathbf{0}_{4,4}$, which gives:

$$
\begin{aligned}
\bar{i}_{h,\mathrm{dq}}(t) &= -[(R_{\mathrm{D}} + jX_{\mathrm{D}})^{-1} + (G_{\mathrm{D}} + jB_{\mathrm{D}})]\,\bar{v}_{h,\mathrm{dq}}(t) \\
&= -\bar{Y}_{\mathrm{D},o}\,\bar{v}_{h,\mathrm{dq}}(t) \;.
\end{aligned}
\tag{2.67}
$$

2.4.3. Transmission System

The high-voltage transmission system is mainly composed of overhead transmission lines and transformers. While detailed dynamic models of these devices can take an entire bookshelf, this section focuses exclusively on lumped, balanced, simplified EMT and QSS models that are commonly used in the dynamic analysis of power systems. The concepts of *admittance matrix* and *extended admittance matrix*, which are prerequisites to understand Chapter 5, are also defined.

2.4.3.1. Transmission Lines

The most detailed model of a transmission line is based on distributed parameters and considers partial differential equations. A first approximation involves transforming partial derivatives of voltages and currents into delayed variables [99]. For sufficiently short lines, a lumped-parameter Π-model is adequate. This is the model shown in Figure 2.8.

The dynamic per-unit equations of a transmission line are as follows:

$$\bar{v}_{h,\mathrm{dq}}(t) = \bar{v}_{k,\mathrm{dq}}(t) + (R_{\mathrm{L}} + \bar{p}_1(t)X_{\mathrm{L}})\,\bar{i}_{\mathrm{L,dq}}(t)\,,$$

$$\bar{i}_{h,\mathrm{dq}}(t) = \bar{i}_{\mathrm{L,dq}}(t) + \frac{1}{2}(G_{\mathrm{L}} + \bar{p}_1(t)B_{\mathrm{L}})\,\bar{v}_{h,\mathrm{dq}}(t)\,,\tag{2.68}$$

$$\bar{i}_{k,\mathrm{dq}}(t) = -\bar{i}_{\mathrm{L,dq}}(t) + \frac{1}{2}(G_{\mathrm{L}} + \bar{p}_1(t)B_{\mathrm{L}})\,\bar{v}_{k,\mathrm{dq}}(t)\,,$$

or, using the notation of (2.15) and dropping for simplicity the dependency on time:

$$\boldsymbol{x}^{\dagger} = [i_{\mathrm{L,d}},\ i_{\mathrm{L,q}},\ v_{h,\mathrm{d}},\ v_{h,\mathrm{q}},\ v_{k,\mathrm{d}},\ v_{k,\mathrm{q}}]\,,$$

$$\boldsymbol{y} = \varnothing\,,$$

$$\boldsymbol{u}^{\dagger} = [i_{h,\mathrm{d}},\ i_{h,\mathrm{q}},\ i_{k,\mathrm{d}},\ i_{k,\mathrm{q}}]\,,$$

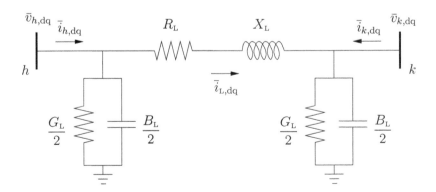

Figure 2.8 Single-phase equivalent Π-model of a transmission line.

$$
f = \begin{bmatrix}
v_{h,\mathrm{d}} - v_{k,\mathrm{d}} - R_{\mathrm{L}}i_{\mathrm{L},\mathrm{d}} + X_{\mathrm{L}}i_{\mathrm{L},\mathrm{q}} \\
v_{h,\mathrm{q}} - v_{k,\mathrm{q}} - R_{\mathrm{L}}i_{\mathrm{L},\mathrm{q}} - X_{\mathrm{L}}i_{\mathrm{L},\mathrm{d}} \\
i_{h,\mathrm{d}} - i_{\mathrm{L},\mathrm{d}} - \dfrac{G_{\mathrm{L}}}{2}v_{h,\mathrm{d}} + \dfrac{B_{\mathrm{L}}}{2}v_{h,\mathrm{q}} \\
i_{h,\mathrm{q}} - i_{\mathrm{L},\mathrm{q}} - \dfrac{G_{\mathrm{L}}}{2}v_{h,\mathrm{q}} - \dfrac{B_{\mathrm{L}}}{2}v_{h,\mathrm{d}} \\
i_{k,\mathrm{d}} + i_{\mathrm{L},\mathrm{d}} - \dfrac{G_{\mathrm{L}}}{2}v_{k,\mathrm{d}} + \dfrac{B_{\mathrm{L}}}{2}v_{k,\mathrm{q}} \\
i_{k,\mathrm{q}} + i_{\mathrm{L},\mathrm{q}} - \dfrac{G_{\mathrm{L}}}{2}v_{k,\mathrm{q}} - \dfrac{B_{\mathrm{L}}}{2}v_{k,\mathrm{d}}
\end{bmatrix} ,
$$

(2.69)

$$
\boldsymbol{g} = \varnothing ,
$$

$$
\boldsymbol{T} = \omega_o^{-1} \begin{bmatrix}
X_{\mathrm{L}} & 0 & 0 & 0 & 0 & 0 \\
0 & X_{\mathrm{L}} & 0 & 0 & 0 & 0 \\
0 & 0 & B_{\mathrm{L}}/2 & 0 & 0 & 0 \\
0 & 0 & 0 & B_{\mathrm{L}}/2 & 0 & 0 \\
0 & 0 & 0 & 0 & B_{\mathrm{L}}/2 & 0 \\
0 & 0 & 0 & 0 & 0 & B_{\mathrm{L}}/2
\end{bmatrix} ,
$$

$$
\boldsymbol{R} = \varnothing ,
$$

where $\bar{i}_{h,\mathrm{dq}}$ and $\bar{i}_{k,\mathrm{dq}}$ are input variables as they are obtained as the sum of all current injections of the devices connected to buses h and k.

The QSS model of the transmission line is obtained assuming $\boldsymbol{T} = \boldsymbol{0}_{6,6}$ in (2.69), which leads to a purely algebraic model. In QSS models, however, it is unusual to use the expressions given in (2.68). Since the currents entering in the line are inputs, a notation exclusively based on admittances is preferred, as follows:

$$
\begin{bmatrix} \bar{i}_{h,\mathrm{dq}}(t) \\ \bar{i}_{k,\mathrm{dq}}(t) \end{bmatrix} = \begin{bmatrix} \bar{Y}_{\mathrm{L},hk} + \bar{Y}_{\mathrm{L},h} & -\bar{Y}_{\mathrm{L},hk} \\ -\bar{Y}_{\mathrm{L},hk} & \bar{Y}_{\mathrm{L},hk} + \bar{Y}_{\mathrm{L},k} \end{bmatrix} \begin{bmatrix} \bar{v}_{h,\mathrm{dq}}(t) \\ \bar{v}_{k,\mathrm{dq}}(t) \end{bmatrix} ,
$$

(2.70)

where

$$
\bar{Y}_{\mathrm{L},hk} = (R_{\mathrm{L}} + jX_{\mathrm{L}})^{-1} ,
$$

$$
\bar{Y}_{\mathrm{L},h} = \bar{Y}_{\mathrm{L},k} = \frac{1}{2}(G_{\mathrm{L}} + jB_{\mathrm{L}}) .
$$

(2.71)

2.4.3.2. Transformers

Transformer models include a series inductive impedance that models resistive losses and flux leakages of the windings. The magnetization reactance (X_μ) and iron losses (R_{Fe}) are modeled through shunt elements, which, however, are often

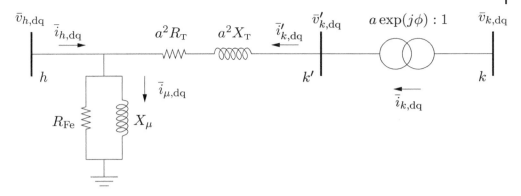

Figure 2.9 Single-phase equivalent circuit a transformer with off-nominal tap ratio and phase shifter.

neglected when studying power systems. Figure 2.9 shows the resulting single-phase equivalent per-unit circuit of the transformer obtained assuming that the magnetization current is small and that primary and secondary winding losses and leakage can be merged together in a single impedance, namely $R_T + jX_T$. Figure 2.9 also shows an ideal (lossless) transformer that takes into account off-nominal tap ratio (a) and phase shift (ϕ) between primary and secondary voltages. The off-nominal tap is assumed to be on the primary winding, hence the need to multiply by a^2 the series impedance of the transformer [113, 127].

The dynamic per-unit equations of a transmission line are as follows:

$$\bar{v}_{h,\mathrm{dq}}(t) = \bar{v}'_{k,\mathrm{dq}}(t) - a^2(R_T + \bar{p}_1(t)X_T)\,\bar{i}'_{k,\mathrm{dq}}(t)\ ,$$

$$\bar{v}_{h,\mathrm{dq}}(t) = \bar{p}_1(t)X_\mu\,\bar{i}_{\mu,\mathrm{dq}}(t)\ ,$$

$$\bar{i}_{h,\mathrm{dq}}(t) = -\bar{i}'_{k,\mathrm{dq}}(t) + \frac{1}{R_{\mathrm{Fe}}}\bar{v}_{h,\mathrm{dq}}(t) + \bar{i}_{\mu,\mathrm{dq}}(t)\ , \tag{2.72}$$

$$\bar{v}'_{k,\mathrm{dq}}(t) = \bar{a}\,\bar{v}_{k,\mathrm{dq}}(t)\ ,$$

$$\bar{i}_{k,\mathrm{dq}}(t) = \bar{a}^*\,\bar{i}'_{k,\mathrm{dq}}(t)\ ,$$

where $\bar{a} = a\exp(j\phi)$ is the complex tap ratio of the transformer. Note that the ideal transformer is power invariant:

$$\bar{v}_{k,\mathrm{dq}}(t)\,\bar{i}^*_{k,\mathrm{dq}}(t) = \bar{a}\,\bar{v}_{k,\mathrm{dq}}(t)\,\frac{\bar{i}^*_{k,\mathrm{dq}}(t)}{(\bar{a}^*)^*} = \bar{v}'_{k,\mathrm{dq}}(t)\,\bar{i}'^*_{k,\mathrm{dq}}(t)\ . \tag{2.73}$$

If iron core magnetization and losses are neglected, as usual in power system studies, $X_\mu \to \infty$, which implies $\bar{i}_{\mu,\mathrm{dq}} = 0$, and $1/R_{\mathrm{Fe}} \to 0$, which leads to $\bar{i}_{h,\mathrm{dq}} = -\bar{i}'_{k,\mathrm{dq}}$.

Using the notation of (2.15) and dropping for simplicity the dependency on time, (2.72) becomes:

$$\boldsymbol{x}^\dagger = [i'_{k,\mathrm{d}},\, i'_{k,\mathrm{q}},\, i_{\mu,\mathrm{d}},\, i_{\mu,\mathrm{q}}]\ ,$$

$$\boldsymbol{y}^\dagger = [v_{h,\mathrm{d}},\, v_{h,\mathrm{q}},\, v_{k,\mathrm{d}},\, v_{k,\mathrm{q}},\, v'_{k,\mathrm{d}},\, v'_{k,\mathrm{q}}],$$

$$\boldsymbol{u}^\dagger = [i_{h,\mathrm{d}},\, i_{h,\mathrm{q}},\, i_{k,\mathrm{d}},\, i_{k,\mathrm{q}}]\ ,$$

$$\boldsymbol{f} = \begin{bmatrix} v'_{k,\mathrm{d}} - v_{h,\mathrm{d}} - a^2 R_\mathrm{T}\, i'_{k,\mathrm{d}} + a^2 X_\mathrm{T}\, i'_{k,\mathrm{q}} \\ v'_{k,\mathrm{q}} - v_{h,\mathrm{q}} - a^2 R_\mathrm{T}\, i'_{k,\mathrm{q}} - a^2 X_\mathrm{T}\, i'_{k,\mathrm{d}} \\ v_{h,\mathrm{d}} + X_\mu\, i_{\mu,\mathrm{q}} \\ v_{h,\mathrm{q}} - X_\mu\, i_{\mu,\mathrm{d}} \end{bmatrix} ,$$

$$\boldsymbol{g} = \begin{bmatrix} i_{h,\mathrm{d}} + i'_{k,\mathrm{d}} - \dfrac{1}{R_\mathrm{Fe}} v_{h,\mathrm{d}} - i_{\mu,\mathrm{d}} \\[4pt] i_{h,\mathrm{q}} + i'_{k,\mathrm{q}} - \dfrac{1}{R_\mathrm{Fe}} v_{h,\mathrm{q}} - i_{\mu,\mathrm{q}} \\[4pt] a\cos(\phi) v_{k,\mathrm{d}} - a\sin(\phi) v_{k,\mathrm{q}} - v'_{k,\mathrm{d}} \\ a\cos(\phi) v_{k,\mathrm{q}} + a\sin(\phi) v_{k,\mathrm{d}} - v'_{k,\mathrm{q}} \\ a\cos(\phi) i'_{k,\mathrm{d}} + a\sin(\phi) i'_{k,\mathrm{q}} - i_{k,\mathrm{d}} \\ a\cos(\phi) i'_{k,\mathrm{q}} - a\sin(\phi) i'_{k,\mathrm{d}} - i_{k,\mathrm{q}} \end{bmatrix} ,$$

$$\mathbf{T} = \omega_o^{-1} \begin{bmatrix} a^2 X_\mathrm{T} & 0 & 0 & 0 \\ 0 & a^2 X_\mathrm{T} & 0 & 0 \\ 0 & 0 & X_\mu & 0 \\ 0 & 0 & 0 & X_\mu \end{bmatrix} ,$$

$$\mathbf{R} = \mathbf{0}_{6,4}\ .$$

(2.74)

Similarly to the model of the transmission line, the QSS model of the transformer is obtained by assuming $\mathbf{T} = \mathbf{0}_{4,4}$ in (2.74), which leads to a purely algebraic model. And, again, in the same vein as the model of transmission lines, a notation based on admittances is preferred. Neglecting iron core magnetization and losses, the currents injected into the transformer are linked to the bus voltages as follows:

$$\begin{bmatrix} \bar{i}_{h,\mathrm{dq}}(t) \\ \bar{i}_{k,\mathrm{dq}}(t) \end{bmatrix} = \bar{Y}_\mathrm{T} \begin{bmatrix} 1/a^2 & -1/\bar{a}^* \\ -1/\bar{a} & 1 \end{bmatrix} \begin{bmatrix} \bar{v}_{h,\mathrm{dq}}(t) \\ \bar{v}_{k,\mathrm{dq}}(t) \end{bmatrix} ,$$

(2.75)

where

$$\bar{Y}_\mathrm{T} = (R_\mathrm{T} + jX_\mathrm{T})^{-1}\ .$$

(2.76)

2.4.3.3. Admittance Matrix

The QSS models of transmission lines and transformers enables the merger of these two devices in a compact formulation based on the matrix notation of (2.70) and (2.75). The result is the so-called *admittance matrix* $\bar{\mathbf{Y}}_{\text{bus}}$, that links current injections and bus voltages at all network buses:

$$\bar{i}_{\text{dq}}(t) = \bar{\mathbf{Y}}_{\text{bus}}\, \bar{v}_{\text{dq}}(t) = (\mathbf{G}_{\text{bus}} + j\mathbf{B}_{\text{bus}})\, \bar{v}_{\text{dq}}(t) \ , \tag{2.77}$$

where, according to the assumptions of the QSS model, the elements of $\bar{\mathbf{Y}}_{\text{bus}}$ are effectively constant as they depend exclusively on transmission line and transformer parameters, on the system impedance and frequency bases (see also equation (2.18) and the discussion of the per-unit time derivative of the Park vector thereafter).

The rules to build the admittance matrix are provided in many books, but it is probably worth repeating such rules here for completeness. For every pair of buses h and k connected by one or more branches, the elements of the admittance matrix are given by:

$$
\begin{aligned}
\bar{Y}_{hh} &= \sum_{m \in \mathbb{L}_{hk}} \left(\bar{Y}^m_{\text{L},hk} + \bar{Y}^m_{\text{L},h} \right) + \sum_{n \in \mathbb{T}_{hk}} \frac{1}{a_n^2} \bar{Y}^n_{\text{T}} \ , \\
\bar{Y}_{hk} &= - \sum_{m \in \mathbb{L}_{hk}} \bar{Y}^m_{\text{L},hk} - \sum_{n \in \mathbb{T}_{hk}} \frac{1}{\bar{a}_n^*} \bar{Y}^n_{\text{T}} \ , \\
\bar{Y}_{kh} &= - \sum_{m \in \mathbb{L}_{hk}} \bar{Y}^m_{\text{L},hk} - \sum_{n \in \mathbb{T}_{hk}} \frac{1}{\bar{a}_n} \bar{Y}^n_{\text{T}} \ , \\
\bar{Y}_{kk} &= \sum_{m \in \mathbb{L}_{hk}} \left(\bar{Y}^m_{\text{L},hk} + \bar{Y}^m_{\text{L},k} \right) + \sum_{n \in \mathbb{T}_{hk}} \bar{Y}^n_{\text{T}} \ ,
\end{aligned}
\tag{2.78}
$$

where $\mathbb{L}_{hk}$ and $\mathbb{T}_{hk}$ are the sets of transmission lines and transformers, respectively, connecting buses h and k, and line and transformer parameters are defined in (2.71) and (2.76), respectively.

For example, the admittance matrix of the Western Systems Coordinating Council (WSCC) system is shown in Table 2.4.[5] Note that, since the WSCC system does not include transformers with off-nominal tap ratios or phase shifts, $\bar{\mathbf{Y}}_{\text{bus}}$ is symmetrical.

In classical transient stability analysis studies, it is often defined an *extended* admittance matrix, which also includes the internal admittances of the synchronous machines connected to the network. The resulting expression links machine current injections and internal EMFs with bus current injections and

---

[5]The data of the WSCC system is given in Appendix A.2.

Table 2.4 Admittance matrix ($\bar{\mathbf{Y}}_{\text{bus}}$) of the WSCC system.

$\mathbf{G}_{\text{bus}} = \text{Re}\{\bar{\mathbf{Y}}_{\text{bus}}\}$								
0	0	0	0	0	0	0	0	0
0	0	0	0	0	0	0	0	0
0	0	0	0	0	0	0	0	0
0	0	0	3.307	−1.365	−1.942	0	0	0
0	0	0	−1.365	2.553	0	−1.188	0	0
0	0	0	−1.942	0	3.224	0	0	−1.282
0	0	0	0	−1.188	0	2.804	−1.617	0
0	0	0	0	0	0	−1.617	2.772	−1.155
0	0	0	0	0	−1.282	0	−1.155	2.437

$\mathbf{B}_{\text{bus}} = \text{Im}\{\bar{\mathbf{Y}}_{\text{bus}}\}$								
−17.361	0	0	17.361	0	0	0	0	0
0	−16.000	0	0	0	0	16.000	0	0
0	0	−17.065	0	0	0	0	0	17.065
17.361	0	0	−39.309	11.604	10.511	0	0	0
0	0	0	11.604	−17.338	0	5.975	0	0
0	0	0	10.511	0	−15.841	0	0	5.588
0	16.000	0	0	5.975	0	−35.446	13.698	0
0	0	0	0	0	0	13.698	−23.303	9.784
0	0	17.065	0	0	5.588	0	9.784	−32.154

voltages, as follows:

$$\begin{bmatrix} \bar{i}_{\text{G,dq}}(t) \\ \bar{i}_{\text{B,dq}}(t) \end{bmatrix} = \bar{\mathbf{Y}}_{\text{ext}} \begin{bmatrix} \bar{e}_{\text{G,dq}}(t) \\ \bar{v}_{\text{B,dq}}(t) \end{bmatrix} = \begin{bmatrix} \bar{\mathbf{Y}}_{\text{GG}} & \bar{\mathbf{Y}}_{\text{GB}} \\ \bar{\mathbf{Y}}_{\text{BG}} & \bar{\mathbf{Y}}_{\text{BB}} \end{bmatrix} \begin{bmatrix} \bar{e}_{\text{G,dq}}(t) \\ \bar{v}_{\text{B,dq}}(t) \end{bmatrix}, \tag{2.79}$$

where $\bar{e}_{\text{G,dq}} = \mathbf{e}_{\text{G}} \exp(j(\delta_{\text{G}} + \pi/2))$ is the vector of the internal EMFs of the synchronous machines, as defined in Figure 2.6. Note that $\bar{\mathbf{Y}}_{\text{BB}} \neq \bar{\mathbf{Y}}_{\text{bus}}$ as the latter does not include in its diagonal elements the internal admittances of the synchronous machines. For example, the four matrices that form $\bar{\mathbf{Y}}_{\text{ext}}$, namely, $\bar{\mathbf{Y}}_{\text{GG}}$, $\bar{\mathbf{Y}}_{\text{GB}}$, $\bar{\mathbf{Y}}_{\text{BG}}$, and $\bar{\mathbf{Y}}_{\text{BB}}$, for the WSCC system are shown in Table 2.5.

Note that $\bar{\mathbf{Y}}_{\text{GG}}$ is diagonal and, since the machines of the WSCC model are defined through a one d- and one q- axis model with null armature resistance, the internal impedances are given by $-j2/(X'_{\text{d}} + X'_{\text{q}})$ as per equation (2.50). Moreover, since the equivalent circuit of the generators does not include shunt elements, the sum of the elements of each row of $[\bar{\mathbf{Y}}_{\text{GG}} \ \bar{\mathbf{Y}}_{\text{GB}}]$ and $[\bar{\mathbf{Y}}_{\text{GG}} \ \bar{\mathbf{Y}}^{\dagger}_{\text{BG}}]$

Table 2.5 Matrices that compose the extended admittance matrix ($\bar{\mathbf{Y}}_{\text{ext}}$) of the WSCC system.

				$\bar{\mathbf{Y}}_{\text{GG}}$				
			$-j12.682$	0	0			
			0	$-j6.315$	0			
			0	0	$-j4.637$			

				$\bar{\mathbf{Y}}_{\text{GB}}$				
$j12.682$	0	0	0	0	0	0	0	0
0	$j6.315$	0	0	0	0	0	0	0
0	0	$j4.637$	0	0	0	0	0	0

				$\bar{\mathbf{Y}}_{\text{BG}}^{\dagger}$				
$j12.682$	0	0	0	0	0	0	0	0
0	$j6.315$	0	0	0	0	0	0	0
0	0	$j4.637$	0	0	0	0	0	0

			$\mathbf{G}_{\text{BB}}$	$= \text{Re}\{\bar{\mathbf{Y}}_{\text{BB}}\}$				
0	0	0	0	0	0	0	0	0
0	0	0	0	0	0	0	0	0
0	0	0	0	0	0	0	0	0
0	0	0	3.307	-1.365	-1.942	0	0	0
0	0	0	-1.365	2.553	0	-1.188	0	0
0	0	0	-1.942	0	3.224	0	0	-1.282
0	0	0	0	-1.188	0	2.804	-1.617	0
0	0	0	0	0	0	-1.617	2.772	-1.155
0	0	0	0	0	-1.282	0	-1.155	2.437

			$\mathbf{B}_{\text{BB}}$	$= \text{Im}\{\bar{\mathbf{Y}}_{\text{BB}}\}$				
-30.043	0	0	17.361	0	0	0	0	0
0	-22.315	0	0	0	0	16.000	0	0
0	0	-22.702	0	0	0	0	0	17.065
17.361	0	0	-39.309	11.604	10.511	0	0	0
0	0	0	11.604	-17.338	0	5.975	0	0
0	0	0	10.511	0	-15.841	0	0	5.588
0	16.000	0	0	5.975	0	-35.446	13.698	0
0	0	0	0	0	0	13.698	-23.303	9.784
0	0	17.065	0	0	5.588	0	9.784	-32.154

is zero and $\bar{\mathbf{Y}}_{ext}$ has the following property:

$$\bar{Y}_{hh} = \sum_{k\in\mathbb{G},k\neq h} -\bar{Y}_{kh} = \sum_{k\in\mathbb{G},k\neq h} -\bar{Y}_{hk} \ , \tag{2.80}$$

where $\mathbb{G}$ is the set of generator indexes.

Another common approximation of conventional transient stability studies is to assume that the loads are constant impedances, as follows:

$$\bar{i}_{B,dq}(t) = -\bar{\mathbf{Y}}_{BD}\,\bar{v}_{B,dq}(t) \ , \tag{2.81}$$

where $\bar{\mathbf{Y}}_{BD}$ is a diagonal matrix whose elements are the equivalent impedances of the loads, typically computed as in equation (2.64). For example, the matrix $\bar{\mathbf{Y}}_{BD}$ calculated at the power flow solution of the WSCC system is shown in Table 2.6. It is important to note that the absolute values of the elements of $\mathbf{B}_{BD}$ are about an order of magnitude smaller than the elements of $\mathbf{G}_{BD}$

Table 2.6 Equivalent load admittance matrix $(\bar{\mathbf{Y}}_{BD})$ of the WSCC system.

$\mathbf{G}_{BD} = \mathrm{Re}\{\bar{\mathbf{Y}}_{BD}\}$								
0	0	0	0	0	0	0	0	0
0	0	0	0	0	0	0	0	0
0	0	0	0	0	0	0	0	0
0	0	0	0	0	0	0	0	0
0	0	0	0	1.260	0	0	0	0
0	0	0	0	0	0.878	0	0	0
0	0	0	0	0	0	0	0	0
0	0	0	0	0	0	0	0.969	0
0	0	0	0	0	0	0	0	0

$\mathbf{B}_{BD} = \mathrm{Im}\{\bar{\mathbf{Y}}_{BD}\}$								
0	0	0	0	0	0	0	0	0
0	0	0	0	0	0	0	0	0
0	0	0	0	0	0	0	0	0
0	0	0	0	0	0	0	0	0
0	0	0	0	-0.504	0	0	0	0
0	0	0	0	0	-0.293	0	0	0
0	0	0	0	0	0	0	0	0
0	0	0	0	0	0	0	-0.339	0
0	0	0	0	0	0	0	0	0

and these are an order of magnitude smaller than the absolute values of the elements of $\mathbf{B}_{\text{BB}}$. Hence:

$$\bar{\mathbf{Y}}_{\text{BB}} + \bar{\mathbf{Y}}_{\text{BD}} \approx \bar{\mathbf{Y}}_{\text{BB}} \ . \tag{2.82}$$

Moreover, in transmission systems the ratio $R_{\text{L}}/X_{\text{L}} \ll 1$. Similarly, for synchronous machines, the ratio $R_{\text{a}}/X_{\text{d}}' \leq R_{\text{a}}/X_{\text{d}}'' \ll 1$. These inequalities lead to the following condition for the elements of $\bar{\mathbf{Y}}_{\text{ext}}$:

$$G_{hk} = \text{Re}\{\bar{Y}_{hk}\} \ll |\text{Im}\{\bar{Y}_{hk}\}| = |B_{hk}| \ , \tag{2.83}$$

and, hence:

$$\bar{\mathbf{Y}}_{\text{ext}} \approx j\text{Im}\{\bar{\mathbf{Y}}_{\text{ext}}\} = j\mathbf{B}_{\text{ext}} \ . \tag{2.84}$$

The approximation (2.84) has been exploited in several applications, e.g. in fault analysis, to determine approximated expressions of short-circuit currents [7].

2.4.3.4. Power Flow Equations

The power flow equations are relevant expressions that can be derived from the admittance matrix, as follows. The net power injections at network buses are defined as:

$$\bar{\boldsymbol{s}}_{\text{dq}}(t) = \bar{\boldsymbol{v}}_{\text{dq}}(t) \circ \bar{\boldsymbol{i}}_{\text{dq}}^{*}(t) \ , \tag{2.85}$$

where "$\circ$" denotes the Hadamard product, i.e. the element-wise vector multiplication and the subscript B has been dropped for simplicity. Recalling the expression of the current given in (2.77), one obtains:

$$\bar{\boldsymbol{s}}_{\text{dq}}(t) = \bar{\boldsymbol{v}}_{\text{dq}}(t) \circ \left(\bar{\mathbf{Y}}_{\text{bus}}^{*} \, \bar{\boldsymbol{v}}_{\text{dq}}^{*}(t)\right) \ , \tag{2.86}$$

which are the sought power flow equations. The expression (2.86) is valid in time as it is deduced from the definition of complex power and Kirchhoff's current law. Very often, however, the power flow equations are used in steady-state to solve the well-known power flow problem.

Before defining such a problem, it is convenient to expand (2.86) into its real and imaginary components. Dropping the dependency on time for compactness, one obtains:

$$p_h = \sum_{k \in \mathbb{B}} \left[v_{h,\text{d}}(G_{hk}\,v_{k,\text{d}} - B_{hk}\,v_{k,\text{q}}) + v_{h,\text{q}}(G_{hk}\,v_{k,\text{q}} + B_{hk}v_{k,\text{d}})\right], \ h \in \mathbb{B} \ ,$$

$$q_h = \sum_{k \in \mathbb{B}} \left[v_{h,\text{q}}(G_{hk}\,v_{k,\text{d}} - B_{hk}\,v_{k,\text{q}}) - v_{h,\text{d}}(G_{hk}\,v_{k,\text{q}} + B_{hk}v_{k,\text{d}})\right], \ h \in \mathbb{B} \ ,$$

$$\tag{2.87}$$

where $\mathbb{B}$ is the set of network buses. These equations can be conveniently rewritten in polar form:

$$p_h = v_h \sum_{k \in \mathbb{B}} v_k \big(G_{hk} \cos(\theta_{hk}) + B_{hk} \sin(\theta_{hk})\big), \quad h \in \mathbb{B} ,$$

$$\tag{2.88}$$

$$q_h = v_h \sum_{k \in \mathbb{B}} v_k \big(G_{hk} \sin(\theta_{hk}) - B_{hk} \cos(\theta_{hk})\big), \quad h \in \mathbb{B} ,$$

where

$$v_h = \sqrt{v_{h,\mathrm{d}}^2 + v_{h,\mathrm{q}}^2} , \tag{2.89}$$

$$\theta_h = \angle \bar{v}_{h,\mathrm{dq}} = \mathrm{atan}\left(-\frac{v_{h,\mathrm{d}}}{v_{h,\mathrm{q}}}\right) , \tag{2.90}$$

and $\theta_{hk} = \theta_h - \theta_k$.

The conventional power flow problem is set up by defining three kinds of buses:

- *Slack bus*, where the voltage magnitude v and the phase angle θ are known and the active and reactive power injections, p and q, are unknown.

- *PV bus*, where the voltage magnitude v and the active power injection p are known and the voltage phase angle θ and reactive power injection q are unknown.

- *PQ bus*, where the active and reactive power injections, p and q, are known and the voltage magnitude v and phase angle θ are unknown.

Several examples of power flow data and solutions are given in Appendix A.

2.4.3.5. DC Power Flow Equations

This section provides the formulation of the DC power flow equations, which are relevant for the developments discussed in Chapter 6. The basic assumption of the DC power flow model is that the reactive powers and voltage magnitude variations are negligible. In particular, the hypotheses are:

1. All voltage magnitudes are assumed constant and equal to 1.0 pu(kV), and reactive powers are neglected.

2. Line resistances and charging are neglected when computing the simplified admittance matrix, hence:

$$\bar{\mathbf{Y}}_{\mathrm{bus}} \approx j\mathbf{B}_{\mathrm{bus}} . \tag{2.91}$$

When computing the matrix $\mathbf{B}_{bus}$, all transformer tap ratios and phase shifter angles are 1 pu(kV)/pu(kV) and 0 rad, respectively.

3. Bus voltage phase angles are considered *small*, so that $\sin(\theta_{hk}) \approx \theta_{hk}$ and $\cos(\theta_{hk}) \approx 1.0$.

The resulting system of equations are:

$$p = \mathbf{B}_{bus}\,\boldsymbol{\theta}\,,\tag{2.92}$$

where p are the active power injections at buses and $\boldsymbol{\theta}$ are the bus voltage phase angles.[6] Table 2.7 shows the result of the DC power flow analysis for the WSCC system and compares it with the exact solution of the equations (2.88). While approximated, the values of the phase angles obtained with the DC power flow analysis show relatively small errors with respect to the exact solution. Finally, note that (2.92) is, in turn, a simplification of (2.85) and is thus valid in time, i.e.:

$$p(t) = \mathbf{B}_{bus}\,\boldsymbol{\theta}(t)\,.\tag{2.93}$$

The latter is the expression that is utilized in Chapter 6.

2.4.4. Example

This example compares of the EMT and QSS models and illustrates the effect of the utilization of the CoI reference on the trajectories of the variables for

Table 2.7 Results of the DC power flow analysis for the WSCC system.

Bus #	Base case (exact) solution				DC power flow solution			
	v_h [pu(kV)]	θ_h [rad]	p_h [pu(MW)]	q_h [pu(MVAr)]	v_h [pu(kV)]	θ_h [rad]	p_h [pu(MW)]	q_h [pu(MVAr)]
1	1.0400	0	0.7164	0.2705	1	0	0.67	0
2	1.0250	0.1620	1.63	0.0665	1	0.1710	1.63	0
3	1.0250	0.0814	0.85	−0.1086	1	0.0883	0.85	0
4	1.0258	−0.0387	0	0	1	−0.0386	0	0
5	0.9956	−0.0696	−1.25	−0.50	1	−0.0709	−1.25	0
6	1.0127	−0.0644	−0.90	−0.30	1	−0.0652	−0.90	0
7	1.0258	0.0649	0	0	1	0.0691	0	0
8	1.0159	0.0127	−1.00	−0.35	1	0.0144	−1.00	0
9	1.0324	0.0343	0	0	1	0.0385	0	0

---

[6]In (2.92), the power shifts introduced by phase shifting transformers are ignored as these are constant and, thus, immaterial for the developments discussed in Chapter 6.

the same system. Simulations are based on the IEEE 14-bus system, whose data are given in Appendix A.3.

The grid includes two d- and two q- axis models of synchronous machines but no primary controllers. The EMT models of machines, loads, transmission lines, and transformers are those described in Section 2.4 including all fast and

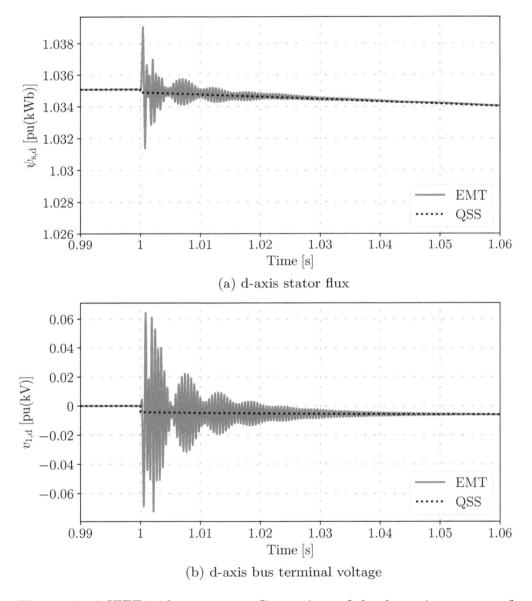

(a) d-axis stator flux

(b) d-axis bus terminal voltage

Figure 2.10 IEEE 14-bus system – Comparison of the dynamic response of the EMT and QSS models: d-axis components of stator flux and bus terminal voltage of the synchronous machine connected to bus 1.

slow dynamics. For the QSS models, fast dynamics of each model are neglected, as discussed in Sections 2.4.1 through 2.4.3.

Figure 2.10 shows the dynamic behavior of the d-axis components of the stator flux and the terminal bus voltage of the generator connected to bus 1 following the outage of the line connecting buses 2 and 4 at $t = 1$ s. Similarly to the results shown in Figure 2.4, the effect of electromagnetic dynamic transients lasts for only a few tens of milliseconds and can be thus safely neglected when focusing on electromechanical transients.

Figure 2.11 compares the trajectories of relevant variables of the synchronous machine connected to bus 1 using either ω_o or ω_{CoI} as reference angular frequency. The rotor speed and the dq-axis components of the internal variables of the machine, such as the d-axis stator flux shown in the figure, are

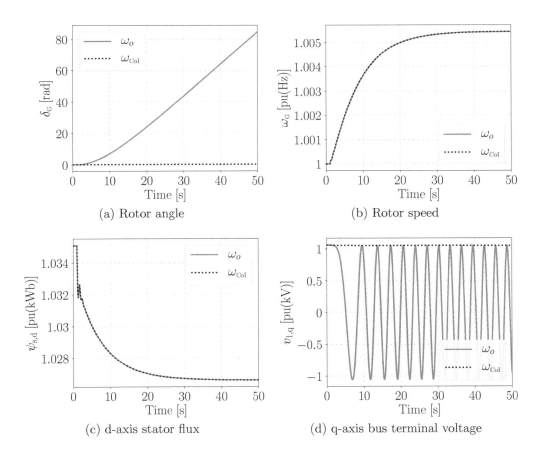

(a) Rotor angle

(b) Rotor speed

(c) d-axis stator flux

(d) q-axis bus terminal voltage

Figure 2.11 IEEE 14-bus system – Comparison of the dynamic response of the QSS model using ω_o or ω_{CoI} as reference frequency for the synchronous machines. All variables refers to the synchronous machine connected to bus 1.

invariant with respect to the reference frequency of the system. This is because dq-axis components of the rotor and stator variables *within* the machine are also referred to the reference frame rotating with angular speed $\omega_{G,1}$. The only variable of the machine that is affected by the choice of the reference frame is the rotor angle $\delta_{G,1}$.

On the other hand, the voltages of the network buses vary greatly. Due to the lack of regulation, following the line outage, the frequency of the machines do not recover the nominal synchronous speed and, in stationary conditions, rotor speeds show a constant frequency variation with respect to the reference ω_o. This explains the drift of the rotor angle $\delta_{G,1}$ and the periodic variations of the terminal bus voltage shown in Figure 2.11.[7] If ω_{CoI} is used as reference frequency, machine rotor angles and the network voltages are constant in stationary conditions. These results have also a relevant impact on the performance of the simulation. The property of CoI-based models to avoid the drift of generator angles can be exploited to increase the integration step and thus improve efficiency [66].

In the remainder of the book, all simulations, unless indicated otherwise, are solved considering QSS models and using ω_{CoI} as reference angular frequency.

---

[7]See also the example in Section 1.4.4.4 for a discussion on the periodic variations of the dq-axis components of the Park transform.

Chapter 3
Dynamic State Estimation

3.1. Basic Concepts

State estimation is a basic area of power system analysis and a fundamental tool for transmission and distribution system operators. While its formulation is relatively simple and, in turn, reduces to an optimization problem, several practical issues, such as the identification of bad data, are challenging and have generated a vast literature.

This section only describes the basic formulation and provides relevant definitions that are functional to the discussions given in Chapter 8. The reader can refer to [143] and, more recently, to [2], and in the literature reviews therein, for a comprehensive introduction to state estimation.

State estimation consists in the determination of the "state" of the system given a set of measurements, as follows:

$$\text{minimize:} \quad J(\varepsilon) = \frac{1}{2} \varepsilon^\dagger \mathbf{U}^{-1} \varepsilon \qquad (3.1)$$
$$\scriptsize (\varepsilon, z)$$

$$\text{subject to:} \quad \mathbf{0}_{n_\eta, 1} = \boldsymbol{\eta}(z) - \tilde{\boldsymbol{\eta}} + \varepsilon \quad : \boldsymbol{\mu}\,, \qquad (3.2)$$

where z ($z \in \mathbb{R}^{n_z}$) are the states to be estimated; $\tilde{\boldsymbol{\eta}}$ ($\tilde{\boldsymbol{\eta}} \in \mathbb{R}^{n_\eta}$) are the measurements; $\boldsymbol{\eta}$ ($\boldsymbol{\eta} : \mathbb{R}^{n_z} \mapsto \mathbb{R}^{n_\eta}$) are nonlinear functions that relate the system states to the measurements; ε ($\varepsilon \in \mathbb{R}^{n_\eta}$) are the measurement errors (often also called *residuals*); $\boldsymbol{\mu}$ ($\boldsymbol{\mu} \in \mathbb{R}^{n_\eta}$) are the dual variables associated with the equality constraints; and $\mathbf{U}$ ($n_\eta \times n_\eta$) is the covariance matrix whose inverse is the weight matrix of the measurement errors.

It is worth noticing that, in state estimation, the vector of states z includes both the state variables (x) and the algebraic variables (y) that appear in the set of DAEs given in Section 2.3. Hence, $z = (x, y)$. Since in the semi-implicit formulation (2.15), algebraic variables can be interpreted as states with infinitely fast dynamics, the notation of (2.15) and (3.2) is formally consistent.

Conventional, although not always necessarily verified assumptions are:

- Each element of the vector of measurement errors ε has null expectation, namely $\mathcal{E}(\varepsilon_i) = 0$, $\forall i = 1, 2, \ldots, n_\eta$.

Frequency Variations in Power Systems: Modeling, State Estimation, and Control, First Edition.
Federico Milano and Álvaro Ortega Manjavacas.
© 2020 John Wiley & Sons Ltd. Published 2020 by John Wiley & Sons Ltd.

- The measurement errors follow a Gaussian distribution and are independent, i.e. $\mathcal{E}(\varepsilon_i \varepsilon_j) = 0$ for $i \neq j$.

- The covariance matrix is diagonal, i.e. $\mathbf{U} = \text{diag}\{\sigma_1^2, \sigma_2^2, \ldots, \sigma_{n_\eta-1}^2, \sigma_{n_\eta}^2\}$, where $\sigma_i^2 = \mathcal{E}(\varepsilon_i^2)$ is the variance of the i-th measurement error ε_i.

While not strictly necessary for the developments described in Chapter 8, the assumptions above are also implied in this book.

The Lagrangian function $\mathcal{L}(z, \varepsilon, \mu)$ associated with the optimization problem (3.1) and (3.2) is:

$$\mathcal{L}(z, \varepsilon, \mu) = J(\varepsilon) - \mu^\dagger \left[\eta(z) - \tilde{z} + \varepsilon \right] . \tag{3.3}$$

Considering the optimal solution $(\check{z}, \check{\varepsilon}, \check{\mu})$ of problem of (3.1) and (3.2), the first order KKT optimality conditions are:

$$\mathbf{0}_{n_z, 1} = \left. \frac{\partial \mathcal{L}}{\partial z} \right|_{(\check{z}, \check{\varepsilon}, \check{\mu})} = \mathbf{H}_z^\dagger(\check{z}) \, \check{\mu} \ , \tag{3.4}$$

$$\mathbf{0}_{n_\eta, 1} = \left. \frac{\partial \mathcal{L}}{\partial \varepsilon} \right|_{(\check{z}, \check{\varepsilon}, \check{\mu})} = \mathbf{U}^{-1} \check{\varepsilon} - \check{\mu} \ , \tag{3.5}$$

$$\mathbf{0}_{n_\eta, 1} = \left. \frac{\partial \mathcal{L}}{\partial \mu} \right|_{(\check{z}, \check{\varepsilon}, \check{\mu})} = \eta(\check{z}) - \tilde{\eta} + \check{\varepsilon} \ . \tag{3.6}$$

3.1.1. Weighted Least Squares Problem

In general, the vector of measurement functions η is nonlinear and the Jacobian matrix $\mathbf{H}_z$ has nonconstant elements. Thus, the solution of (3.4)–(3.6) requires an iterative technique, such as the Newton method or more sophisticated techniques, such as the interior-point method [126]. However, the absence of inequalities enables the simplification of the solution of the optimization problem (3.1) and (3.2) without the need to involve dual variables.

The optimality conditions (3.4)–(3.6) can be rewritten in the following compact form:

$$\mathbf{0}_{n_z, 1} = \mathbf{H}_z^\dagger(\check{z}) \, \mathbf{U}^{-1} \left[\tilde{\eta} - \eta(\check{z}) \right] , \tag{3.7}$$

which is obtained by substituting the expression of $\check{\varepsilon}$ from (3.6) into (3.5) and then substituting the obtained expression for $\check{\mu}$ into (3.4).

Equation (3.7) is the well-known weighted least squares (WLS) state estimation problem and can be conveniently solved using an iterative method, such as the Newton method. Since the factorization of the Jacobian matrix of equation (3.7) can be numerically challenging due to the large diagonal elements of $\mathbf{U}^{-1}$, some robust factorization technique is often used, e.g. the QR factorization. The reader can refer to [2] for details on the numerical challenges of solving (3.7).

The special case of linear measurement functions is relevant for the state estimation problem discussed in Chapter 8:

$$\eta(z) = \mathbf{H}\, z \;, \tag{3.8}$$

where $\mathbf{H}$ has constant elements. Then the optimality conditions (3.7) become:

$$\mathbf{0}_{n_z,1} = \mathbf{H}^\dagger \mathbf{U}^{-1}\left[\tilde{\eta} - \mathbf{H}\,\check{z}\right] \;, \tag{3.9}$$

which can be solved with a direct calculation:

$$\check{z} = \left[\mathbf{H}^\dagger \mathbf{U}^{-1}\mathbf{H}\right]^{-1}\left[\mathbf{H}^\dagger \mathbf{U}^{-1}\tilde{\eta}\right] \;. \tag{3.10}$$

Assuming that the diagonal elements of $\mathbf{U}$ are equal, i.e. all measurements are retrieved using devices and/or techniques of the same technology and configuration, the solution of (3.10) reduces to that of a linear least square problem:

$$\check{z} = \left[\mathbf{H}^\dagger \mathbf{H}\right]^{-1}\mathbf{H}^\dagger \tilde{\eta} = \mathbf{H}^+ \tilde{\eta} \;, \tag{3.11}$$

where $\mathbf{H}^+$ is the Moore-Penrose pseudo-inverse, or *left inverse*, of $\mathbf{H}$, which is unique if $\mathbf{H}$ has rank n_η.

3.1.1.1. Example

Table 3.1 shows a set of measurements for the WSCC system. All measurements have a standard deviation $\sigma_i = 0.01$ in per unit with respect to system bases, except for the active and reactive power measurements at the transit buses 4, 7 and 9, which are "pseudo-measurements," i.e. quantities known *a priori*. For these quantities, the standard deviation is $\sigma_i = 10^{-5}$ pu.

Table 3.2 shows the solution obtained by solving problem (3.7) for the WSCC with the data of Table 3.1 and assuming that the phase angle of bus 1 is the reference. The actual solution of the power flow problem is also included in Table 3.2 for comparison.

3.1.2. Detection of Bad Data

An important area of research in state estimation is the definition of robust algorithms that are able to identify bad data and topological errors. With this aim, the largest normalized residual and the χ^2 (*chi squared*) tests are simple and commonly used techniques.

Table 3.1 Measurements for the WSCC system.

Bus #	$\tilde{v}_h$ [pu(kV)]	$\tilde{p}_h$ [pu(MW)]	$\tilde{q}_h$ [pu(MVAr)]	Line #	$\tilde{p}_{hk}$ [pu(MW)]
1	1.0424	0.7221	0.2747	9-8	0.2442
2	1.0203	1.6261	0.0710	7-8	0.7622
3	1.0291	0.8599	−0.0895	9-6	0.6160
4	1.0197	0	0	7-5	0.8786
5	0.9975	−1.2687	−0.4967	5-4	−0.4027
6	1.0206	−0.8874	−0.3003	6-4	−0.3044
7	1.0379	0	0	2-7	1.6241
8	1.0032	−0.9908	−0.3233	3-9	0.8350
9	1.0214	0	0	1-4	0.7202

Table 3.2 Results of the state estimation problem for the WSCC system.

	Base case solution				State estimation solution			
Bus #	v_h [pu(kV)]	θ_h [rad]	p_h [pu(MW)]	q_h [pu(MVAr)]	$\breve{v}_h$ [pu(kV)]	$\breve{\theta}_h$ [rad]	$\breve{p}_h$ [pu(MW)]	$\breve{q}_h$ [pu(MVAr)]
1	1.0400	0	0.7164	0.2705	1.0351	0	0.7187	0.2692
2	1.0250	0.1620	1.63	0.0665	1.0215	0.1630	1.6298	0.0636
3	1.0250	0.0814	0.85	−0.1086	1.0231	0.0819	0.8493	−0.0977
4	1.0258	−0.0387	0	0	1.0209	−0.0392	0	0
5	0.9956	−0.0696	−1.25	−0.50	0.9906	−0.0711	−1.2642	−0.5033
6	1.0127	−0.0644	−0.90	−0.30	1.0081	−0.0646	−0.8958	−0.3068
7	1.0258	0.0649	0	0	1.0225	0.0653	0	0
8	1.0159	0.0127	−1.00	−0.35	1.0137	0.0131	−0.9907	−0.3318
9	1.0324	0.0343	0	0	1.0299	0.0346	0	0

Largest Normalized Residual Test

The normalized residue vector is calculated at the optimal solution of the optimization problem (3.1)–(3.2) as:

$$\breve{\varepsilon}_N = \frac{\breve{\varepsilon}}{\sqrt{\mathbf{S}(\breve{z})}} \ , \tag{3.12}$$

where $\breve{\varepsilon}$ is the residue vector:

$$\breve{\varepsilon} = \tilde{\eta} - \eta(\breve{z}) \ , \tag{3.13}$$

and $\mathbf{S}(\breve{z})$ is the residue covariance matrix:

$$\mathbf{S}(\breve{z}) = \mathbf{U} - \mathbf{H}_z(\breve{z}) \left[\mathbf{H}_z^\dagger(\breve{z})\mathbf{U}^{-1}\mathbf{H}_z(\breve{z}) \right]^{-1} \mathbf{H}_z^\dagger(\breve{z}) . \tag{3.14}$$

The test consists in verifying the condition:

$$\max\{\breve{e}_N\} > \varepsilon_{\text{thres}} . \tag{3.15}$$

If true, the measurement associated with the maximum value of $\breve{e}_N$ is discarded and the problem (3.1)–(3.2) is solved again. A common choice for the threshold is $\varepsilon_{\text{thres}} = 3$.

Chi Squared Test

The test is based on the scalar performance index defined as:

$$\text{SPI} = \breve{z}^\dagger\mathbf{U}^{-1}\breve{z} , \tag{3.16}$$

which is compared with the threshold value $\text{SPI}_{\text{thres}}$ returned by the percent point function of the χ^2 distribution with a given percentage, e.g. 95%, and shape factor given by $n_\eta - n_z$. If $\text{SPI} > \text{SPI}_{\text{thres}}$, then the test fails and some bad data is present in the set of measurements.

3.1.2.1. Example

Introducing an erroneous measurements in Table 3.1, say $\tilde{v}_9 = 1.3$ pu, the tests discussed above give:

$$\max\{\breve{e}_N\} = \breve{e}(\tilde{v}_9) = 25.8 > 3 = \varepsilon_{\text{thres}} ,$$

$$\text{SPI} = 685.25 > 25 = \text{SPI}_{\text{thres}} ,$$

where $\varepsilon_{\text{thres}} = 3$ is a conventional common choice for the maximum residue and $\text{SPI}_{\text{thres}} = 25$ is obtained from (3.16). Once the measurement is removed from the measurement set, the tests return the following values:

$$\max\{\breve{e}_N\} = 2.23 < 3 = \varepsilon_{\text{thres}} ,$$

$$\text{SPI} = 17.89 < 25 = \text{SPI}_{\text{thres}} .$$

3.1.3. Phasor Measurement Units

Phasor measurement units (PMUs) have led to a sort of small revolution in power system metering as, for the first time, not only the magnitudes of phasors but also the phase angles can be measured. This is possible thanks to the global

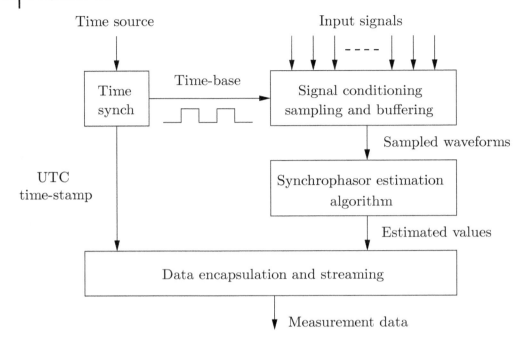

Figure 3.1 Main components of a PMU [135].

positioning system (GPS) owned by the United States government, operated by the United States Air Force and, since the 1980s, made available for civilian use. The GPS allows any device to have a precise and unique time reference, which is crucial to define the reference phase of phasors in a power system. Figure 3.1 shows the main components of a PMU (see Chapter 3 of [135]).

With the advent of the PMU technology, the state estimation has gained the ability to acquire the measurements of bus voltage and current phase angles. This has led to a variety of new state-estimation models, mostly coupled with the extended Kalman filter (EKF) [74, 75, 92, 140]. In [67], frequency measurements from PMUs are also utilized.

The ability to measure the phase angle of phasors comes also with the ability to estimate the frequency of such phasors. This is the feature that is mostly relevant in this book, as the measurement of the frequency can be utilized for both dynamic state estimation (DSE) and control and is thoroughly discussed in Section 3.3. The remainder of this section exclusively and briefly discusses the utilization of PMUs for static state estimation and, in particular, their ability to measure phase angles.

In an ideal scenario where PMUs are installed at all system buses, one has the ability to define a set of measurements that lead to a *linear* state estimation problem. This is true, of course, only if PMU measurements are precise enough, which is often not the case and is currently the main issue that operators face when dealing with PMU measurements. A vast literature is dedicated to the definition of robust techniques to filter bad data and/or topology errors. Relevant works from where the interested reader can start an interesting journey into the world of PMU-based state estimation are [82, 185, 214, 226, 227].

3.1.3.1. Example

The state estimation of the WSCC system discussed in Section 3.1.1.1 is revisited in this example by assuming that a PMU is installed at every bus of the system. The PMUs measure both the magnitude and the phase angle of the bus voltage. Same measurements as those given in Table 3.1 are assumed plus the phase angle measurements indicated in Table 3.3. The standard deviation of the PMU phase angle measurements is $\sigma = 0.01$ rad of the actual phase angle values except for θ_1, which is the phase reference and is thus a pseudo-measurement with standard deviation 10^{-6} rad.

The results of the state estimation problem with inclusion of PMU data are given in Table 3.4. Note that the inclusion of additional measurements does not improve the results of the state estimation problem with respect to the results shown in Table 3.2. This is because the errors on the phase angles impact consistently on the solution. The normalized residual and the χ^2 tests return the following values:

$$\max\{\breve{\varepsilon}_N\} = 2.1889 < 3 = \varepsilon_{\text{thres}} \ ,$$

$$\text{SPI} = 21.89 < 35.17 = \text{SPI}_{\text{thres}} \ .$$

Table 3.3 PMU measurements for the WSCC system.

Bus #	$\tilde{\theta}_h$ [rad]
1	0
2	0.1698
3	0.0855
4	−0.0445
5	−0.0658
6	−0.0690
7	0.0588
8	0.0206
9	0.0217

Table 3.4 Results of the state estimation problem with PMU measurements for the WSCC system.

Bus #	Base case solution				State estimation solution			
	v_h [pu(kV)]	θ_h [rad]	p_h [pu(MW)]	q_h [pu(MVAr)]	$\breve{v}_h$ [pu(kV)]	$\breve{\theta}_h$ [rad]	$\breve{p}_h$ [pu(MW)]	$\breve{q}_h$ [pu(MVAr)]
1	1.0400	0	0.7164	0.2705	1.0350	0	0.7189	0.2692
2	1.0250	0.1620	1.63	0.0665	1.0213	0.1630	1.6299	0.0636
3	1.0250	0.0814	0.85	−0.1086	1.0230	0.0819	0.8492	−0.0975
4	1.0258	−0.0387	0	0	1.0208	−0.0392	0	0
5	0.9956	−0.0696	−1.25	−0.50	0.9905	−0.0711	−1.2641	−0.5032
6	1.0127	−0.0644	−0.90	−0.30	1.0079	−0.0646	−0.8961	−0.3066
7	1.0258	0.0649	0	0	1.0223	0.0653	0	0
8	1.0159	0.0127	−1.00	−0.35	1.0135	0.0131	−0.9907	−0.3317
9	1.0324	0.0343	0	0	1.0298	0.0346	0	0

3.2. Introducing Dynamics

DSE is a fundamental tool of energy management systems and control centers of transmission system operators. DSE can also be useful for short-term transient stability predictions and has become even more important and challenging with the recent development of the smart grid. This, in fact, typically requires faster and system-wider controls than traditional power systems. With the introduction of PMUs, which have a high sampling rate – up to 120 measurements per second – and accurate synchronization, has made possible the implementation of a fast and accurate DSE.

There is a vast literature on DSE. Traditional methods are based on a set of nonlinear DAEs that models the machines and the controllers of the system [77]. The nonlinearity of the model is a major challenge, as the solution of the state estimation problem is iterative and requires the calculation and factorization of the Jacobian matrix of the DAEs [91]. Another important aspect of state estimation and, hence, also of DSE, is how to process bad data. With this aim, several techniques, mostly based on the Kalman filter, have been proposed [13, 93]. Other filtering techniques include, for example, particle filter [232] and mean squared estimator [18]. Finally, decentralization of the state estimation is also important, especially for large systems [230].

A comprehensive treatise on DSE is beyond the scope of this chapter. However, it is useful to outline the basic formulation that enables the solution of the DSE problem. First, the DAE model of the system to be estimated is assumed to be known, as follows:

$$\Theta \frac{d}{dt} z(t) = l(z(t), u(t)) , \qquad (3.17)$$

where (3.17) represents a compact notation of (2.15) with $l = (f, g)$ and:

$$\Theta = \begin{bmatrix} \mathbf{T} & \mathbf{0}_{n_x, n_y} \\ \mathbf{R} & \mathbf{0}_{n_y, n_y} \end{bmatrix} . \qquad (3.18)$$

Then, the DAE in (3.17) is formulated through a discrete-time approximation, in the form:

$$\mathbf{0} = \lambda(z_k, z_{k-1}, u_k, u_{k-1}) , \qquad (3.19)$$

where z_k and u_k are the state and input variables, respectively, at the current time t_k; and z_{k-1} and u_{k-1} are the state and input variables, respectively, at the time $t_{k-1} = t_k - \Delta t$, where Δt is the time interval of the discretization.

Equation (3.19) can be set up using any numerical integration scheme. For example, applying the implicit trapezoidal method, one obtains:

$$\lambda(z_k, z_{k-1}, u_k, u_{k-1}) = \Theta (z_k - z_{k-1}) - \frac{1}{2} \Delta t (l_k - l_{k-1}) , \qquad (3.20)$$

where $l_k = l(z_k, u_k)$ and $l_{k-1} = l(z_{k-1}, u_{k-1})$.

Similarly to the standard state estimation, the DSE problem can be formulated as an optimization problem to be solved at every time step t_k, as follows:

$$\begin{aligned}
&\underset{(\xi_k, \varepsilon_k, z_k)}{\text{minimize:}} && J(\xi_k, \varepsilon_k) = \frac{1}{2} \left(\xi_k^\dagger \mathbf{Q}_k^{-1} \xi_k + \varepsilon_k^\dagger \mathbf{U}_k^{-1} \varepsilon_k \right) && (3.21) \\
&\text{subject to:} && \mathbf{0} = \lambda(z_k, z_{k-1}, u_k, u_{k-1}) + \xi_k && : \nu_k , && (3.22) \\
&&& \mathbf{0} = \eta(z_k) - \tilde{\eta}_k + \varepsilon_k && : \mu_k , && (3.23)
\end{aligned}$$

where (3.22) defines the *process* and (3.23) defines the *observations*; ν_k and μ_k are the process and observation dual variables, respectively; and ξ_k and ε_k are the process and observation noises, respectively, with covariance matrices $\mathbf{Q}_k$ and $\mathbf{U}_k$.

3.2.1. Extended Kalman Filter

Problem (3.21)–(3.23) can be solved with any nonlinear programming technique. However, by linearizing (3.22) and (3.23) around an estimate of the current mean and covariance, one obtains the well known EKF, which provides sub-optimal solutions. The global optimum can be obtained only for linear l and η functions, for which the EKF becomes the classical Kalman filter (and problem (3.21)–(3.23) becomes a linear quadratic estimation, thus showing a formal similarity with the χ^2 merit function approach).

For their relevance for DSE, it is useful to provide the steps to solve the EKF problem that aims at minimizing the effect of the noise. The starting point is the system:

$$0 = \lambda(z_k, z_{k-1}, u_k, u_{k-1}) + \xi_k \; , \tag{3.24}$$

$$0 = \eta(z_k) - \tilde{\eta}_k + \varepsilon_k \; . \tag{3.25}$$

Let $\check{z}_{n|m}$ represent the estimate of z after n given observations at time $m \le n$. Then the solution of the EKF consists in two steps: *predict* and *update*, as follows.

Predict

- Predicted state estimate: Determine $\check{z}_{k|k-1}$ by solving (3.20) with initial condition $\check{z}^0_{k|k-1} = \check{z}_{k-1|k-1}$ and applying an iterative solver, e.g. the Newton method, one has:

$$
\begin{aligned}
z^i_k &= z^{i-1}_k - [\Lambda^i_{z,k}]^{-1} \lambda(z^i_k, z_{k-1}, u_k, u_{k-1}) \\
&= z^{i-1}_k - \big[\Theta - \frac{1}{2}\Delta t L_z(z^i_k, u_k)\big]^{-1} \lambda(z^i_k, z_{k-1}, u_k, u_{k-1}) \; ,
\end{aligned}
\tag{3.26}
$$

where L_z is the Jacobian matrix of function l_k, and for $i = 1, 2, \ldots, i_{\max}$, with initial condition $z^0_k = z_{k-1}$ and stopping if $|z^i_k - z^{i-1}_k| < \epsilon$, with ϵ a given tolerance or, in case of nonconvergence, if $i > i_{\max}$.

- Predicted covariance estimate: $\Pi_{k|k-1} = \Lambda_{z,k}\Pi_{k-1|k-1}\Lambda^\dagger_{z,k} + Q_k$, where $\Lambda_{z,k}$ is the Jacobian matrix of function λ calculated at $(\check{z}_{k|k-1}, u_k)$.

Update

- Residue covariance: $S_k = H_k\Pi_{k|k-1}H^\dagger_k + U_k$.

- Near-optimal Kalman gain: $K_k = \Pi_{k|k-1}H^\dagger_k S^{-1}_k$.

- Updated state estimate: $\check{z}_{k|k} = \check{z}_{k|k-1} + K_k(\tilde{\eta} - \eta(\check{z}_{k|k-1}))$.

- Updated covariance estimate: $\Pi_{k|k} = (I_{n_z} - K_kH_k)\Pi_{k|k-1}$.

The continuous version of the EKF merges together the predict and update steps in a set of DAEs, as follows:

$$\Theta \frac{d}{dt}\check{z}(t) = l\big(\check{z}(t), u(t)\big) + \mathbf{K}(t)\big(\tilde{\eta}(t) - \eta(\check{z}(t))\big) ,$$

$$\frac{d}{dt}\Pi(t) = -\mathbf{\Lambda}_{\boldsymbol{z}}(t)\Pi(t) - \Pi(t)\mathbf{\Lambda}_{\boldsymbol{z}}(t)^\dagger - \mathbf{K}(t)\mathbf{H}_{\boldsymbol{z}}(t)\Pi(t) + \mathbf{Q}(t) , \qquad (3.27)$$

$$\mathbf{K}(t) = \Pi(t)\mathbf{H}_{\boldsymbol{z}}(t)^\dagger \mathbf{U}(t)^{-1} .$$

The DSE problem (3.21)–(3.23) and, hence, also the EKF have the same hypothesis and, additionally, require that the initial state z_o as well as the noises ε_k and ξ_k are mutually independent. The linearization around a single point is also a well-known limitation of the EKF, which can be solved using *sigma points* $\mathcal{X}$ and the unscented Kalman filter [219]. However, the most relevant limit of the whole Kalman filtering approach is that its estimation is model dependent. The set of equations and parameters that describe the process, in fact, have to be defined *a priori*.

3.2.1.1. Example

The classical model of a synchronous machine is (see also Section 2.4.1):

$$\omega_o^{-1}\frac{d}{dt}\delta_{\mathrm{G}}(t) = \omega_{\mathrm{G}}(t) - 1 ,$$

$$M\frac{d}{dt}\omega_{\mathrm{G}}(t) = p_{\mathrm{m}}(t) - p_{\mathrm{e}}(t) - D\big(\omega_{\mathrm{G}}(t) - \omega_o\big) , \qquad (3.28)$$

where p_{e} is the electromechanical power generated by the machine.

The owner of the power plant is able to measure the rotor speed $\tilde{\omega}_{\mathrm{G}}$, the mechanical power $\tilde{p}_{\mathrm{m}}$, and the generator power $\tilde{p}_{\mathrm{e}}$. The parameters of the machine, namely M and D, as well as the reference angular speed ω_o are known and constant. Then, using the notation of (3.17) and considering the continuous version of the EKF, the state vector is $z = (\delta_{\mathrm{G}}, \omega_{\mathrm{G}})$, the input vector is $u = (\tilde{p}_{\mathrm{m}}, \tilde{p}_{\mathrm{e}})$, the process is:

$$\Theta = \begin{bmatrix} \omega_o^{-1} & 0 \\ 0 & M \end{bmatrix} , \quad l = \begin{bmatrix} \omega_{\mathrm{G}}(t) - 1 \\ \tilde{p}_{\mathrm{m}}(t) - \tilde{p}_{\mathrm{e}}(t) - D\big(\omega_{\mathrm{G}}(t) - \omega_o\big) \end{bmatrix} , \qquad (3.29)$$

and the observation is:

$$0 = \omega_{\mathrm{G}}(t) - \tilde{\omega}_{\mathrm{G}}(t) + \varepsilon(t) , \qquad (3.30)$$

where the observation equation is $\eta = \omega_{\mathrm{G}}$ and the measurement is $\tilde{\eta} = \tilde{\omega}_{\mathrm{G}}$.

Since both the process and the observation are linear, the EKF returns an optimal estimation and the matrices $\mathbf{\Lambda}_z$ and $\mathbf{H}_z$ in (3.27) are constant:

$$\mathbf{\Lambda}_z = \begin{bmatrix} \omega_o^{-1} & -0.5\,\Delta t \\ 0 & M + 0.5\,D\,\Delta t \end{bmatrix}, \quad \mathbf{H}_z = \begin{bmatrix} 0 & 1 \end{bmatrix}. \tag{3.31}$$

Figure 3.2 shows the dynamic response of the continuous EKF discussed above, which has been set up to estimate the angle position and the angular speed of the rotor of the synchronous machine connected to bus 1 of the WSCC system. The contingency is the loss of load connected to bus 5 at $t = 1$ s and the reconnection of the load after 100 ms.

The inputs $\tilde{p}_{m,1}$ and $\tilde{p}_{e,1}$ as well as the measurement of the rotor speed $\tilde{\omega}_{G,1}$ are assumed to be affected by independent errors with Gaussian distribution, zero mean and standard deviation $\sigma = 0.02$ pu. The same value is used for the standard deviations to calculate the elements of the matrices $\mathbf{Q}$ and $\mathbf{U}$. Note that, since $\mathbf{\Lambda}_z$, $\mathbf{H}_z$, $\mathbf{Q}$, and $\mathbf{U}$ are constant, so are $\mathbf{\Pi}$ and $\mathbf{K}$, whose equations can be thus solved at the equilibrium point and removed from the set of DAEs that defines the continuous EKF. In this example, $\mathbf{K} = [0 \quad 0.01057]^{\dagger}$.

Overall, the error between the estimated frequency $\breve{\omega}_{G,1}$ and the actual rotor speed $\omega_{G,1}$ is of the order of 10^{-5} pu. The EKF is thus able to return an estimation of the rotor angular speed about three orders of magnitude more precise than the available rotor speed measurement. Note also that the original model of the machine is of fourth order, while the process implemented in the EKF is a second order classical model and that the value of the damping D has been introduced with an error of 0.1 in the EKF process.

The catch of this example is the assumption that the measurements of the internal mechanical power as well as of the rotor angular speed are available.

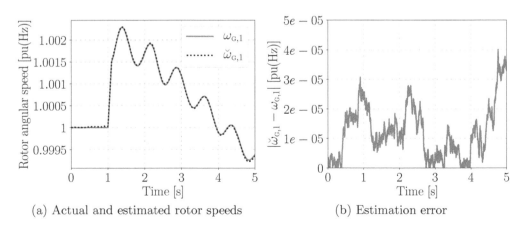

(a) Actual and estimated rotor speeds

(b) Estimation error

Figure 3.2 WSCC system – Dynamic response of the continuous EKF that estimates the rotor angle and speed of the synchronous machine at bus 1.

These measurements can be obtained only by the owner of the power plant. Chapter 8 shows how the rotor speed of the machine can be estimated without any knowledge of the machine model or internal measurements, based on the theoretical developments given in Part II.

3.3. Estimation of Bus Frequencies

The conventional power system model discussed in Chapter 2 is based on the assumption of quasi-steady-state (QSS) phasors for voltages and currents. The crucial hypothesis on which such a model is defined is that the frequency required to define all phasors and system parameters is constant and equal to its reference value ω_o. Trying to determine local frequency variations in the QSS model, thus, appears somewhat "inconsistent" as, in such a model, the frequency is assumed to be constant everywhere in network.

Nevertheless, using an electromagnetic transients (EMT) model instead of the QSS one, while certainly retains *all* information, does not appear as a great solution either. It is true that, in real-world applications, measurements are closer to the EMT quantities than to the phasors of QSS models, but it is also true that such an information has to be extracted and properly filtered. Defining local frequencies is thus not any simpler for the EMT model than for the QSS one.

Despite its intrinsic inconsistencies, the QSS model has been and still is very successful to carry out angle and voltage stability analyses. Until very recently, in fact, synchronous machines were the only devices available to regulate the system frequency through standard primary and secondary frequency regulators. Hence, the only frequency variations of interest in power system models were those of the rotor speeds of synchronous machines and these are properly taken into account in the QSS model. In recent years, however, an increasing number of devices other than synchronous machines are expected to provide frequency regulation. These include, among others, distributed energy resources such as wind and solar generation [6, 15, 125, 145, 176]; flexible loads providing load demand response [124, 183]; high-voltage direct current (HVDC) transmission systems [28, 139, 204]; and energy storage devices [119, 132, 200, 225].

Whenever these devices include a frequency control there is the need, both in practice and in simulations, to define the local frequency at the point of connection of the device with the grid. This section discusses the techniques that are currently available to estimate such frequencies. With this aim, three approaches are considered: (i) the numerical derivative of the phase angle; (ii) phase-locked loop (PLL) devices; and (iii) methods based on the DFT. These methods can be used both in real-world applications and in simulations.

3.3.1. Numerical Derivative of Phase Angles

The time derivative of the phase angle of a QSS phasor is an estimation of the frequency variations of such a phasor. In simulations, the phase angle is either directly available if using polar coordinates or can be easily deduced from the dq-axis components of the Park vectors. In real-world applications, there exist PMUs that calculate the frequency through the numerical derivative of the phase angle [171]. Two simple approaches are discussed below.

In the following, the determination of the frequency variations is assumed to be an *indirect measurement* rather than an estimation, as it is not the result of an optimization problem, hence the notation "$\Delta\tilde{\omega}$."

3.3.1.1. Rough Numerical Derivative

The most intuitive and simple method to calculate per-unit frequency variations based of a voltage phasor at a bus h is to approximate it by means of the time derivative with finite differences:

$$\Delta\tilde{\omega}_h[n\Delta t] \approx \omega_o^{-1} \frac{\theta_h[n\Delta t] - \theta_h[(n-1)\Delta t]}{\Delta t} , \qquad (3.32)$$

where Δt is the sampling time of θ_h or the time step of the numerical integration; ω_o is the reference angular frequency in rad/s; and $\Delta\tilde{\omega}_h$ is the sought frequency variation in pu(rad/s). It is worth recalling that, in the QSS model (2.15), the phase angles θ_h of voltage phasors are shifted with respect to a reference. Since the reference is the frequency of the CoI (see the discussion at the end of Section 2.4.4), using the notation of (1.2) and (2.59), one has:

$$\vartheta_h(t) = \delta_{\text{CoI}}(t) + \theta_h(t) , \qquad (3.33)$$

and, hence:

$$\tilde{\omega}_h[n\Delta t] \approx \omega_{\text{CoI}}[n\Delta t] + \Delta\tilde{\omega}_h[n\Delta t] . \qquad (3.34)$$

The calculation of (3.32) is straightforward if a discrete time series of θ_h is available, e.g. in numerical time domain integration. This, however, is quite a rough approach as noise and sudden variations of the phase generate "spikes" that are not physical variations of the frequency. While noise can be removed in mathematical models such as (2.15), jumps are unavoidable as θ_h is an algebraic variable and, thus, it shows sudden variations after contingencies and discrete events.

Figures 3.3.a and 3.3.b show the trajectory of the voltage phase angle at bus 5 of the WSCC system and its numerical derivative calculated as in (3.32) following a disconnection of the load at the same bus at $t = 1$ s and its reconnection after 100 ms. As expected, the phase angles show two discontinuities at $t = 1$ s and $t = 1.1$ s. These lead to the spikes of the measured frequency $\Delta\tilde{\omega}_5$.

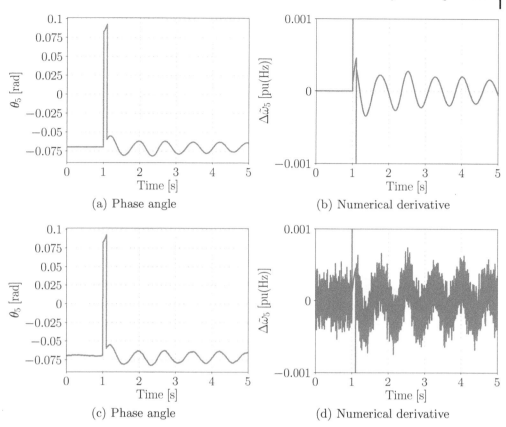

(a) Phase angle

(b) Numerical derivative

(c) Phase angle

(d) Numerical derivative

Figure 3.3 WSCC system – Phase angle of the voltage at bus 5 and its rough numerical derivative: (a) and (b) connection and disconnection of the load at bus 5; and (c) and (d) connection and disconnection of the load at bus 5 and stochastic perturbation of the load.

The amplitude of these spikes depends on the time step of the simulation. For example, for a time step $\Delta t = 0.001$, the spikes have an amplitude of about 100 rad/s. Figures 3.3.c and 3.3.d show the same simulation with inclusion of a stochastic perturbation of the power consumption of the load at bus 5. The noise, while almost negligible in the signal of the phase angle, greatly affects the calculation of its numerical derivative.

3.3.1.2. Washout Filter

A simple approach to mitigate the issues of the brute-force numerical derivative discussed above is the utilization of a washout filter (WF). This approach was first discussed in [97] along with the CoI model, and is commonly used in proprietary software tools for power system simulation, e.g. [48]. The scheme of the WF with inclusion of a low-pass filter (LPF) is shown in Figure 3.4.

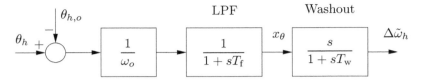

Figure 3.4 Scheme of low-pass and washout filters to calculate the numerical derivative of a bus voltage phase angle.

The differential equations of the WF are as follows:

$$T_\text{f}\frac{d}{dt}x_\theta(t) = \omega_o^{-1}\big(\theta_h(t) - \theta_{h,o}\big) - x_\theta(t) \,, \tag{3.35}$$

$$T_\text{w}\frac{d}{dt}\Delta\tilde{\omega}_h(t) - \frac{d}{dt}x_\theta(t) = -\Delta\tilde{\omega}_h(t) \,, \tag{3.36}$$

where $\theta_{h,o}$ is the initial bus voltage phase angle, e.g. the phase angle as obtained with the power flow analysis; ω_o is the system reference frequency in rad/s and is included to make the output in per unit; T_f and T_w are the time constants of the LPF and the WF, respectively; x_θ is the filtered phase angle; and $\Delta\tilde{\omega}_h$ is the estimated frequency variation at bus h.

If using dq-axis coordinates, the bus voltage phase angle θ_h has to be defined first. Instead of computing directly θ_h, which might lead to numerical issues, one can define two fictitious state variables, namely $s_{\theta,h}$ and $c_{\theta,h}$, whose dynamics are defined as follows [48]:

$$T_\text{f}\frac{d}{dt}c_{\theta,h}(t) = \frac{v_{h,\text{q}}(t)}{v_h(t)} - c_{\theta,h}(t) \,, \tag{3.37}$$

$$T_\text{f}\frac{d}{dt}s_{\theta,h}(t) = \frac{v_{h,\text{d}}(t)}{v_h(t)} + s_{\theta,h}(t) \,, \tag{3.38}$$

where $v_h = \sqrt{v_{h,\text{d}}^2 + v_{h,\text{q}}^2}$ and, according to the notation of the Park transform used in Section 1.4.4.4:

$$v_{h,\text{d}}(t) = -v_h(t)\sin\big(\theta_h(t)\big) = -v_h(t)s_{\theta,h}(t) \,,$$
$$v_{h,\text{q}}(t) = v_h(t)\cos\big(\theta_h(t)\big) = v_h(t)c_{\theta,h}(t) \,. \tag{3.39}$$

Then, $\Delta\tilde{\omega}_h$ is obtained from:

$$T_\text{w}\frac{d}{dt}\Delta\tilde{\omega}_h(t) = \begin{cases} \dfrac{1}{\omega_o\,c_{\theta,h}(t)}\dfrac{d}{dt}s_{\theta,h}(t) - \Delta\tilde{\omega}_h(t) \,, & \text{if } |c_{\theta,h}| > |s_{\theta,h}| \,, \\[2mm] \dfrac{-1}{\omega_o\,s_{\theta,h}(t)}\dfrac{d}{dt}c_{\theta,h}(t) - \Delta\tilde{\omega}_h(t) \,, & \text{otherwise.} \end{cases} \tag{3.40}$$

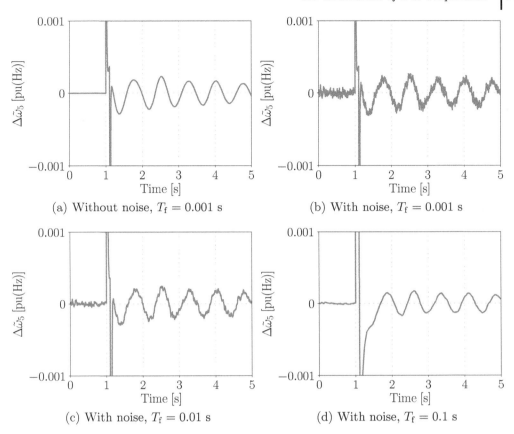

(a) Without noise, $T_f = 0.001$ s

(b) With noise, $T_f = 0.001$ s

(c) With noise, $T_f = 0.01$ s

(d) With noise, $T_f = 0.1$ s

Figure 3.5 WSCC system – Numerical derivative of the phase angle of the voltage at bus 5 through low-pass and washout filters with and without noise and with $T_w = 3/\omega_o$ s and different values of the LPF time constant T_f.

Figure 3.5 shows the output of the WF for the same system and contingency considered in Figure 3.3. The LPF mitigates the numerical issues due to noise and jumps of the phase angle but it also introduces a delay. A trade-off between accuracy and numerical efficiency of the WF is obtained with $T_f = 0.05$ s and $T_w = 3/\omega_o$ s. These are the values used in the simulations of this book.

The rationale of the trade-off between accuracy and effectiveness to filter the noise of the WF can be formally justified as follows. Deriving equation (3.35) with respect to time[1] and substituting the expression of x_θ from (3.36)

---

[1]The phase angle θ_h is an algebraic variable and, as illustrated in Section 3.3.1.1, it is not smooth for all t. It is smooth, however, almost everywhere, except for the instants at which a discrete event occurs. Its derivative can be thus calculated everywhere except for a finite number of points.

into (3.35) leads to:

$$\omega_o^{-1}\frac{d}{dt}\theta_h(t) = \Delta\tilde{\omega}_h(t) + (T_w + T_f)\frac{d}{dt}\Delta\tilde{\omega}_h(t) + T_w T_f\frac{d^2}{dt^2}\Delta\tilde{\omega}_h(t) . \qquad (3.41)$$

Equation (3.41) indicates that the ability of the WF to estimate frequency variations highly depends on the values of T_w and T_f. The smaller these time constants, the closer is (3.41) is to the actual time derivative of the phase angle, but also, the higher the effect of noise and numerical issues. On the other hand, increasing the values of T_w and T_f reduces the effect of numerical spikes but, necessarily, decreases the accuracy of the estimation of the frequency variations by introducing delays.

3.3.2. Phase-Locked Loop

The main purpose of a PLL is the synchronization of power electronic converters connected to a three-phase AC grid. Since, in turn, a PLL is a closed-loop controller with inclusion of a filter, its implementation is not unique. In the literature, the main focus so far has been to propose and test PLL designs that properly filter harmonics, compensate unbalanced conditions and reduce the delay and the error with which the phase is tracked.

Recent publications have recognized the impact of PLLs in the regulation provided by nonsynchronous devices [90, 221], but also the potential instabilities that these devices can cause to electronic converters [19, 81, 233]. Reference [161] shows how fast flux dynamics and the delays introduced by PLLs can affect the ability of nonsynchronous devices to properly regulate the frequency. On the other hand, the noise of frequency signals can be filtered and is, generally, less harmful [60].

The basic scheme of the fundamental-frequency model of a PLL is shown in Figure 3.6. The main components are as follows.

- The phase detector (PD) measures the three-phase voltage vector at the bus of connection h, $\boldsymbol{v}_{h,abc}$, and calculates the phase angle difference, ϵ_θ, between the bus voltage phase angle, θ_h, and the phase angle of the dq

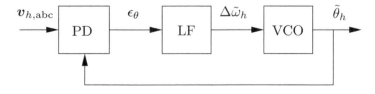

Figure 3.6 Basic scheme of a PLL.

frame obtained with the PLL, $\tilde{\theta}_h$. The voltage is converted from "abc" into "$\alpha\beta\gamma$" first and then into "dqo" components (see Sections 1.4.4.3 and 1.4.4.4, and [54]).

- The loop filter (LF) filters the error ϵ_θ. While there exist several configurations of the LF, they are generally based on a perfect tracking controller that, in steady state, imposes $\epsilon_\theta = 0$. Remarkably, the output of the LF is an indirect measurement of the frequency variation at the bus of connection, namely, $\Delta\tilde{\omega}_h$.

- The voltage controlled oscillator (VCO) takes the bus frequency variation, $\Delta\tilde{\omega}_h$, and provides the estimation of the dq-frame phase angle $\tilde{\theta}_h$. The VCO typically consists of an integrator that, in steady state, leads to $\Delta\tilde{\omega}_h = 0$.

In the conventional TS model, $\boldsymbol{v}_{h,\text{abc}}$ is not available, as the model consists in a balanced, fundamental frequency and QSS phasor representation of network branches and devices. The transformations operated within the PD are thus not explicitly implemented. Depending on the network model, however, either the components $v_{h,\text{d}}$ and $v_{h,\text{q}}$ or the polar representation in terms of voltage magnitude $v_h = |v_{h,\text{d}} + jv_{h,\text{q}}|$ and phase angle $\theta_h = \angle(v_{h,\text{d}} + jv_{h,\text{q}}) + \pi/2$ are available at the point of connection of the converter (see also equations (2.51) and (3.39)). In the following, polar coordinates for bus voltage phasors are used. But any component of the dq-frame representation of the grid voltage, namely $v_{h,\text{d}}$ or $v_{h,\text{q}}$, would work equally well. Note also that most PLL configurations are perfect tracking the reference phase, i.e. $\epsilon_\theta = 0$ in steady state, which, in turn, implies $v_{h,\text{d}} = 0$.

There are several PLL solutions specifically designed for power electronic converters. The synchronous reference frame PLL (SRF-PLL) is likely the simplest and the most commonly utilized scheme [150]. Other configurations are aimed at improving the SRF-PLL to reduce noise, distortions, and internal parameter uncertainties. Relevant implementations are the Lag PLL [29], the low-pass filter PLL [88], the enhanced PLL [102], and the second-order generalized integrator with frequency-locked loop [36, 177, 178]. The interested reader can find a comparison of the aforementioned PLL implementations in Chapter 7 of [142] and in [157].

A drawback of efficient noise filtering is the introduction of delays in the measurement of the phase and, consequently, of the frequency variations. In this regard, alternative solutions to PLLs for grid synchronization aimed at improving the trade-off between speed and noise filtering have been proposed in the literature [24]. In this reference, the authors propose an algorithm in the discrete-domain based on the well-known Kalman filter. The main drawback of these solutions is their generally high computational burden, which justifies the wide utilization of *conventional* PLLs.

The remainder of this section describes the SRF-PLL, which is widely used in the simulations of Parts II and III, and a continuous-time version of the PLL described in [24], which is a didactic application of the EKF discussed in Section 3.2.1. In the following, the SRF-PLL returns an indirect measurement, hence the notation $\Delta\tilde{\omega}_h$ for the output of such a device. On the other hand, the EKF-PLL is, *de facto*, a dynamic state estimator and, hence the notation $\Delta\breve{\omega}_h$ for its output.

3.3.2.1. Synchronous Reference Frame PLL

Figure 3.7 shows the commonly used SRF-PLL, where the PD includes a constant delay τ_θ; and the LF consists of an LPF and a proportional integral (PI) regulator.

The equations that describe the SRF-PLL shown in Figure 3.7 are:

$$T_{\mathrm{f}}\frac{d}{dt}x_{\mathrm{f}}(t) = \epsilon_\theta(t) - x_{\mathrm{f}}(t) = \theta_h(t - \tau_\theta) - \tilde{\theta}_h(t) - x_{\mathrm{f}}(t) , \qquad (3.42)$$

$$T_{\mathrm{I}}\frac{d}{dt}x_{\mathrm{I}}(t) = x_{\mathrm{f}}(t) , \qquad (3.43)$$

$$\Delta\tilde{\omega}_h(t) = x_{\mathrm{I}}(t) + K_{\mathrm{P}}\,x_{\mathrm{f}}(t) , \qquad (3.44)$$

$$\frac{d}{dt}\tilde{\theta}_h(t) = \omega_o\,\Delta\tilde{\omega}_h(t) . \qquad (3.45)$$

In turn, the SRF-PLL takes the phase angle of the measured voltage phasor as reference frame of the local dq-axis. In transient conditions, such a reference is tracked with error ϵ_θ:

$$v_{h,\mathrm{d}}(t) + jv_{h,\mathrm{q}}(t) = v_h(t)\left(-\sin(\epsilon_\theta(t)) + j\cos(\epsilon_\theta(t))\right) . \qquad (3.46)$$

The fidelity of the SRF-PLL can be readily shown. Assuming $T_{\mathrm{f}} \approx 0$, deriving (3.44) with respect to time, and substituting $\frac{d}{dt}x_{\mathrm{I}}$ and the expression of

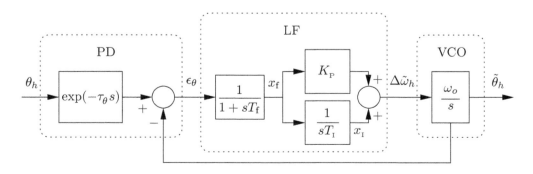

Figure 3.7 Scheme of an SRF-PLL.

$\frac{d}{dt}\tilde{\theta}_h$ in (3.45) into (3.42), one obtains:

$$w_o^{-1}\frac{d}{dt}\theta_h(t-\tau_\theta) = \Delta\tilde{w}_h(t) + \frac{w_o^{-1}}{K_\mathrm{P}}\frac{d}{dt}\Delta\tilde{w}_h(t) - \frac{w_o^{-1}}{K_\mathrm{P}T_\mathrm{I}}\,\epsilon_\theta(t)\,, \tag{3.47}$$

from which it follows that, apart from the delay τ_θ, the fidelity of the estimation of the frequency variations by means of the SRF-PLL is a trade-off. Assuming that the product $K_\mathrm{P}T_\mathrm{I}$ is constant, the higher K_P, the faster is the response of the SRF-PLL, but also the more sensitive the frequency estimation with respect to noise.

Note that, in the literature, the conventional SRF-PLLs model does not include an LPF. However, the LPF is often used due to its ability to filter noise. In the remainder of this book, unless otherwise indicated, it is assumed that $T_\mathrm{f} = 0$.

3.3.2.2. Enhanced Kalman Filter PLL

Assuming that one can measure $\tilde{\theta}_h$, the continuous version of the EKF can be written as a set of linear DAEs, which has thus gain matrix $\mathbf{K}$ – as well as constant covariance matrix $\mathbf{\Pi}$ – as follows:

$$\mathbf{\Theta}\frac{d}{dt}\breve{z}(t) = \mathbf{L}\,\breve{z}(t) + \mathbf{K}(t)\Big(\tilde{\eta}(t) - \eta\big(\breve{z}(t)\big)\Big)\,, \tag{3.48}$$

where $\breve{z} = (\breve{\theta}_h, \Delta\breve{w}_h)$, $\tilde{\eta} = \tilde{\theta}_h$, $\eta = \breve{\theta}_h$, $\mathbf{\Theta} = \mathbf{I}_2$ and

$$\mathbf{L} = \begin{bmatrix} 0 & w_o \\ 0 & 0 \end{bmatrix}, \quad \mathbf{K} = -\begin{bmatrix} \sqrt{2\,k_o\,w_o} \\ k_o \end{bmatrix},$$

where $k_o = \sigma_w/\sigma_\eta$ and σ_w and σ_η are the standard deviations of the noise of the process and the observation, respectively.

An advantage of the EKF with respect to conventional PLLs is that its order can be arbitrarily augmented by including higher order derivatives. For example, including the RoCoF, say $\alpha_h = \frac{d}{dt}\Delta w_h$, in the model of the process, one has: $\breve{z} = (\breve{\theta}_h, \Delta\breve{w}_h, \breve{\alpha}_h)$, $\tilde{\eta} = \tilde{\theta}_h$, $\eta = \breve{\theta}_h$, $\mathbf{\Theta} = \mathbf{I}_3$ and:

$$\mathbf{L} = \begin{bmatrix} 0 & w_o & 0 \\ 0 & 0 & 1 \\ 0 & 0 & 0 \end{bmatrix}, \quad \mathbf{K} = -\begin{bmatrix} \sqrt{2\,k_o\,w_o} \\ k_o \\ (\sigma_w\,\sigma_\alpha)^{-1} \end{bmatrix},$$

where σ_α is the standard deviation of the noise of the RoCoF equation.

Another advantage of the EKF is that the optimal gains are unequivocally determined, as opposed to the SRF-PLL where the parameters of the LF have to the tuned. However, the EKF approach requires the knowledge of the standard deviation of the process and observations.

3.3.2.3. Example

Figure 3.8 shows the output of the SRF-PLL and the EKF-PLL for different parameters and for the same system and contingency considered in Figures 3.3 and 3.5. Only the second order EKF-PLL is considered in this example as the third order model shows effectively same dynamic response. Similarly to the WF, the PLLs are able to filter the noise of the load, provided that their parameters are properly tuned. The trade-off is, as expected, between filtering too much thus introducing delays, and filtering too little and then having to deal with noisy signals. Ultimately, all PLL implementations show relatively similar behaviors. For simplicity, the SRF-PLL is used in all simulation with parameters $K_P \in [0.1, 0.2]$ and $T_I = 20$ s. This setup provides an acceptable trade-off between accuracy and numerical efficiency.

3.3.3. Enhanced DFT Techniques

The DFT is an effective and widely used technique to study stationary signals (see Section 1.4.2). For time-varying signals and, in particular, for signals with a time-varying frequency, the Fourier approach can still be applied but it requires some adjustments. The main idea is that if one takes a sliding window of samples of a given signal, then it is possible to follow the spectrum of variations of the signal in time. The result is a matrix, rather than a vector of harmonics. Each column of the matrix refers to a given window of samples and each row to the components of the spectrum of the DFT.

While there exist PMUs that calculate the frequency through the numerical derivative of the phase angle [171], most PMUs use some variant of the DFT to estimate the frequency. Before abandoning for good the Fourier approach, it is worth outlining the basic DFT-based technique that is utilized in practice to estimate frequency variations of QSS phasors. The principle is relatively simple but there are several issues that need to be solved when applying the DFT to a sliding window. These issues are briefly discussed below.

Aliasing

The number of samples in a given time period is a crucial parameter of the DFT to reconstruct the original signal. The reconstruction is accurate only if the bandwidth of the signal is smaller than the half of the frequency of the sampling rate of the signal itself. The rationale for this statement finds its roots in the well-known Nyquist-Shannon theorem [155]. For nonperiodic signals and, thus, also frequency-varying signals, this condition cannot be satisfied and, as a consequence, the DFT will inevitable distort the original signal. The choice of the sampling rate is thus crucial for the accuracy of the frequency estimation. In practice, this choice is a trade-off between computational burden, hardware limitations and required accuracy.

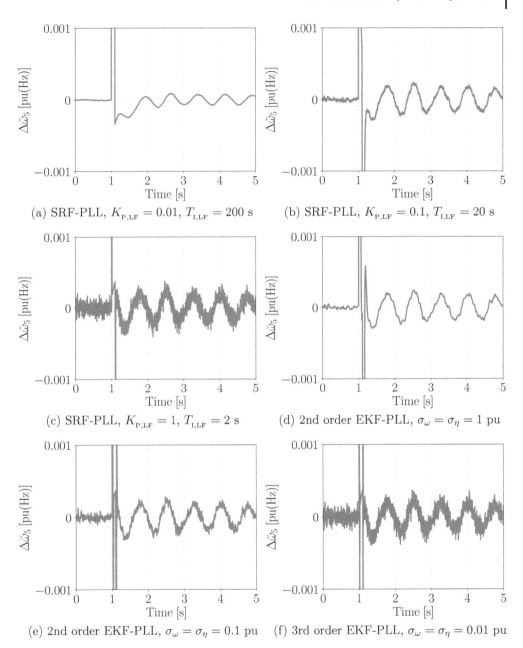

Figure 3.8 WSCC system – Performance of the SRF-PLL and EKF-PLL with different parameters.

Spectral Leakage

The sliding window with which a signal, say u, is sampled has a crucial role in the estimation process. This is also known as the *windowing* problem. In turn, one has to choose a function, say w, through which the original signal is clustered in a limited region of time. The DFT is applied not to the whole signal, but to its windowed version:

$$u_w(t) = w(t)\, u(t) \ . \tag{3.49}$$

The simplest function – but not the best – that one can use is a rectangular window. Applying the DFT to (3.49), one has:

$$\hat{u}\{k\omega_s\} = \frac{1}{W} \sum_{n=0}^{N-1} u_w[n\Delta t]\exp(-j\,k\,\omega_s\,n\Delta t), \quad k = 0, 1, 2, \ldots, N - 1 \ , \tag{3.50}$$

where:

$$W = \sum_{n=0}^{N-1} w[n\Delta t] \ , \tag{3.51}$$

and N is the number of samples of the window of length T, $n\Delta t$ is the time-step index and k is the DFT-bin index.

Spectral leakage appears when the chosen window does not contain an integer number of periods of the input signal. Since signals with time-varying frequency do not have a fixed period, spectral leakage is unavoidable for such signals. The length of the window is a crucial choice for the estimation of the frequency. The longer the window the more precise is the estimation for nonperiodic signals. On the other hand, the longer the window the fewer fast frequency variations of the signal can be captured. The window length has thus to be chosen carefully. It appears that the window profile can help and several window shapes (typically with decaying lobes) have been proposed, e.g. the Hann window.

Computational Burden

The DFT is the most computationally demanding part of the software included in the PMUs and can be the bottleneck for real-time DSE applications. Many studies have thus focused on techniques to reduce the number of computations required to determine the spectral content of the windowed signal u_w. With this aim, there are several approaches: short-time Fourier transform (STFT), modulated sliding DFT, interpolated DFT, enhanced interpolated DFT, etc. The starting point of each of these techniques is that the DFT can be updated at every new sample $n\Delta t$ using the information of the previous $N - 1$ samples.

For example, the STFT of (3.50) is:

$$\hat{u}_k\{k\omega_s\} = \sum_{m=0}^{N-1} u[(\nu + m)\Delta t] \exp(-j\,k\,\omega_s\,m\Delta t) , \qquad (3.52)$$

where $\nu = n - N + 1$.

3.3.3.1. Example

Consider the following voltage signal:

$$v(t) = \sqrt{2}V \sin\left(\omega_o t + \Delta\theta(t)\right) , \qquad (3.53)$$

where $\Delta\theta(t)$ is defined by the test function (1.15). According to the definition of frequency as the time derivative of the phase angle given in Chapter 1, the signal (3.53) has frequency:

$$w(t) = \omega_o + \Delta\omega(t) , \qquad (3.54)$$

where $\Delta\omega$ is defined in (1.69) and recalled below for clarity:

$$\Delta\omega(t) = \frac{d}{dt}\Delta\theta(t) = \Delta\theta_o\big[\omega_\theta \sin(\omega_\theta t) + \lambda_\theta\big(\cos(\omega_\theta t) - 1\big)\big]\exp(-\lambda_\theta t) . \quad (3.55)$$

Figure 3.9 shows the result of applying the Scipy function **spectrogram** – which, in turn, implements an STFT algorithm – with default parameters and the Hann window to the signal (3.53) with 50% window overlapping. The analytical expression (3.54) is also shown in Figure 3.9.d for comparison. The sampling rate of the measurements is 50 kHz and the parameters of the phase modulation function $\Delta\theta$ are $\Delta\theta_o = 2\pi 0.1$ rad, $\omega_\theta = 2\pi$ rad/s and $\lambda_\theta = 1$ Hz. The match between the STFT results and the analytical expression of the frequency increases as the number of samples per window increases, but so does the computational burden and the delay with which the measurement is available. As a byproduct, these results also confirm that the frequency of a signal *is* the derivative of its time-varying phase – see also the discussion in Section 1.1.

The STFT provides the main components of the DFT of the signal at discrete times. Depending on the sampling rate and the size of the window, however, results can be blurry due to the aliasing phenomenon. This can be corrected by means of some advanced technique, such as the interpolated DFT. Such an analysis goes beyond the scope of this example. The interested reader can refer to Chapter 3 of [135] for an in-depth discussion of advanced techniques to estimate time-varying frequencies based on the DFT approach.

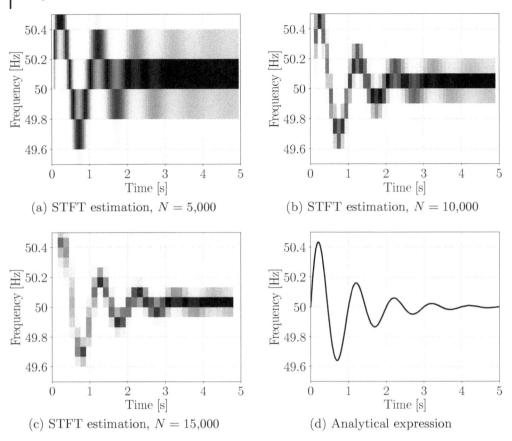

(a) STFT estimation, $N = 5,000$

(b) STFT estimation, $N = 10,000$

(c) STFT estimation, $N = 15,000$

(d) Analytical expression

Figure 3.9 Performance of the STFT for the determination of the time-varying frequency of the signal defined by the test function (1.15).

Even considering more sophisticated techniques than the STFT, the limitation of the Fourier approach is apparent (see also the discussion in Section 1.4.2). The Fourier transform essentially assumes "stationary" conditions and returns a "snapshot" of the harmonic content of a signal at a given time. The sliding window "fakes" such stationarity by moving the snapshot along the signal. Nevertheless, for a signal with time-varying frequency, aliasing and spectral leakage are unavoidable and only approximated "patches" can be found.

Finally, it is worth mentioning that, in general, not only the frequency but also the amplitude of the signal varies in time (see also footnote 4 in Chapter 1). With this kind of signals, the DFT and its variants do not give good results. An option, of course, is to look for other transforms, such as the Hilbert transform, which can achieve a higher accuracy than the DFT [71]. However, it is important to note that, any approach that attempts to reconstruct the harmonic contents of a general time-varying signal is arbitrary as one has to define *a priori* the mathematical expression of the signal to be reconstructed and then fit it with the measurements. Thus, independently from the transform, the fitting inevitably ends up being a sort of Procrustes' approach (see also the discussion in Section 1.1).

The approach discussed in Part II is not based on any frequency-domain analysis or transform and, as it appears, removes the arbitrariness of the definition of frequency variations in transient conditions.

Chapter 4
Frequency Control

4.1. Introduction

Despite their natural tendency to synchronize and maintain the system power balance by varying their kinetic energy, additional regulation is required to help synchronous machines recover an acceptable steady-state operating point following a contingency. Conventional power plants, in fact, are allowed to connect to the grid only if they provide frequency and voltage regulation.

The need for frequency control is twofold. The machines alone cannot recover synchronism following a large contingency that creates a power unbalance larger than the available inertia. Even if the power balance can be recovered, without control, the steady state following a contingency is characterized by a frequency variation with respect to the reference frequency and, the more severe the contingency, the bigger such a variation. Primary and secondary frequency controllers of synchronous machines cope with these issues and are discussed in Section 4.2.

Another intrinsic feature of synchronous machines is their oscillatory response following power unbalances. The electromechanical equations of the machine, in fact, resemble those of a *pendulum*. These oscillations affect the magnitude of electrical quantities as well as the variations of the frequency with respect to the reference value. Since the damping of the machines is necessarily small to reduce mechanical losses, such oscillations can last for several seconds unless proper control is implemented. In most cases, the primary control of the machines is enough to damp electromechanical oscillations. However, there are cases for which the automatic voltage regulator (AVR) and the electromechanical modes of the machine become resonant. This resonance is triggered by a high gain of the primary voltage control of the machines or, alternatively, by high loading conditions [218]. This is a well-known phenomenon that can be corrected using a power system stabilizer (PSS). This controller measures the rotor speed of the machine and modifies the reference voltage of the AVR to emulate a higher damping of the electromechanical modes. Section 4.3 discusses the dynamic response of AVRs and PSSs and their impact on frequency oscillations.

Frequency Variations in Power Systems: Modeling, State Estimation, and Control, First Edition. Federico Milano and Álvaro Ortega Manjavacas.

A decade ago, the controllers above would have been enough to complete a chapter on frequency control and damping of frequency oscillations. At most, one could have mentioned power oscillation dampers for flexible AC transmission system (FACTS) devices as an alternative solution to PSSs. In recent years, however, the penetration of nonsynchronous generation has forced system operators to look for alternative ways to provide primary frequency control. It appears that there are a plethora of available options: renewable generation, especially wind and photo-voltaic solar power plants; energy storage systems; and controllable loads. Section 4.4 outlines the control schemes of the nonsynchronous devices that are considered in the examples of Parts II and III.

4.2. Power Balancing

The main purpose of frequency regulation is to balance active power generation, load consumption, and network losses. In the first seconds after a contingency, the balance is attained by synchronous machines by means of their inertial response. The machines function as *flywheels*, and exchange or store kinetic energy with the system to counteract the shortage or excess of generation. This behavior can be observed in the simulation results discussed in Section 2.4.4.

Then, the primary frequency control (PFC) takes over the dynamic behavior of the system. The PFC consists in regulating the active power input of the synchronous machines through the turbine governor (TG) which, in turn, increases or decreases the volumetric flow of steam, water, or gas into the turbine connected to the shaft of the synchronous machine.

The TG acts in the range of a few seconds to a few tens of seconds and modifies the flow of the thermodynamic fluid in the turbine utilizing the *spinning reserve* of the turbines. The PFC, however, does not substantially change the power set point of the generators. This is done by the secondary frequency control (SFC), often called automatic generation control (AGC), which coordinates the power plants that belong to a certain area. The SFC is slower than the PFC by at least an order of magnitude, i.e. its action requires from a few tens of seconds to a few minutes.

The time scales and capacity of the inertial response of the synchronous machines, the PFC and the SFC is qualitatively illustrated in Figure 4.1.

While there exist several implementations, the purpose and basic dynamic behavior of the primary and secondary frequency regulators is common for all power plant technologies. The scope of this section is not to provide a comprehensive description of the models of existing primary and secondary frequency controllers. Rather, it describes simple conventional control schemes that are relevant for the examples discussed in the remainder of the book.

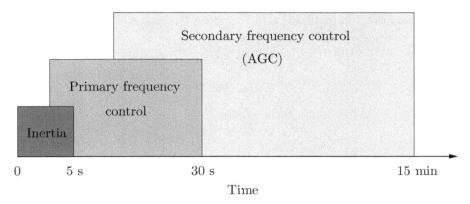

Figure 4.1 Time scales of inertial response and frequency control of conventional power plants based on synchronous machines. The area of the rectangles qualitatively indicates the relative amount of energy involved.

4.2.1. Primary Frequency Control

The PFC is effectively the TG model of synchronous machines. The PFC is mandatory for large conventional generators connected to a transmission system. This controller utilizes the power reserve of the machines to recover the power balance after a contingency and to reduce the frequency variation.

Figure 4.2 shows a typical, simplified scheme of the PFC and turbine of a steam power plant. It includes a governor, a servomotor and a reheater block. The set of per-unit DAEs that describes this model is:

$$T_g \frac{d}{dt} x_g(t) = p_{in}(t) - x_g(t) \ ,$$

$$T_{sm} \frac{d}{dt} x_c(t) - T_t \frac{d}{dt} x_g(t) = x_g(t) - x_c(t) \ ,$$

$$T_{rh} \frac{d}{dt} x_{rh}(t) = x_c(t) - x_{rh}(t) \ ,$$

$$-\kappa_{rh} T_{rh} \frac{d}{dt} x_{rh}(t) = x_{rh}(t) - \tau_m(t) \ ,$$

(4.1)

where p_{in} is the output of windup limiter, as follows:

$$p_{in}(t) = \mathcal{P}_{in}(\hat{p}_{in}(t)) = \begin{cases} \hat{p}_{in}(t) & \text{if } p^{min} \leq \hat{p}_{in}(t) \leq p^{max} \ , \\ p^{max} & \text{if } \hat{p}_{in}(t) > p^{max} \ , \\ p^{min} & \text{if } \hat{p}_{in}(t) < p^{min} \ , \end{cases}$$

(4.2)

where

$$\hat{p}_{in}(t) = p_G^{ref}(t) + \frac{1}{\mathcal{R}} \left(\omega_o - \omega_G(t) \right) \ .$$

(4.3)

Using the notation of (2.15) and dropping the dependency on time, the set (4.1)–(4.3) becomes:

$$\boldsymbol{x}^\dagger = [x_{\mathrm{g}}, x_{\mathrm{c}}, x_{\mathrm{rh}}] \ ,$$

$$\boldsymbol{y}^\dagger = [\tau_{\mathrm{m}}, \hat{p}_{\mathrm{in}}, p_{\mathrm{in}}] \ , \tag{4.4}$$

$$\boldsymbol{u}^\dagger = [\omega_{\mathrm{G}}, p_{\mathrm{G}}^{\mathrm{ref}}] \ ,$$

$$\boldsymbol{f} = \begin{bmatrix} p_{\mathrm{in}} - x_{\mathrm{g}} \\ x_{\mathrm{g}} - x_{\mathrm{c}} \\ x_{\mathrm{c}} - x_{\mathrm{rh}} \end{bmatrix} \ , \qquad \boldsymbol{g} = \begin{bmatrix} x_{\mathrm{rh}} - \tau_{\mathrm{m}} \\ p_{\mathrm{G}}^{\mathrm{ref}} + \dfrac{1}{\mathcal{R}}(\omega_o - \omega_{\mathrm{G}}) - \hat{p}_{\mathrm{in}} \\ \mathcal{P}_{\mathrm{in}}(\hat{p}_{\mathrm{in}}) - p_{\mathrm{in}} \end{bmatrix} \ , \tag{4.5}$$

$$\mathbf{T} = \begin{bmatrix} T_{\mathrm{g}} & 0 & 0 \\ -T_{\mathrm{t}} & T_{\mathrm{sm}} & 0 \\ 0 & 0 & T_{\mathrm{rh}} \end{bmatrix} \ , \qquad \mathbf{R} = \begin{bmatrix} 0 & 0 & -\kappa_{\mathrm{rh}} T_{\mathrm{rh}} \\ 0 & 0 & 0 \\ 0 & 0 & 0 \end{bmatrix} \ , \tag{4.6}$$

where $p_{\mathrm{G}}^{\mathrm{ref}}$ is the power set point of the TG as defined by the electricity market and adjusted by the AGC (see also Section 4.2.2); and other variables are defined in Figure 4.2. Table 4.1 defines the parameters of the turbine and TG.

All dynamics of the TG are relevant for the conventional QSS model and are generally retained. However, the TG model can be further simplified by setting $T_{\mathrm{g}} = T_{\mathrm{sm}} = T_{\mathrm{t}} = 0$, thus neglecting the dynamics of the governor and the servo motor and leaving only the lag compensator block that describes the reheater. When such a simplification is implemented, typical values for the lag compensator parameters are $T_{\mathrm{rh}} \approx 15$ s and $\kappa_{\mathrm{rh}} \approx 0.3$.

A lag compensator with similar values for T_{rh} and κ_{rh} is also suitable for modeling, as a very first approximation, hydroelectric TGs. Gas turbines, on the other hand, are more complex due to the control design to improve performance, increase efficiency and lower emissions. Their essential dynamic response, however, can be still be captured by a droop controller similar to that shown in Figure 4.2. The interested reader is referred to the IEEE Report [96] for a recent review of current TG modeling practices in industry.

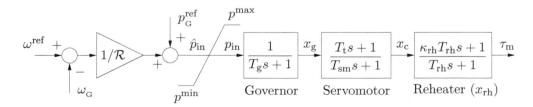

Figure 4.2 Scheme of a simple steam turbine governor model.

Table 4.1 Parameters of the steam turbine governor model.

Parameter	Description	Unit
p^{max}	Maximum turbine output	pu(MW)
p^{min}	Minimum turbine output	pu(MW)
$\mathcal{R}$	Droop of the turbine governor	pu(Hz/MW)
T_{g}	Governor time constant	s
T_{rh}	Reheater time constant	s
T_{sm}	Servomotor time constant	s
T_{t}	Transient gain time constant	s
κ_{rh}	Reheater fraction	–

Regardless of the complexity of the model, the common feature of all TGs is that they are based on a *droop control* that introduces a steady-state error in the system frequency after a disturbance. This happens on purpose, as the droop control enables the sharing of the power unbalance among all power plants proportionally to the capacity of their relative machines,[1] thus optimizing the utilization of the available spinning reserve. This result cannot be obtained if the PFC tracked perfectly the reference frequency, e.g. by means of PI controllers. Note also that, even if PI controllers were used in practice, the frequency error $\omega^{\mathrm{ref}} - \omega_{\mathrm{G}}$ is passed through a dead band, whose typical value is 0.0006 pu(Hz), to reduce the stress of the mechanical parts that compose the TG. Hence, PFC always introduces a steady-state frequency error which requires to be corrected. This is achieved with the SFC, whose model is described next.

4.2.2. Automatic Generation Control

The secondary frequency control, often called automatic generation control (AGC), coordinates the power plants of an area of the system. Its purpose is to reinstate the spinning reserve of the system after a contingency by redispatching the power set point of the turbine governors of the synchronous machines.

The AGC is generally perfectly tracking the reference frequency, i.e., it eliminates the steady-state frequency error introduced by the TGs after a disturbance. The AGC is a regional controller that measures the frequency of a pilot bus or an average regional frequency, compares it with a reference frequency and sends control signals to each TG of the synchronous machines. To be stable, the AGC has to be an order of magnitude slower than the PFC.

---

[1]This is true only if the values of $\mathcal{R}$ in per unit with respect to machine bases are similar across all power plants of the system.

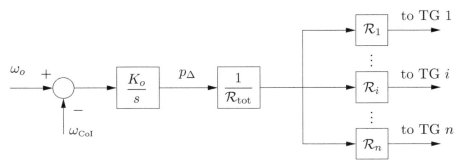

Figure 4.3 Simplified automatic generation control scheme.

The AGC scheme utilized in the examples of this book is depicted in Figure 4.3. The AGC includes an integral controller to guarantee that the frequency error is zero in steady state. The controller measures the angular frequency of the CoI (see Section 2.4.1.4) and compares it with the reference ω_o, whose value is 1 in per unit if ω_o is also the frequency base:

$$\frac{d}{dt} p_\Delta(t) = K_o \left(1 - \omega_{\mathrm{CoI}}(t) \right) , \tag{4.7}$$

where K_o is the integral gain. In the scheme of Figure 4.3, the AGC coordinates n TGs. Then:

$$\mathcal{R}_{\mathrm{tot}} = \sum_{i=1}^{n} \mathcal{R}_i , \tag{4.8}$$

where $\mathcal{R}_i$ is the droop of TG i. The signals $\left(\mathcal{R}_i / \mathcal{R}_{\mathrm{tot}} \right) p_\Delta$ are added to the $p_{\mathrm{G}}^{\mathrm{ref}}$ of the turbine governor. For example, if using an AGC, equation (4.3) becomes:

$$\hat{p}_{\mathrm{in}}(t) = p_{\mathrm{G}}^{\mathrm{ref}}(t) + \frac{1}{\mathcal{R}} \left(1 - \omega_{\mathrm{G}}(t) \right) + \underbrace{\frac{\mathcal{R}}{\mathcal{R}_{\mathrm{tot}}} p_\Delta(t)}_{\text{AGC signal}} , \tag{4.9}$$

where the index i has been omitted for simplicity and p_Δ has to be included in the vector $\boldsymbol{u}$ of the inputs of the DAE model (4.4)–(4.6) of the TG.

Using the notation of (2.15) and dropping the dependency on time, (4.7) and (4.9) become:

$$\boldsymbol{x} = [p_\Delta] , \quad \boldsymbol{y} = \emptyset, \quad \boldsymbol{u} = [\omega_{\mathrm{CoI}}] , \tag{4.10}$$

$$\boldsymbol{f} = \left[K_o (1 - \omega_{\mathrm{CoI}}) \right] , \quad \boldsymbol{g} = \emptyset , \tag{4.11}$$

$$\mathbf{T} = [1] , \quad \mathbf{R} = \emptyset . \tag{4.12}$$

The time response of the SFC is slow with respect to the machine inertial response and the PFC (see Figure 4.1). For short-term simulations – e.g. up to 20 or 30 s – it is not necessary to take into account the effect of the SFC. In this case, the model of the AGC reduces to $p_\Delta = 0$.

4.2.3. Examples

The examples presented in this section aim to illustrate the inertial response of the synchronous machines and the effect of their primary and secondary frequency controls following a contingency. The WSCC system described in Appendix A.2 is utilized. In all examples, the contingency is the loss of 25 MW of the load at bus 5.

Example 1: Inertial Response

No PFC nor AGC are considered in this example. Figure 4.4.a shows how all machines follow the same trend following the loss of the load. Figures 4.4.b and 4.4.c show the rotor speed of the synchronous machine at bus 1 for different values of the inertia and damping coefficients of equation (2.26). As expected, higher inertia values lead to smaller rates of change of the rotor speed for the same power unbalance. Nonzero values of the damping, on the other hand, lead to eventual steady-state conditions, thus preventing the rotor angular speed to drift permanently from its reference value as observed from the case when $D = 0$. The higher the value of the damping, the smaller the steady-state error with respect to the synchronous frequency. However, damping needs to be small to minimize mechanical losses.

Example 2: Primary Frequency Control

Turbine governors as that shown in Figure 4.2 are included in the power plants of the WSCC system. The behavior of the rotor speed of the machine at bus 1, and the mechanical power delivered by each machine for different values of the droop $\mathcal{R}_i$, $i = 1, 2, 3$, are shown in Figure 4.5. All machines have the same $\mathcal{R}_i = \mathcal{R}$ for each case. Smaller values of $\mathcal{R}$ lead to faster PFC responses, reducing the frequency zenith due to the loss of the load, and obtaining a steady-state frequency closer to the synchronous one. Different values of droop affect the speed of response of the machines, given by the slope of the variations of the machine active power outputs. Steady-state conditions are then reached when the power balance is met. This occurs for the same machine power outputs, but at different rotor angular speeds depending on the value of $\mathcal{R}$.

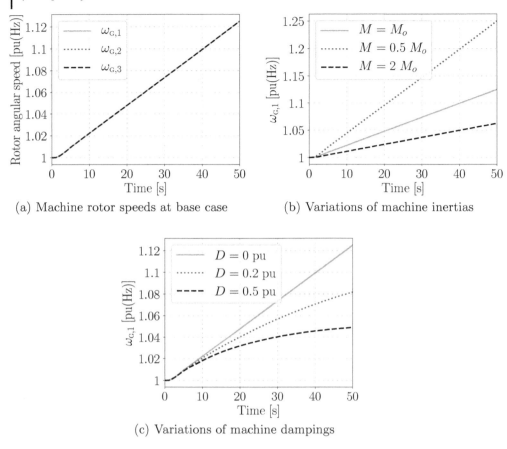

(a) Machine rotor speeds at base case

(b) Variations of machine inertias

(c) Variations of machine dampings

Figure 4.4 WSCC system – Response of the machine rotor speeds for different values of the inertia and damping when no PFC nor SFC are installed.

A second PFC example is shown in Figure 4.6, where different $\mathcal{R}_i$ are considered for the TGs of the machines. As expected, different relative $\mathcal{R}_i$ lead to different contributions from each machine to balance the power after the loss of the load. This does not appear to affect the response of the angular speed of the rotor of machine 1 of Figure 4.6.a, as machines at buses 1 and 3 have similar power ratings.

Note that the case where one or more droops are equal to zero is not practical, as in this case, the corresponding machine(s) will try to meet the power balance on its (their) own. In practice, all $\mathcal{R}_i$ are similar and usually have values around 4–5%. This necessarily implies that rotor angular speeds

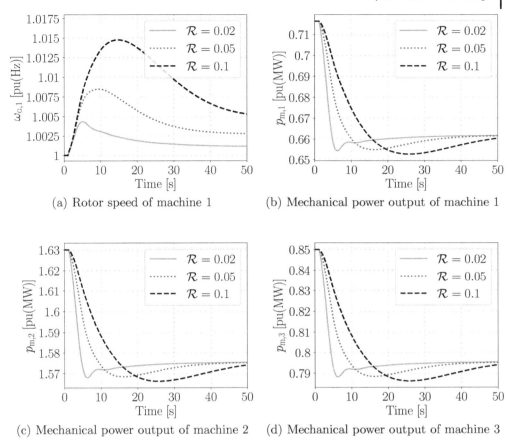

(a) Rotor speed of machine 1

(b) Mechanical power output of machine 1

(c) Mechanical power output of machine 2

(d) Mechanical power output of machine 3

Figure 4.5 WSCC system – Response of the rotor speed of machine at bus 1 and the mechanical power outputs of the three machines for different values of the droop $\mathcal{R}$ of the PFC.

and bus frequencies will differ from the reference frequency of the system in steady state. To bring the system frequency back to the reference value, SFC must be implemented. This is illustrated in the following example.

Example 3: Secondary Frequency Control

A common example of SFC is the AGC described in Section 4.2.2, and its dynamic behavior is illustrated in Figure 4.7. The AGC scheme of Figure 4.3 is included in the system and different values of the integral gain K_o are compared.

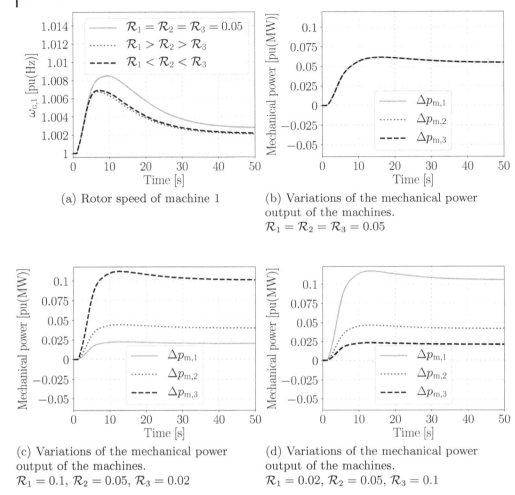

(a) Rotor speed of machine 1

(b) Variations of the mechanical power output of the machines. $\mathcal{R}_1 = \mathcal{R}_2 = \mathcal{R}_3 = 0.05$

(c) Variations of the mechanical power output of the machines. $\mathcal{R}_1 = 0.1$, $\mathcal{R}_2 = 0.05$, $\mathcal{R}_3 = 0.02$

(d) Variations of the mechanical power output of the machines. $\mathcal{R}_1 = 0.02$, $\mathcal{R}_2 = 0.05$, $\mathcal{R}_3 = 0.1$

Figure 4.6 WSCC system – Response of the rotor speed of machine at bus 1 and the mechanical power outputs of the three machines for different values of the droop $\mathcal{R}$ of the PFC.

Smaller values of K_o lead to faster responses of the AGC, reducing the frequency zenith after the loss of load, and reducing also the time during which the frequency is above the maximum threshold for a secure operation of the system (in this example, a threshold of ± 0.006 pu(Hz) has been considered for illustration). However, smaller values of K_o also imply oscillations that can lead to instabilities, e.g. for $K_o = 80$. The integral gain of the AGC must thus be tuned carefully and a compromise should be met to avoid on one hand the coupling between the SFC and the PFC and, on the other hand, to prevent a slow recovery of the frequency to its reference value after a contingency.

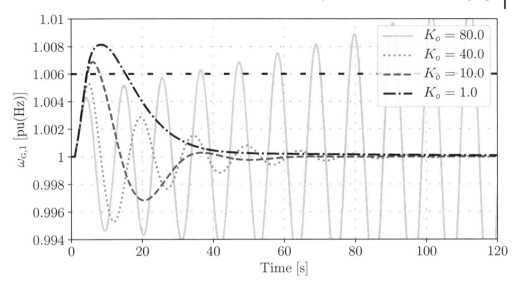

Figure 4.7 WSCC system – Response of the rotor speed of machine at bus 1 for different values of the integral gain K_o of the SFC.

4.3. Power Oscillation Damping

Similarly to the PFC, the primary voltage regulation, generally called automatic voltage regulator (AVR), of synchronous machines is mandatory for conventional power plants connected at the transmission system level. AVRs often include several auxiliary controllers and limiters. In the context of this book, the most relevant of such controllers is the power system stabilizer (PSS).

Both AVR and PSS are widely discussed in the literature and a huge variety of models for these controllers exists. The purpose of this section is not to provide a comprehensive overview of these models but, rather, to discuss the role that AVRs and PSSs play in the birth and damping of electromechanical oscillations.

4.3.1. Automatic Voltage Regulator

The main function of the AVR is to keep the bus terminal voltage of a synchronous machine equal to a reference value by regulating the reactive power produced by the machine itself. AVRs are fundamental to the grid as they prevent the voltage magnitudes of the system from becoming too high or too low and help prevent voltage instability.

Three main types of AVRs are identified based on the technology used for the excitation system: DC generator, AC generator with rectifier, and

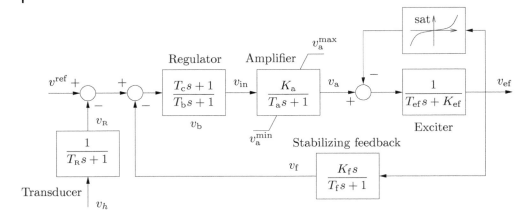

Figure 4.8 Control diagram of the IEEE Type DC1 exciter [98].

transformer with rectifier [98]. Figure 4.8 shows a very common and widely used control scheme, namely the IEEE Type DC1 exciter.

The AVR depicted in Figure 4.8 is described by the following equations:

$$T_R(t)\frac{d}{dt}v_R(t) = v_h(t) - v_R(t) \ ,$$

$$T_b\frac{d}{dt}v_b(t) = v^{\text{ref}} - v_R(t) - v_f(t) - v_b(t) \ ,$$

$$z_1(t)\, z_2(t)\, T_a\frac{d}{dt}v_a(t) = z_1(t)\, z_2(t)\, K_a v_{\text{in}}(t) + \hat{z}_1(t)\, v_a^{\text{max}}$$
$$+ \hat{z}_2(t)\, v_a^{\text{min}} - v_a(t) \ ,$$

$$T_f\frac{d}{dt}v_f(t) - K_f\frac{d}{dt}v_{\text{ef}}(t) = -v_f(t) \ ,$$

$$T_{\text{ef}}\frac{d}{dt}v_{\text{ef}}(t) = v_a(t) - \Big(K_{\text{ef}} + \text{sat}(t)\Big)v_{\text{ef}}(t) \ ,$$

(4.13)

where v_h is the regulated voltage at the generator terminal bus h, and v_{in} is the amplifier input signal, which, for the IEEE Type DC1 is:

$$v_{\text{in}}(t) = T_c\frac{d}{dt}v_b(t) + v_b(t) \ ,$$

(4.14)

and the anti-windup limiter of the amplifier is implemented as:

$$z_1(t) = \mathcal{P}_1\big(v_a(t), \tfrac{d}{dt}v_a(t)\big) \ ,$$
$$z_2(t) = \mathcal{P}_2\big(v_a(t), \tfrac{d}{dt}v_a(t)\big) \ ,$$
$$\hat{z}_1(t) = 1 - z_1(t) \ ,$$
$$\hat{z}_2(t) = 1 - z_2(t) \ ,$$

(4.15)

where $\mathcal{P}_1$ and $\mathcal{P}_2$ are antiwindup limiter functions:

$$\mathcal{P}_1\left(v_a(t), \tfrac{d}{dt}v_a(t)\right) = \begin{cases} 0, & \text{if } v_a(t) \geq v_a^{\max} \text{ and } \dfrac{d}{dt}v_a(t) \geq 0 , \\ 1, & \text{otherwise} , \end{cases} \quad (4.16)$$

$$\mathcal{P}_2\left(v_a(t), \tfrac{d}{dt}v_a(t)\right) = \begin{cases} 0, & \text{if } v_a(t) \leq v_a^{\min} \text{ and } \dfrac{d}{dt}v_a(t) \leq 0 , \\ 1, & \text{otherwise} . \end{cases} \quad (4.17)$$

The ceiling function approximates the saturation of the excitation winding:

$$\text{sat}(t) = A_{\text{ef}} \exp\left(B_{\text{ef}}|v_{\text{ef}}(t)|\right) . \quad (4.18)$$

Further details on the definition of the ceiling function can be found in [186] and [127]. Table 4.2 defines the parameters that appear in (4.13)–(4.18).

Using the notation of (2.15) and dropping the dependency on time, the set (4.13)–(4.18) becomes:

$$\boldsymbol{x}^\dagger = [v_R,\ v_b,\ v_a,\ v_f,\ v_{\text{ef}}] ,$$

$$\boldsymbol{y}^\dagger = [v_{\text{in}},\ z_1,\ z_2,\ \hat{z}_1,\ \hat{z}_2], \quad (4.19)$$

$$\boldsymbol{u} = [v_h] ,$$

$$\boldsymbol{f} = \begin{bmatrix} v_h - v_R \\ v^{\text{ref}} - v_R - v_f - v_b \\ z_1 z_2 K_a v_{\text{in}} + \hat{z}_1 v_a^{\max} + \hat{z}_2 v_a^{\min} - v_a \\ -v_f \\ v_a - \left(K_{\text{ef}} + \text{sat}\right) v_{\text{ef}} \end{bmatrix} , \quad \boldsymbol{g} = \begin{bmatrix} v_b - v_{\text{in}} \\ \mathcal{P}_1\left(v_a, \tfrac{d}{dt}v_a\right) - z_1 \\ \mathcal{P}_2\left(v_a, \tfrac{d}{dt}v_a\right) - z_2 \\ 1 - z_1 - \hat{z}_1 \\ 1 - z_2 - \hat{z}_2 \end{bmatrix} , \quad (4.20)$$

$$\boldsymbol{T} = \begin{bmatrix} T_R & 0 & 0 & 0 & 0 \\ 0 & z_1 z_2 T_a & 0 & 0 & 0 \\ 0 & 0 & T_b & 0 & 0 \\ 0 & 0 & 0 & T_f & -K_f \\ 0 & 0 & 0 & 0 & T_{\text{ef}} \end{bmatrix} , \quad \boldsymbol{R} = \begin{bmatrix} 0 & -T_c & 0 & 0 & 0 \\ 0 & 0 & 0 & 0 & 0 \\ 0 & 0 & 0 & 0 & 0 \\ 0 & 0 & 0 & 0 & 0 \\ 0 & 0 & 0 & 0 & 0 \end{bmatrix} . \quad (4.21)$$

The dynamics of the IEEE Type DC1 exciter overlap the time scale of the inertial response of the machine. In some cases, however, faster dynamics can be neglected. The minimal AVR model that retains its basic dynamic behavior is obtained by setting $T_c = T_b = T_m = T_f = K_f = A_{\text{ef}} = 0$ and $K_{\text{ef}} = 1$.

Table 4.2 Parameters of the IEEE Type DC1 exciter model.

Parameter	Description	Unit
A_{ef}	First ceiling coefficient	–
B_{ef}	Second ceiling coefficient	$pu(kV)^{-1}$
K_a	Amplifier gain	–
K_{ef}	Field circuit integral deviation	–
K_f	Stabilizer gain	–
T_a	Amplifier time constant	s
T_b	Pole of the regulator inherent dynamic	s
T_c	Zero of the regulator inherent dynamic	s
T_{ef}	Field circuit time constant	s
T_f	Stabilizer time constant	s
T_R	Measurement time constant	s
v_a^{max}	Maximum regulator voltage	$pu(kV)$
v_a^{min}	Minimum regulator voltage	$pu(kV)$

4.3.2. Power System Stabilizer

The PSS is a very common auxiliary controller of synchronous machines aimed at damping electromechanical oscillations. As discussed in the introduction of this section, the PSS takes as input the measurement of the rotor speed of the synchronous machine and utilizes this signal to modify the reference voltage of the AVR. If properly tuned, the effect of the PSS is to increase the damping of electromechanical oscillations of the rotor, without actually increasing mechanical losses. The energy is in fact exchanged with the field winding of the machine, which functions as an inductive storage.

A typical PSS scheme with a washout filter and two lead-lag blocks is shown in Figure 4.9, and its parameters are defined in Table 4.3. The set of DAEs that describes such a PSS model is:

$$T_w \frac{d}{dt} v_1(t) = K_w \, \omega_G - v_1(t) \, ,$$

$$T_2 \frac{d}{dt} v_2(t) - T_w \frac{d}{dt} v_1(t) = -v_2(t) \, ,$$

$$T_4 \frac{d}{dt} v_3(t) - T_1 \frac{d}{dt} v_2(t) = v_2(t) - v_3(t) \, ,$$

$$-T_3 \frac{d}{dt} v_3(t) = v_3(t) - \hat{v}_s(t) \, ,$$

(4.22)

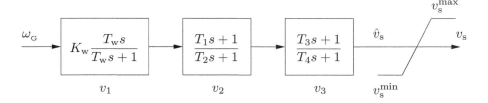

Figure 4.9 PSS with washout filter and two lead-lag compensators.

Table 4.3 Parameters of the PSS model.

Parameter	Description	Unit
K_w	Stabilizer gain	–
T_1	First zero time constant	s
T_2	First pole time constant	s
T_3	Second zero time constant	s
T_4	Second pole time constant	s
T_w	Washout time constant	s
$v_\mathrm{s}^\mathrm{max}$	Max stabilizer output signal	pu(kV)
$v_\mathrm{s}^\mathrm{min}$	Min stabilizer output signal	pu(kV)

where $\hat{v}_\mathrm{s}$ is subject to a windup limiter $\mathcal{P}_\mathrm{s}$ that constraints the output v_s of the PSS:

$$v_\mathrm{s}(t) = \mathcal{P}_\mathrm{s}(\hat{v}_\mathrm{s}(t)) \,, \qquad (4.23)$$

where the expression of the limiter is as follows:

$$\mathcal{P}_\mathrm{s}(\hat{v}_\mathrm{s}(t)) = \begin{cases} v_\mathrm{s}^\mathrm{min} & \text{if } \hat{v}_\mathrm{s}(t) < v_\mathrm{s}^\mathrm{min} \,, \\ v_\mathrm{s}^\mathrm{max} & \text{if } \hat{v}_\mathrm{s}(t) > v_\mathrm{s}^\mathrm{max} \,, \\ \hat{v}_\mathrm{s}(t) & \text{otherwise} \,. \end{cases} \qquad (4.24)$$

Finally, the output of the PSS has to be added to the reference voltage of the AVR. Recalling the second equation of (4.13), one has:

$$T_\mathrm{b}\frac{d}{dt}v_\mathrm{b}(t) = v^\mathrm{ref} + \underbrace{v_\mathrm{s}(t)}_{\text{PSS signal}} - v_\mathrm{m}(t) - v_\mathrm{f}(t) - v_\mathrm{b}(t) \,, \qquad (4.25)$$

and v_s has to be included in the vector $\boldsymbol{u}$ of the inputs of the DAE model (4.19)–(4.21) of the AVR.

Using the notation of (2.15) and dropping the dependency on time, the set (4.22)–(4.24) becomes:

$$\boldsymbol{x}^{\dagger} = [v_1, \, v_2, \, v_3] \, ,$$

$$\boldsymbol{y}^{\dagger} = [\hat{v}_{\mathrm{s}}, \, v_{\mathrm{s}}] \, , \tag{4.26}$$

$$\boldsymbol{u} = [\omega_{\mathrm{G}}] \, ,$$

$$\boldsymbol{f} = \begin{bmatrix} K_{\mathrm{w}} \, \omega_{\mathrm{G}} - v_1 \\ -v_2 \\ v_2 - v_3 \end{bmatrix} , \qquad \boldsymbol{g} = \begin{bmatrix} v_3 - \hat{v}_{\mathrm{s}} \\ \mathcal{P}_{\mathrm{s}}(\hat{v}_{\mathrm{s}}) - v_{\mathrm{s}} \end{bmatrix} , \tag{4.27}$$

$$\mathbf{T} = \begin{bmatrix} T_{\mathrm{w}} & 0 & 0 \\ -T_{\mathrm{w}} & T_2 & 0 \\ 0 & -T_1 & T_4 \end{bmatrix} , \qquad \mathbf{R} = \begin{bmatrix} 0 & 0 & -T_3 \\ 0 & 0 & 0 \end{bmatrix} . \tag{4.28}$$

The most important part of the control block diagram of the PSS is the washout filter, which can be viewed as an approximated derivative of the input signal (see also Section 3.3.1.2). Thus, the PSS effectively estimates the RoCoF and varies accordingly the reference voltage of the AVR.[2]

Since the PSS is aimed at damping oscillations, taking the derivative of the rotor speed implies that the PSS uses a signal that is shifted by about $\pi/2$ rad with respect to the actual phase of the rotor speed oscillation. Moreover, using a washout filter implies that the PSS output is null in steady state. The lead compensators that are included in the control diagram of the PSS improve its dynamic response but, in a first approximation, can be neglected. Hence, the minimal PSS model can be defined assuming $T_1 = T_2 = T_3 = T_4 = 0$.

Two remarks are relevant:

- The signal utilized as input of the PSS does not need to be a local measurement of the machine rotor speed. Frequency signals from remote buses/areas can be utilized as part of a wide-area stabilizing control (WASC) to damp inter-area oscillations [117, 173, 184]. The main difficulty of this kind of wide-area controllers is to cope with communication delays.

- The ability of damping electromechanical oscillations through a proper modulation of a voltage controller is not exclusive of the AVRs of synchronous machines. Any fast enough voltage controller, in principle, can be coupled with a PSS. In this case, the PSS is often renamed power oscillation damper (POD) and is often coupled to FACTS devices [78, 191], HVDC transmission systems [101] or, even, converter-interfaced generation [189].

---

[2]PSSs have been utilized for decades in power systems. However, the term RoCoF has become popular only in recent years due to the high penetration of nonsynchronous generation.

4.3.3. Examples

The examples presented in this section illustrate the dynamic behavior, following a disturbance, of the AVR and the PSS. The impact of load models on the dynamic response of the system is also discussed. The examples below considers the outage of the line connecting buses 2 and 4 of the IEEE 14-bus system described in Appendix A.3.

Example 1: Primary Voltage Control

Figure 4.10 shows the trajectories of the voltage magnitudes at the generator bus 2 and at the load bus 14 as well as the rotor angular speed of generator 2 as obtained with and without the inclusion of AVRs in the system. Loads are modeled as constant impedances ($\gamma_p = \gamma_q = 2$).

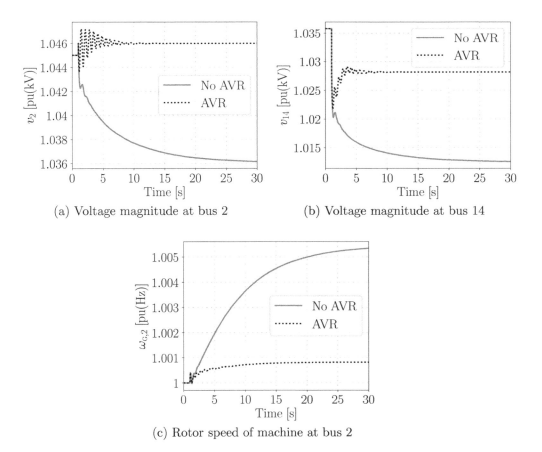

(a) Voltage magnitude at bus 2

(b) Voltage magnitude at bus 14

(c) Rotor speed of machine at bus 2

Figure 4.10 IEEE 14-bus system – Response of the voltage magnitude at buses 2 and 14, and rotor angular speed of the synchronous machine at bus 2 with and without AVRs in the machines. Loads are modeled as constant impedances.

The AVRs help maintain the voltage magnitude at the machine bus close to its initial value, albeit oscillations are observed in the first seconds after the line outage. At the load bus, the steady-state deviation with respect to the initial value is about 1.8% and 0.6% without and with AVRs, respectively. The effect of the AVR is also seen in the rotor angular speeds of the machines and, consequently, on the system frequency. The speed deviations are about 0.6% and 0.1% for the cases without and with AVRs, respectively.

Figure 4.11 shows the impact of the gains of the AVR model, namely K_a and K_f. $K_{a,o} = 200$ and $K_{f,o} = 0.0012$ are used for the machine at bus 1, and $K_{a,o} = 20$ and $K_{f,o} = 0.001$ are used for all other machines.

High values of the amplifier gain K_a increase the speed of response of the AVRs, leading to steady-state values of the voltage magnitude that are closer to its initial value. However, high K_a values also lead to bigger oscillations, which can lead to unstable conditions such as the limit cycle observed when K_a is twice the base-case value. Steady-state values of voltage magnitudes closer to the reference voltage also lead to steady-state frequencies that are closer to the reference system frequency.

Higher values of the stabilizer gain K_f lead to better damped voltage profiles with same steady-state values. Therefore, a common strategy followed in practice is to vary simultaneously both gains, in such a way that the ratio $K_{a:f} = K_a/K_f$ is kept close to a constant value, as shown in Figures 4.11.e and 4.11.f, where higher values of $K_{a:f}$ provide the best response.

Example 2: Load Modeling

Figure 4.12 shows the same voltage magnitudes and machine rotor angular speed as in Figure 4.10 of Example 1 above, but considering constant power consumption of the loads ($\gamma_p = \gamma_q = 0$).

Without AVRs, the voltage magnitudes show a greater drop after the line outage. Moreover, the oscillations introduced by the AVR last longer than when using constant impedance loads. Interestingly, the rotor angular speeds of the machines vary with the opposite trend to that seen in Example 1, being such variations smaller if the AVRs are installed. The rationale of this result is as follows. If loads impose a constant power consumption, after the line outage, the same active power has to be delivered through less transmission lines, meaning higher currents and, thus, higher losses. Therefore, more energy needs to be delivered by the generators, leading to negative frequency deviations. If loads are modeled as constant impedances, their power consumption varies quadratically with the voltage. Lower voltages thus lead to lower power consumption from the loads. Such a power unbalance translates into a positive frequency variation.

The discussion above motivates the question "which is the right load model?" As discussed in Section 2.4.2, a common assumption is that, for

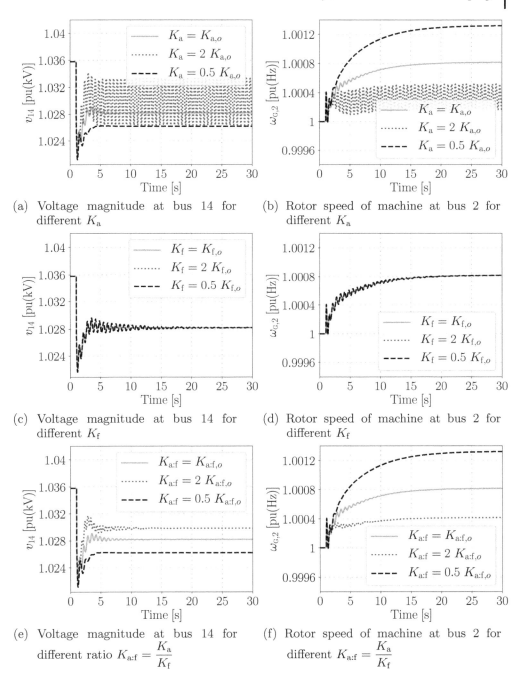

(a) Voltage magnitude at bus 14 for different K_a

(b) Rotor speed of machine at bus 2 for different K_a

(c) Voltage magnitude at bus 14 for different K_f

(d) Rotor speed of machine at bus 2 for different K_f

(e) Voltage magnitude at bus 14 for different ratio $K_{a:f} = \dfrac{K_a}{K_f}$

(f) Rotor speed of machine at bus 2 for different $K_{a:f} = \dfrac{K_a}{K_f}$

Figure 4.11 IEEE 14-bus system – Response of the voltage magnitude at bus 14, and rotor angular speed of the synchronous machine at bus 2 for different values of K_a and K_f of the AVRs in the machines.

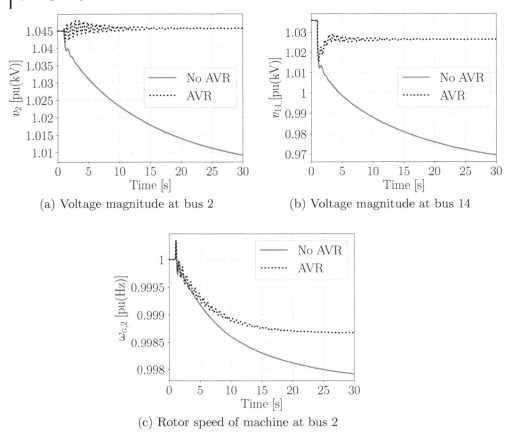

(a) Voltage magnitude at bus 2

(b) Voltage magnitude at bus 14

(c) Rotor speed of machine at bus 2

Figure 4.12 IEEE 14-bus system – Response of the voltage magnitude at buses 2 and 14, and rotor angular speed of the synchronous machine at bus 2 with and without AVRs in the machines. Loads are modeled as constant power consumption.

transient stability analyses which consider time scales of up to several seconds, constant impedance loads are a good approximation. For studies characterized by longer time scales (from tens of seconds to few minutes), constant power loads are commonly utilized.

Example 3: Power System Stabilizer

The effect of installing a PSS in the system is shown in Figure 4.13. The PSS filters out the oscillations caused by the AVRs, providing a much damper response of both voltage magnitudes and machine rotor angular speeds. The main drawback of the PSS is the slower recovery of the voltage magnitude at the machine bus back to its reference value.

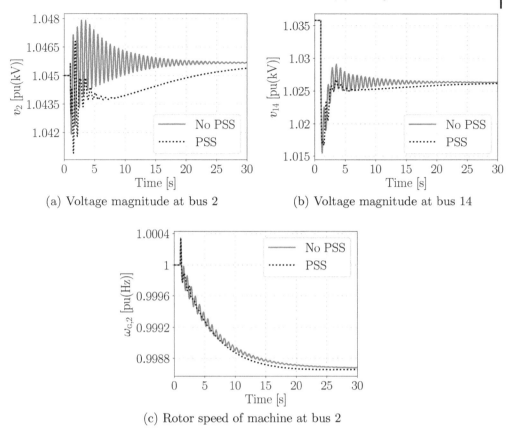

(a) Voltage magnitude at bus 2

(b) Voltage magnitude at bus 14

(c) Rotor speed of machine at bus 2

Figure 4.13 IEEE 14-bus system – Response of the voltage magnitude at buses 2 and 14, and rotor angular speed of the synchronous machine at bus 2 with and without a PSS. AVRs are included in all machines.

4.4. Nonsynchronous Devices

There exists a vast literature on the provision of frequency regulation from renewable energy sources, HVDC systems, energy storage system (ESS), and flexible loads. This section briefly outlines a variety of nonconventional frequency control schemes that complement the primary frequency regulation provided by synchronous machines.

4.4.1. Distributed Energy Resources

There is an ongoing debate on the role of the frequency controllers of converter-interfaced DERs such as converter-interfaced generation (CIG) and ESSs.

In recent years, transmission system operators (TSOs) have recognized the need for new ancillary services, such as the fast responding regulation service (FRRS) defined by Electric Reliability Council of Texas (ERCOT) already in 2002 [62], or the fast frequency response (FFR) defined by Australian Energy Market Operator (AEMO) [137]. A common feature of these definitions is their attempt to make a neat distinction between synchronous inertia response and frequency droop control.

CIG can respond to RoCoF and frequency variations, provided that they are operated with sufficient power reserve [31, 154, 180]. However, it is also recognized that, due to their intermittent nature, most CIGs have a limited frequency control capability. In this regard, ESSs become relevant thanks to their capability to provide fast active and reactive power regulation simultaneously [132]. Depending on the response times[3] and capacities of the storage technology, ESSs can provide either RoCoF control or FFR, or both at the same time. The solution of combining CIG, characterized by limited reserve, with flexible ESSs is thus becoming more and more common.

FFR of DERs is conceptually similar to the droop control of synchronous machines. The main difference lies in the time scale of the response of the regulation. The response time of the turbine governors of synchronous machines spans up to several seconds. DERs on the other hand typically require from a few tens of milliseconds to several hundred milliseconds to start regulating.

The general scheme of the fast frequency and RoCoF control of DERs is shown in Figure 4.14. For the sake of clarity and completeness, the voltage control of DERs is also illustrated in the scheme. The dq representation of the power converter and its inner and outer current controls enables fully decoupled voltage and frequency regulations, at least in stationary conditions [132, 228].

The frequency control loop of Figure 4.14 typically takes, as input, the error between the measured frequency at the bus of connection, $\tilde{\omega}_h$, and the reference frequency, ω^{ref}, which is generally 1 pu. This error is then passed through dead band and LPF blocks, which reduce the sensitivity of the frequency control to small and/or high-frequency perturbations.

4.4.1.1. Frequency Control Block

There exists a large number of possibilities to implement the frequency control transfer function, say $H_{\text{fc}}(s)$, of DERs. Until recently, the most common configuration consisted in a single droop control, as depicted in Figure 4.15. For CIGs, the control is typically composed of a droop gain $\mathcal{R}_{\text{fc}}$ (see Figure 4.15.a) [6, 201]. For ESSs, on the other hand, the frequency control is commonly

---

[3]*Response time* refers to the time required for a specific device to vary its active power generation/consumption from its reference value to the maximum/minimum value allowed.

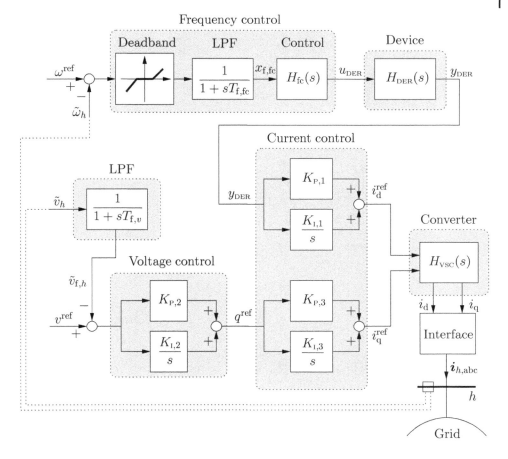

Figure 4.14 Scheme of fast frequency/RoCoF and voltage controls of DERs.

implemented as a PI control, with integral deviation $\mathcal{R}'_{\text{fc}}$ (see Figure 4.15.b) [115, 156]. The output of the PI control is then passed through a block called "storage input limiter," which is responsible of moderating the effect of reaching energy saturation of the storage device [160].

In recent years, alternative solutions have been proposed to take advantage of the fast response of converter-interfaced DERs to regulate, additionally, the RoCoF at the bus of connection with the grid [31, 59, 125, 163]. This is usually achieved by including an additional channel in the control that computes the derivative of the frequency error. Rudimentary studies proposed assigning different weights to each of the two channels (droop and RoCoF), depending on the requirements of the system [210]. More recent studies demonstrate the inadequacy of this approach, and make use of the fact that the time scales of RoCoF and droop controls are effectively decoupled [159]. This solution enables the allocation, at any time, of 100% of the regulation capability of the DER,

Droop

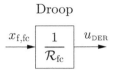

(a) Configuration for CIG

PI control   Storage input limiter

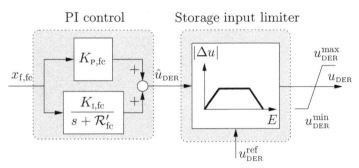

(b) Configuration for ESSs

Figure 4.15 $H_{\text{fc}}(s)$ – Models of the fast frequency control for DERs.

thus optimizing its performance. Therefore, only the latter implementation is discussed below.

As for the case with only droop control, there exist two main implementations depending on the application (CIG or ESS), as illustrated in Figure 4.16.

In both configurations, the additional RoCoF channel consists in a washout filter whose numerator computes the derivative of the filtered frequency error $x_{\text{f,fc}}$. This signal is then added to the output of the droop channel. The resulting signal is the input of the CIG device, or the input of the PI control of the ESS. Note that, if one decides to disable the RoCoF channel, it will suffice to set $K_{\text{rc}} = 0$. Similarly, the droop channel can be disabled if $\mathcal{R}_{\text{fc}} \gg 1$. Note also that, if only the RoCoF channel is enabled for ESSs, the integral deviation coefficient $\mathcal{R}'_{\text{fc}}$ must be set to 0 to avoid nonzero steady-state errors in the output of the controller, thus leading to a steady nonzero slope of the frequency.

4.4.1.2. DER Device

Wind Energy Conversion System

Wind turbines have been used to generate electricity since July 1887 by Scottish academic, James Blyth, and later that year by the American inventor Charles F. Brush [174]. During this almost century and a half, the technology of

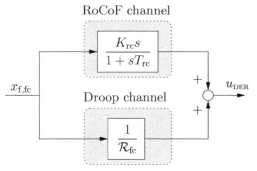

(a) Configuration for CIG

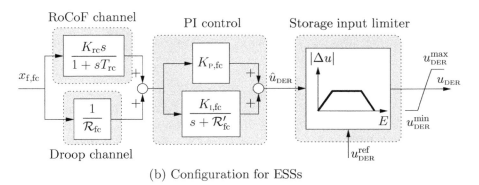

(b) Configuration for ESSs

Figure 4.16 $H_{fc}(s)$ – Models of the fast frequency and RoCoF control for DERs.

wind energy conversion systems (WECSs) has evolved substantially. Currently, there exist a large variety of well-established wind turbine configurations, among which there are noncontrolled-speed wind turbines with constant-speed induction generator (CSWT – Type A), controlled-speed wind turbines with doubly-fed wound-rotor induction generator (DFIG – Type C), and full converter-based wind turbines (Types D-G). The general components of any WECS are the following:

- *Induction machine.* Type A constant-speed wind turbines (CSWTs) usually include a squirrel-cage induction generator connected to the turbine through a gearbox and to the grid in parallel with a capacitor bank. Type C doubly-fed induction generators (DFIGs) on the other hand connect the rotor and stator of their doubly-fed induction machine through an AC/AC power converter. While not strictly necessary due to its symmetrical construction, the Park transform is generally utilized also for the formulation of the model of the induction machine [110].

- *Dynamic shaft.* The shaft links, mechanically, the turbine and the rotor of the generator, and is usually modeled as a two-mass system connected by a spring which represents the stiffness of the shaft. Some references utilize a model with three masses (turbine, machine, and gearbox) and two shafts [4].

- *Turbine.* The mechanical model of the wind turbine is independent from the wind turbine type. Type A wind turbines tend to include fixed-blade turbines, while other types often include variable-pitch angle turbines [84].

- *Pitch angle control.* The conventional pitch angle control prevents the turbine from rotating at speeds that exceed a given limit and, in some cases, can be designed to provide, to a limited extent, frequency control [31].

- *Maximum power point tracking.* The MPPT speed control aims at maximizing the power generated by WECSs, generally based on the speed-power characteristic of the wind turbine. Some works have proposed the design of MPPT techniques that do not require the knowledge of the optimal power characteristics [109].

The interested reader can find in Chapter 4 of [132] detailed TS models of the components of WECSs above that, combined, constitute the set of DAEs required to construct the transfer function $H_{\text{DER}}(s)$. This is the model used in the simulations of this book.

Solar Photovoltaic Generation

The Solar photovoltaic generation (SPVG) model suitable for voltage and transient stability analysis is relatively standard (see e.g. [202] and [132]). The PV array is connected to the grid through the series of a DC/DC and a DC/AC full converter. The voltage regulation of the SPVG and the MPPT are generally located in the DC/DC converter at the output of the array. The changes of solar irradiation levels are considered slow with respect to the time scales of TS analysis, thus the dynamics of the PV array are often neglected in short-term simulations. On the other hand, in TS studies, the MPPT response can be considered instantaneous. The scheme shown in Figure 4.14 includes the current control, which defines the current set points, and the converter $H_{\text{VSC}}(s)$.

Finally, the model of the SPVG $H_{\text{DER}}(s)$ is simply:

$$y_{\text{DER}}(t) = u_{\text{DER}}(t) + p_{\text{MPPT}}(t) , \tag{4.29}$$

where p_{MPPT} is the power output of the MPPT.

Energy Storage System

There exists a large variety of ESS technologies: batteries (lithium-ion, nickel-cadmium, etc.), flywheels, compressed air tanks, electrochemical capacitors, superconducting magnetic coils, etc. Detailed transient stability models as well as more general, highly simplified models have been proposed in the literature for all technologies above [132]. Overly simplified models show low accuracy, as they reduce the behavior of the storage device to a single time constant, the dynamics of the DC circuit and the Voltage-sourced converter (VSC) are neglected, and nonlinearities such as current or energy saturation are generally not properly captured. Detailed transient stability models, on the other hand, are often cumbersome and require accurate parameter values. To overcome the issues above, the generalized ESS model proposed in [132, 162] is used in the case studies discussed in the remainder of this book.

4.4.1.3. Voltage-sourced Converter

In Figure 4.14, the block $H_{\mathrm{VSC}}(s)$ represents a two-input, two-output model of the VSC used to interface the DER with the grid. The scheme of the VSC and its inner current-control in the dq-reference frame is shown in Figure 4.17, where $\bar{i}_{\mathrm{dq}}$ and $\bar{i}_{\mathrm{dq}}^{\mathrm{ref}}$ are the current flow at the AC side of the converter and its reference value, respectively; $\bar{v}_{h,\mathrm{dq}}$, $\bar{v}_{t,\mathrm{dq}}$, and v_{dc} are the voltages at the bus of connection with the grid and at the AC- and DC-side terminals of the VSC, respectively; $\bar{m}_{\mathrm{dq}}$ is the modulating signal of the pulse-width modulation (PWM) [141]; R_{VSC} and L_{VSC} are the series resistance and inductance, respectively, of

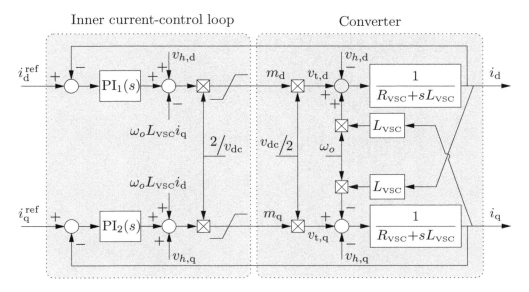

Figure 4.17 $H_{\mathrm{VSC}}(s)$ – VSC and inner current-control in the dq-reference frame.

the converter transformer and phase reactor; ω_o is the synchronous reference frequency; and $\text{PI}_1(s)$ and $\text{PI}_2(s)$ are two PI regulators, each with a zero canceling the pole of the converter, $R_{\text{VSC}} + sL_{\text{VSC}}$.

The interested reader can find more details on VSC models used for power system applications in [32, 132, 228].

4.4.2. Thermostatically Controlled Loads

thermostatically controlled loads (TCLs) are dynamic loads with temperature control [85]. These can be air conditioning systems, industrial refrigerators or heating systems. In most cases, the reference temperature is fixed to an assigned value. There are, however, prototypes of TCLs that measure the system frequency and vary the reference temperature in order to reduce frequency variations [112, 124, 195]. The control scheme of the TCL is depicted in Figure 4.18, where Θ_{TCL} is the load temperature (lumped model); Θ_{a} is the ambient temperature; G_{TCL} is the equivalent load conductance; v_h is the load terminal voltage; and p_h is the consumed active power.

The gain $K_{1,\text{TCL}}$ and the maximum conductance $G_{\text{TCL}}^{\text{max}}$ are determined based on the initial values of the voltage, $v_{h,o}$, and the active power, $p_{h,o}$, as follows:

$$
\begin{aligned}
K_{1,\text{TCL}} &= \frac{\Theta_{\text{TCL}}^{\text{ref}} - \Theta_{\text{a}}}{p_{h,o}} \ , \\
G_{\text{TCL}}^{\text{max}} &= K_{\text{L,TCL}} G_{\text{TCL},o} = K_{\text{L,TCL}} \frac{p_{h,o}}{v_{h,o}^2} \ ,
\end{aligned}
\tag{4.30}
$$

where $K_{\text{L,TCL}}$ is the ceiling conductance output ratio ($K_{\text{L,TCL}} < 1$ for cooling systems and $K_{\text{L,TCL}} > 1$ for heating systems).

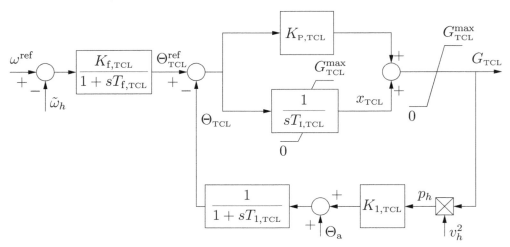

Figure 4.18 Thermostatically controlled load with frequency control.

4.4.3. Example

The performance of the fast frequency control provided by a WECS when its input signal is provided by an SRF-PLL is illustrated with an example based on the WSCC system described in Appendix A.2.

The synchronous machine at bus 3 is replaced with a wind power plant of the same capacity, composed of 50 variable-speed wind turbines modeled with a fifth-order DFIG model [193]. The simulation data of each DFIG that composes the wind plant have been obtained from Appendix D of [132]. In the long term, the stochastic process applied to the wind follows a Weibull distribution or some other nonsymmetrical distribution [229]. However, in the short term, one can approximate wind fluctuations with a Gaussian process [100].

In particular, the noise is modeled as a zero-average Ornstein-Uhlenbeck process, also known as *mean-reverting* process [79, 134, 170]. This process is characterized by a bounded standard deviation, as follows:

$$\frac{d}{dt} x_{\text{OU},h}(t) = -a_{\text{OU},h}(t)\, x_{\text{OU},h}(t) + b_{\text{OU},h}\, \xi_{\text{OU},h}(t) \;, \tag{4.31}$$

where $x_{\text{OU},h}$ is the stochastic state variable; $a_{\text{OU},h}$ and $b_{\text{OU},h}$ are the drift and the diffusion, respectively, of the stochastic process; and $\xi_{\text{OU},h}$ is the white noise. The drift is the mean reversion speed of the process itself. An individual uncorrelated process $x_{\text{OU},h}$ is generated per each variable to be perturbed and then added to such a variable (additive noise). The interested reader can find a detailed description of this model and the numerical scheme to integrate it in [134].

The contingency is the outage of the line connecting buses 5 and 7 at $t = 30$ s. Figure 4.19 shows the response of the WSCC system for the cases with and without WECS frequency regulation, whose parameters are given in Table 4.4.

The input signal of the controller is depicted in Figure 4.19.a, which clearly shows the characteristic spikes and noise of the PLL measurement. The result-

Table 4.4 Values of the parameters of the WECS controller.

Parameter	Value	Unit
K_{rc}	80.0	–
$\mathcal{R}_{\text{fc}}$	0.05	–
$T_{\text{f,fc}}$	0.5	s
T_{rc}	4.0	s

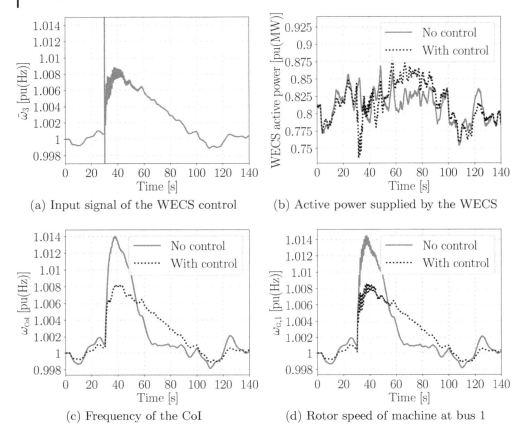

(a) Input signal of the WECS control

(b) Active power supplied by the WECS

(c) Frequency of the CoI

(d) Rotor speed of machine at bus 1

Figure 4.19 WSCC system – Response following the outage of the line 5–7 with and without frequency control implemented in the WECS of bus 3.

ing active power output of the WECS is shown in Figure 4.19.b. The fast response of the WECS, which quickly drops its power generation during the occurrence of the line outage from 0.85 to 0.74 pu, helps reduce the frequency zenith by over 40%, and smooths the frequency recovery in the next tens of seconds, reducing also the stress of the synchronous machines, as observed from Figures 4.19.c and 4.19.d.

The control is not capable of reducing the oscillations of the synchronous machines in the few seconds after the line outage. This is due, to a great

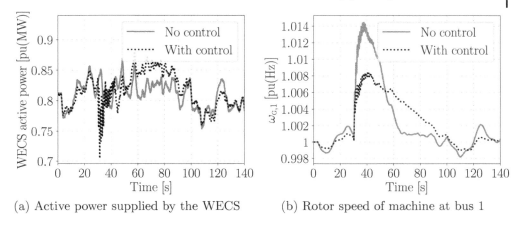

(a) Active power supplied by the WECS

(b) Rotor speed of machine at bus 1

Figure 4.20 WSCC system – Response following the outage of the line 5–7 with and without frequency control implemented in the WECS at bus 3 ($T_{\text{f,fc}} = 0.1$ s).

extent, to the LPF of the frequency error. Lower values of the time constant $T_{\text{f,fc}}$ reduces both the amplitude and duration of the oscillations. However, this also leads to a greater effort from the WECS, as observed from Figure 4.20. A compromise must thus be found for each application.

Part II
Theory

Chapter 5

Frequency Divider Formula

5.1. Rationale

The ENTSO-E Commission Regulation (EU) 2016/631 of 14 April 2016 "Establishing a Network Code on Requirements for Grid Connection of Generators" provides the following definition: *frequency means the electric frequency of the system expressed in hertz that can be measured in all parts of the synchronous area under the assumption of a consistent value for the system in the time frame of seconds, with only minor differences between different measurement locations. Its nominal value is 50 Hz.*

This definition is acceptable for time scales of several seconds. However, in the first seconds of a transient triggered by a large disturbance, e.g. the occurrence of a fault followed by its clearance, synchronous machine rotor speeds oscillate. Each machine shows local nondominant modes and/or interarea oscillation modes common to its coherent cluster [224]. The crucial point is that, in the first seconds after a large disturbance, machine rotor angular speeds are not equal and can show oscillations in counter-phase among clusters.

This behavior is illustrated in Figure 5.1, where a three-phase fault cleared after 70 ms is simulated in the WSCC system of the examples discussed in Section 4.2.3. The rotor speeds of the three machines tend to a steady-state value of about 1.0015 pu(Hz) in about 25 s after the fault clearance. However, during the transient, the machine at bus 1 oscillates in counter-phase with the machines at buses 2 and 3.

The different transient response of rotor angular speeds in the first seconds after a large contingency strongly depends on the inertia of the synchronous machines. As the penetration of nonsynchronous generation based on DERs increases, the overall inertia of the system necessarily decreases, thus leading to higher differences between the oscillatory modes of the remaining synchronous machines of the system. Figure 5.2 shows a comparison of the response of the rotor angular speeds of the machines for the base case condition, and for the case where the inertias of the machines have been decreased by half.

Frequency Variations in Power Systems: Modeling, State Estimation, and Control, First Edition.
Federico Milano and Álvaro Ortega Manjavacas.

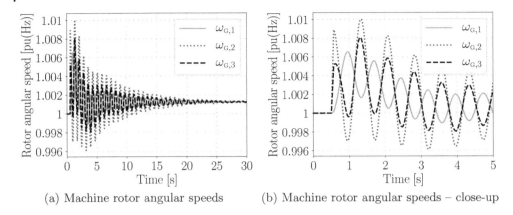

(a) Machine rotor angular speeds

(b) Machine rotor angular speeds – close-up

Figure 5.1 WSCC system – Response of the machine rotor speeds following a three-phase fault.

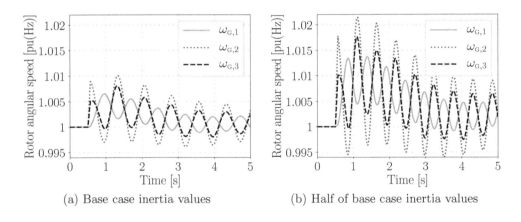

(a) Base case inertia values

(b) Half of base case inertia values

Figure 5.2 WSCC system – Response of the machine rotor speeds following a three-phase fault for different values of the machine inertias.

Each synchronous machine imposes, through its rotor angular speed, the frequency at its bus of connection with the rest of the grid. If the rotor speeds are different, the frequency of the buses and the transmission lines connecting different machines have also to evolve in time. Figure 5.3 illustrates this concepts and shows the frequency measurements from SRF-PLLs devices of the buses located between the synchronous machines 1 and 2, namely buses 1, 2, 4, 5, and 7.

The frequency measurements can be characterized based on the proximity of each bus to each machine. As they impose the frequency in the air gap between stator and rotor, synchronous machines set the boundary conditions

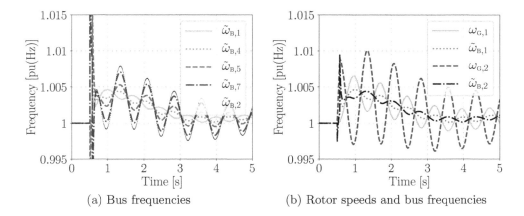

(a) Bus frequencies   (b) Rotor speeds and bus frequencies

Figure 5.3 WSCC system – Frequency measurements at the buses between the synchronous machines 1 and 2.

of the frequencies of all buses located between them. Note also that the rotor speeds of the machines are not exactly the same as the frequencies at terminal buses where the machines are connected (see Figure 5.3.b). These observations are the seeds of the formula derived in the next section.

5.2. Derivation

In [187] and, later, in [188], Semlyen proposed to model the transmission system as a *continuum* where the speeds of synchronous machines are the *boundary conditions* that the frequency has to satisfy.

The starting point of the definition of the frequency divider formula (FDF) is Semlyen's continuum. However, since the time scale of interest is that of electromechanical transients, the transient effects of wave propagation in transmission lines can be safely neglected. The speed of traveling waves, in fact, is given by $1/\sqrt{LC}$, where L and C are the inductance and the capacitance of the transmission line or cable, and is of the order of 10^6 m/s or higher, the upper limit being the speed of light. This speed enables the assumption that wave propagation transients do not overlap local and inter-area oscillation modes.

As a consequence, to compute the spatial variations of the frequency, the problem can be formulated as a steady-state boundary value problem, where boundary conditions are given by synchronous machine rotor speeds. This implies that any variation of machine rotor speeds leads to *instantaneous* variations of the frequencies everywhere in the grid. This assumption is consistent with the formulation of the QSS model described in Chapter 2.

The starting point of the FDF is the extended admittance matrix defined in Section 2.4.3.3, with inclusion of synchronous machine internal impedances as it is commonly defined for fault analysis. Recalling for clarity (2.79) and dropping for simplicity the dq-axis reference frame subscript, one has:

$$
\begin{bmatrix} \bar{i}_G(t) \\ \bar{i}_B(t) \end{bmatrix} = \begin{bmatrix} \bar{Y}_{GG}(t) & \bar{Y}_{GB}(t) \\ \bar{Y}_{BG}(t) & \bar{Y}_{BB}(t) \end{bmatrix} \begin{bmatrix} \bar{e}_G(t) \\ \bar{v}_B(t) \end{bmatrix} , \tag{5.1}
$$

where $\bar{v}_B$ and $\bar{i}_B$ are bus voltages and current injections, respectively, at network buses; $\bar{i}_G$ are generator current injections; $\bar{e}_G$ are generator EMFs behind the internal generator impedance; $\bar{Y}_{BB}$ is the network admittance matrix whose diagonal is augmented with the internal impedances of the synchronous machines at generator buses; and $\bar{Y}_{GG}$, $\bar{Y}_{GB}$, and $\bar{Y}_{BG}$ are admittance matrices obtained using the internal impedances of the synchronous machines (see Section 2.4.1.2 and Figure 2.6). For generality, all quantities in (5.1) depend on time, including the elements of the admittance matrices, even though the conventional QSS model of the transmission system neglects such a dependency (see Section 2.4.3).

In the derivation discussed in this section, bus currents $\bar{i}_B$ are assumed to be negligible. This assumption leads to rewrite (5.1) as:

$$
\begin{bmatrix} \bar{i}_G(t) \\ \mathbf{0} \end{bmatrix} = \begin{bmatrix} \bar{Y}_{GG}(t) & \bar{Y}_{GB}(t) \\ \bar{Y}_{BG}(t) & \bar{Y}_{BB}(t) \end{bmatrix} \begin{bmatrix} \bar{e}_G(t) \\ \bar{v}_B(t) \end{bmatrix} . \tag{5.2}
$$

If loads are linear, their admittances are much smaller than the diagonal elements of the admittance matrix – see (2.81) and (2.82) and the discussion thereafter. The load model, however, is not relevant for the derivation provided below and this is confirmed by the example presented in Section 5.2.1.2. From (5.2), it descends the relation between bus voltages $\bar{v}_B$ and generator EMFs $\bar{e}_G$:

$$
\mathbf{0} = \bar{Y}_{BB}(t)\,\bar{v}_B(t) + \bar{Y}_{BG}(t)\,\bar{e}_G(t) , \tag{5.3}
$$

which, remembering (2.84), can be approximated as:

$$
\mathbf{0} = \mathbf{B}_{BB}(t)\,\bar{v}_B(t) + \mathbf{B}_{BG}(t)\,\bar{e}_G(t) , \tag{5.4}
$$

where

$$
\begin{aligned}
\bar{v}_B(t) &= v_B(t) \circ \exp\big(j\boldsymbol{\theta}_B(t)\big) , \\
\bar{e}_G(t) &= e_G(t) \circ \exp\big(j\boldsymbol{\delta}_G(t)\big) ,
\end{aligned} \tag{5.5}
$$

where "∘" denotes the Hadamard product, i.e. the element-wise vector multiplication; v_B and $\boldsymbol{\theta}_B$ are the vectors of the magnitudes and phase angles, respectively, of the bus voltage phasors; e_G is the vector of the time-dependent

EMFs of the approximated dynamic circuit representation of the synchronous machines as in Figure 2.6; and $\boldsymbol{\delta}_{\mathrm{G}}$ is the vector of rotor angle positions.

The next step consists in deriving (5.4) with respect to time. While the literature on the numerical differentiation of the phase angles of voltage phasors has attempted to obtain precise model-dependent analytical expressions, e.g. [152], or to set up accurate numerical methods, e.g. [89], a more pragmatic approach is needed. The goal is to introduce appropriate simplifications and obtain a simple yet accurate expression.

The total time-derivative of (5.4) gives:

$$
\begin{aligned}
\mathbf{0} = \mathbf{B}_{\mathrm{BB}}(t) &\left(\frac{d}{dt} \boldsymbol{v}_{\mathrm{B}}(t) \circ \exp\left(j\boldsymbol{\theta}_{\mathrm{B}}(t)\right) + j\frac{d}{dt}\boldsymbol{\theta}_{\mathrm{B}}(t) \circ \bar{\boldsymbol{v}}_{\mathrm{B}}(t) \right) \\
+ \mathbf{B}_{\mathrm{BG}}(t) &\left(\frac{d}{dt} \boldsymbol{e}_{\mathrm{G}}(t) \circ \exp\left(j\boldsymbol{\delta}_{\mathrm{G}}(t)\right) + j\frac{d}{dt}\boldsymbol{\delta}_{\mathrm{G}}(t) \circ \bar{\boldsymbol{e}}_{\mathrm{G}}(t) \right) \\
+ \frac{d}{dt}&\mathbf{B}_{\mathrm{BB}}(t)\,\bar{\boldsymbol{v}}_{\mathrm{B}}(t) + \frac{d}{dt}\mathbf{B}_{\mathrm{BG}}(t)\,\bar{\boldsymbol{e}}_{\mathrm{G}}(t)\ ,
\end{aligned}
\tag{5.6}
$$

or, equivalently:

$$
\begin{aligned}
\mathbf{0} = \mathbf{B}_{\mathrm{BB}}(t) &\left(\frac{d}{dt} \boldsymbol{v}_{\mathrm{B}}(t) \circ \left(\boldsymbol{v}_{\mathrm{B}}(t) \div \boldsymbol{v}_{\mathrm{B}}(t)\right) \circ \exp\left(j\boldsymbol{\theta}_{\mathrm{B}}(t)\right) + j\frac{d}{dt}\boldsymbol{\theta}_{\mathrm{B}}(t) \circ \bar{\boldsymbol{v}}_{\mathrm{B}}(t) \right) \\
+ \mathbf{B}_{\mathrm{BG}}(t) &\left(\frac{d}{dt} \boldsymbol{e}_{\mathrm{G}}(t) \circ \left(\boldsymbol{e}_{\mathrm{G}}(t) \div \boldsymbol{e}_{\mathrm{G}}(t)\right) \circ \exp\left(j\boldsymbol{\delta}_{\mathrm{G}}(t)\right) + j\frac{d}{dt}\boldsymbol{\delta}_{\mathrm{G}}(t) \circ \bar{\boldsymbol{e}}_{\mathrm{G}}(t) \right) \\
+ \frac{d}{dt}&\mathbf{B}_{\mathrm{BB}}(t)\,\bar{\boldsymbol{v}}_{\mathrm{B}}(t) + \frac{d}{dt}\mathbf{B}_{\mathrm{BG}}(t)\,\bar{\boldsymbol{e}}_{\mathrm{G}}(t)\ ,
\end{aligned}
\tag{5.7}
$$

or, equivalently:

$$
\begin{aligned}
\mathbf{0} = \mathbf{B}_{\mathrm{BB}}(t) &\left[\left(\boldsymbol{u}_{\mathrm{B}}(t) + j\frac{d}{dt}\boldsymbol{\theta}_{\mathrm{B}}(t) \right) \circ \bar{\boldsymbol{v}}_{\mathrm{B}}(t) \right] + \frac{d}{dt}\mathbf{B}_{\mathrm{BB}}(t)\,\bar{\boldsymbol{v}}_{\mathrm{B}}(t) \\
+ \mathbf{B}_{\mathrm{BG}}(t) &\left[\left(\boldsymbol{u}_{\mathrm{G}}(t) + j\frac{d}{dt}\boldsymbol{\delta}_{\mathrm{G}}(t) \right) \circ \bar{\boldsymbol{e}}_{\mathrm{G}}(t) \right] + \frac{d}{dt}\mathbf{B}_{\mathrm{BG}}(t)\,\bar{\boldsymbol{e}}_{\mathrm{G}}(t)\ ,
\end{aligned}
\tag{5.8}
$$

where

$$
\begin{aligned}
\boldsymbol{u}_{\mathrm{B}}(t) &= \frac{d}{dt}\boldsymbol{v}_{\mathrm{B}}(t) \div \boldsymbol{v}_{\mathrm{B}}(t)\ , \\
\boldsymbol{u}_{\mathrm{G}}(t) &= \frac{d}{dt}\boldsymbol{e}_{\mathrm{G}}(t) \div \boldsymbol{e}_{\mathrm{G}}(t)\ .
\end{aligned}
\tag{5.9}
$$

and "$\div$" indicates the element-wise division.

Assume:

$$
\begin{aligned}
\bar{\boldsymbol{v}}_{\mathrm{B}}(t) &\approx \mathbf{1}_{n_{\mathrm{B}}}\ , \\
\bar{\boldsymbol{e}}_{\mathrm{G}}(t) &\approx \mathbf{1}_{n_{\mathrm{G}}}\ ,
\end{aligned}
\tag{5.10}
$$

with n_B and n_G the number of network buses and generators, respectively. Then, equation (5.8) can be simplified as:

$$0 = \mathbf{B}_{BB}(t)\boldsymbol{u}_B(t) + \frac{d}{dt}\mathbf{B}_{BB}(t) + j\mathbf{B}_{BB}(t)\frac{d}{dt}\boldsymbol{\theta}_B(t)$$
$$+ \mathbf{B}_{BG}(t)\boldsymbol{u}_G(t) + \frac{d}{dt}\mathbf{B}_{BG}(t) + j\mathbf{B}_{BG}(t)\frac{d}{dt}\boldsymbol{\delta}_G(t) .$$
(5.11)

It is important to note that (5.10) is applied only after calculating the time derivative of (5.4). It is thus a second-order approximation, which is almost always acceptable in transient conditions but during faults, which necessarily last only a few tens of milliseconds. Note also that, when the voltage magnitude is zero, condition that is verified for a zero-impedance short-circuit, its phase angle is undetermined. It is thus of little interest defining the phase angles and frequencies *during* a fault.

Equaling the imaginary part of (5.11) to zero, one has:

$$0 = \mathbf{B}_{BB}(t)\frac{d}{dt}\boldsymbol{\theta}_B(t) + \mathbf{B}_{BG}(t)\frac{d}{dt}\boldsymbol{\delta}_G(t) .$$
(5.12)

Finally, recovering the assumption of constant elements of the admittance matrices and according to the discussion on the time derivative of Park vectors and phase angles given in Section 1.4.5 and Section 3.3, respectively, (5.12) can be rewritten as:

$$\boxed{0 = \mathbf{B}_{BB}\,\Delta\boldsymbol{\omega}_B(t) + \mathbf{B}_{BG}\,\Delta\boldsymbol{\omega}_G(t)} ,$$
(5.13)

which is the sought *frequency divider formula*.

If $\mathbf{B}_{BB}$ is full rank, which is a condition commonly satisfied for conventional transmission grids,[1] bus frequencies can be determined explicitly as:

$$\Delta\boldsymbol{\omega}_B(t) = \mathbf{D}\,\Delta\boldsymbol{\omega}_G(t) ,$$
(5.14)

where $\mathbf{D}$ is the *frequency divider matrix* and is formally defined as:

$$\mathbf{D} = -\mathbf{B}_{BB}^{-1}\mathbf{B}_{BG} .$$
(5.15)

From the computational point of view, (5.14) is not the most convenient expression to implement. In fact, while $\mathbf{B}_{BB}$ and $\mathbf{B}_{BG}$ tend to be extremely sparse matrices, $\mathbf{D}$ is dense and its computational burden and memory requirements can be unacceptable for large systems. For these reasons, the use

---

[1] Zero-impedance connections can be easily taken into account in (5.13) as they impose $\omega_h = \omega_k$ at their sending (h) and receiving (k) ends. For idiosyncratic cases for which $\mathbf{B}_{BB}$ is irremediably singular, the Moore-Penrose pseudo-inverse can be used.

of (5.14) is impractical for a computer-based implementation of the FDF and may cause memory errors on common workstations. On the contrary, (5.13) has a negligible computational burden and can be included as is in the set of DAEs (2.15) that describes the power system model. However, **D** has some relevant properties that are thoroughly discussed in Section 5.5.

Equation (5.13) has another interesting property, namely, it is *acasual*. This means that it is agnostic with respect to what variables are inputs and what are outputs. In Chapter 7, (5.13) serves to calculate $\boldsymbol{\omega}_B$ from $\boldsymbol{\omega}_G$, whereas in Chapter 8, the same formula allows determining $\boldsymbol{\omega}_G$ from $\boldsymbol{\omega}_B$. On the other hand, (5.14) suggests causality as $\boldsymbol{\omega}_B$ is expressed as a function of $\boldsymbol{\omega}_G$. The interested reader can find in [30] and [208] extensive discussions on causality and its implications on the modeling and simulation of physical systems.

The simplifications that lead to (5.13) and (5.14), while may appear drastic, are motivated by usual assumptions and typical transient conditions encountered in fault analysis and transient stability analysis. The examples presented in the next section confirm the accuracy of the FDF.

5.2.1. Examples

This section presents a series of examples that serve to illustrate the rationale behind the notation *frequency divider* and to compare the dynamic behavior of (5.13) with respect to the numerical time derivative of the bus voltage phase angles as obtained, for example, through washout filters, as well as to discuss the differences between the local frequencies obtained with the FDF and the frequency of the CoI.

5.2.1.1. Two-Bus System

The simple radial system shown in Figure 5.4 is a good starting point to illustrate the FDF. The lossless connection, with total reactance $X_{hk} = X_{hi} + X_{ik}$, represents the series of the internal reactances of the machines, and series

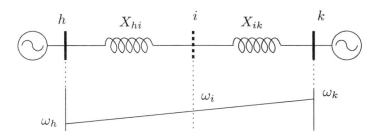

Figure 5.4 Two-bus system.

reactances of the step-up transformers and the transmission line. Hence, the frequencies at buses h and k, say ω_h and ω_k, respectively, are the rotor speeds of the synchronous generators.

The application of (5.13) to the system of Figure 5.4 leads to:

$$0 = \mathbf{B}_{BB} \, \Delta\omega_i(t) + \mathbf{B}_{BG} \begin{bmatrix} \Delta\omega_h(t) \\ \Delta\omega_k(t) \end{bmatrix} ,$$

$$0 = \left[\frac{1}{X_{hi}} + \frac{1}{X_{ik}} \right] \Delta\omega_i(t) + \left[\frac{1}{X_{hi}} \quad \frac{1}{X_{ik}} \right] \begin{bmatrix} \Delta\omega_h(t) \\ \Delta\omega_k(t) \end{bmatrix} ,$$

(5.16)

and solving for $\Delta\omega_i$:

$$\Delta\omega_i(t) = \frac{X_{hi}X_{ik}}{X_{hk}} \left[\frac{1}{X_{hi}} \quad \frac{1}{X_{ik}} \right] \begin{bmatrix} \Delta\omega_h(t) \\ \Delta\omega_k(t) \end{bmatrix}$$

$$= \frac{X_{ik}}{X_{hk}} \, \Delta\omega_h(t) + \frac{X_{hi}}{X_{hk}} \, \Delta\omega_k(t) ,$$

(5.17)

and, since $X_{hi} + X_{ik} = X_{hk}$, the expression (5.17) is valid also for total frequencies, not only for their variations, hence:

$$\omega_i(t) = \frac{X_{ik}}{X_{hk}} \, \omega_h(t) + \frac{X_{hi}}{X_{hk}} \, \omega_k(t) ,$$

(5.18)

where $\omega_{\{i,h,k\}} = \omega^{\text{ref}} + \Delta\omega_{\{i,h,k\}}$, and ω^{ref} is either 1 pu or ω_{CoI} depending on which reference frequency is used in the swing equations of the synchronous machines (see Section 2.4.1).

As a direct consequence of (5.13), the instantaneous frequency ω_i at a generic point i between the boundaries h and k is a *linear interpolation* between ω_h and ω_k (see lower part of Figure 5.4). Such a linear behavior is consistent with the assumption of steady-state conditions in the distribution of the frequency along the transmission line. Moreover, (5.18) has the same formal structure of the well-known *voltage divider* of a resistive circuit where the frequencies are analogous to voltage potentials. Hence the name of (5.13).

For this simple radial system, there exists a physical position i for which the measured frequency ω_i coincides with the frequency of the center of inertia ω_{CoI}. In fact, let M_h and M_k be the inertias of the machine at buses h and k, respectively. Then, the frequency of the center of inertia is given by:

$$\omega_{\text{CoI}}(t) = \frac{M_h}{M_h + M_k} \, \omega_h(t) + \frac{M_k}{M_h + M_k} \, \omega_k(t) ,$$

(5.19)

hence, the position i for which $\omega_i = \omega_{\text{CoI}}$ is given by:

$$X_{hi} = \frac{M_k X_{hk}}{M_h + M_k} , \qquad X_{ik} = \frac{M_h X_{hk}}{M_h + M_k} .$$

(5.20)

Depending on the ratio M_h/M_k, such a point can be anywhere along the transmission line that connects the two generators, the step-up transformers or, even, the internal reactances of a generator. The latter case occurs, for example, if $M_h \gg M_k$, i.e. in the case machine h is an equivalent model of a large network that includes several generators, while machine k models a single power plant.

For meshed networks, the determination of the ω_{CoI} based on the FDF is not straightforward and, in general, does not correspond to any physical point of the grid. The general approach to estimate ω_{CoI} is given in Section 8.2.

5.2.1.2. Three-Bus System

The transient behavior of (5.18) is illustrated through numerical simulations based on the three-bus system shown in Figure 5.5. This system includes two synchronous machines and a load. The impedances of the transmission lines include the step-up transformers and transmission lines ($\bar{Z} = 0.025 + j0.075$ pu).

The conventional QSS model is considered first. In this model, transmission lines are lumped and modeled as constant impedances and generator flux dynamics are neglected. Generators are equal and are modeled as a 6th order synchronous machine, an IEEE Type DC1 automatic voltage controller and a turbine governor with inclusion of servo and re-heater models (see Sections 4.3.1 and 4.2.1, respectively). Loads are modeled as constant admittances. The disturbance is a three-phase fault that occurs at bus 3 at $t = 1$ s and is cleared after 150 ms by opening one of the two lines connecting buses 1 and 3.

Figure 5.6 shows the transient behavior of the synchronous machine rotor speeds, the frequency of the CoI (ω_{CoI}), and the estimated frequency at the load bus using the proposed frequency divider approach. Since the inertias of the machines are equal, oscillations are averaged out from the value of ω_{CoI} as it can be readily deduced by the CoI frequency expression (2.59). The frequency ω_{CoI} can thus be used only as an indication of the overall *trend* of the system frequency. The frequency of the CoI is unable to capture the proximity to any machine of the system. On the other hand, the estimated bus frequency $\omega_{\text{B},3}$ provided by (5.18) shows oscillations in phase with the rotor speed $\omega_{\text{G},1}$, as expected, since the load bus is electrically closer to generator 1 ($X_{13} < X_{32}$).

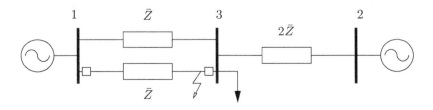

Figure 5.5 Three-bus system.

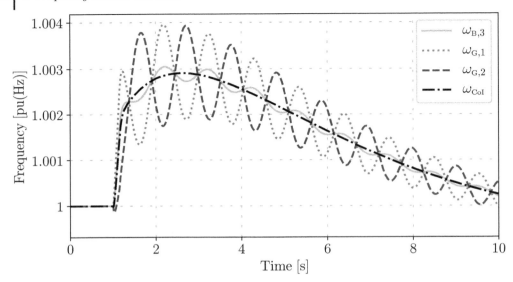

Figure 5.6 Three-bus system – Synchronous machine rotor speeds, CoI frequency, and frequency at bus 3 obtained with the FDF.

It is important to note that the admittance of the load connected to bus 3 is not included in (5.18) and need not to be known to define $\omega_{B,3}$. The load does certainly impact on the transient behavior of the system, as its power consumption affects the dynamic response of the synchronous machines at buses 1 and 2 and the rotor speeds of such machines are required to compute $\omega_{B,3}$. The load is thus *implicitly* taken into account in the FDF. Further discussion on the behavior of loads is provided in Section 6.3.2.3.

Figure 5.7 shows the results obtained with a more detailed model of the system considering eighth-order models of synchronous machines and dq-frame dynamic models of the transmission lines and the load at bus 3. The figures compares the trajectories of the frequency estimation at the load bus for the three-bus system using the FDF and the WF described in Section 3.3.1.2. All parameters are same as in Figure 5.6, which is obtained using a conventional transient stability model. This more accurate model shows that, during the fault, the frequency drops due to the effect of machine fluxes. After the fault occurrence and clearance, the frequency also shows small high-frequency oscillations which are properly captured by (5.18). These oscillations cause severe numerical issues along the entire simulation in the behavior of the WF – see also [89] for an in-depth discussion on this matter – as well as a significant delay of the filter to show the over-frequency after the line disconnection.

As indicated at the beginning of Section 5.2, one of the main assumptions on which the FDF is based is that load currents can be neglected – see

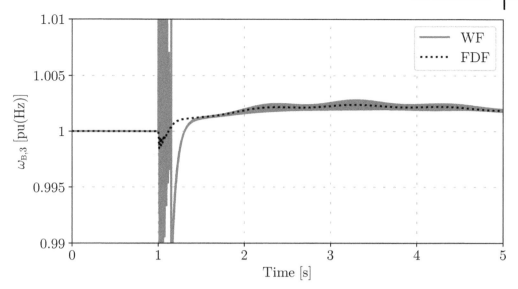

Figure 5.7 Three-bus system – Frequency at bus 3 estimated with the FDF and the WF. The system is simulated using the fully fledged dq-axis model.

equation (5.2). The remainder of this section shows that the effect of loads, including nonlinear and dynamic ones is actually negligible for the calculation of the bus frequencies. The study is based again on the dynamic response of the three-bus system of Figure 5.5 following a short-circuit at bus 3 but the constant admittance load is substituted with (i) a static voltage- and frequency-dependent load as the one described in Section 2.4.2; and (ii) a fifth-order dq-axis model of a squirrel-cage induction motor, which, as it is well known, keeps an almost constant active power consumption when operating close to the nominal speed. Results are shown in Figure 5.8.

In the simulations carried out to obtain Figure 5.8.a, the frequency $\omega_{B,3}$ is estimated using the WF or the proposed frequency divider formula, depending on the model considered. The parameters $p_{D,o}$, $q_{D,o}$, and $v_{h,o}$ are the initial load active and reactive powers and voltage magnitude at bus 3, respectively, determined with the power flow analysis. The parameters γ_p, β_p, γ_q, and β_q resemble those of an aluminum plant and are based on [14]. Finally, the dynamic model of the asynchronous motor is based on [110].

Simulation results confirm that the FDF (5.13) is accurate as it is able to estimate the frequency at the load bus similarly to the WF but avoiding the numerical issues of the latter. It is interesting to note that transient behavior of the frequency in the scenario that considers the induction motor is consistently different from that of the static load model. The load model, in fact, does impact on the overall dynamic behavior of the system and, hence, also on the

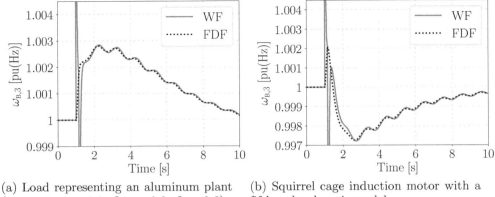

(a) Load representing an aluminum plant $(\gamma_p = 1.8, \gamma_q = 2.2, \beta_p = -0.3, \beta_q = 0.6)$

(b) Squirrel cage induction motor with a fifth-order dq-axis model

Figure 5.8 Three-bus system – Frequency at bus 3 estimated with the FDF and the WF considering nonlinear static and dynamic load models.

variations of rotor speeds of synchronous machines. Since the FDF is based on such variations, load models are indirectly taken into account in (5.13). Similar conclusions can be drawn for other static and dynamic models of loads. The FDF is thus accurate and the approximations discussed in Section 5.2, including that related on load models, are acceptable.

5.2.1.3. WSCC System

This example illustrates the dynamic performance of the FDF through the WSCC system. For clarity, matrices $\mathbf{B}_{BB}$, $\mathbf{B}_{BG}$, and $\mathbf{D}$ are shown in Table 5.1. As expected, $\mathbf{B}_{BB}$ and $\mathbf{B}_{BG}$ are sparse, whereas $\mathbf{D}$ is dense.

The frequency signals estimated through the FDF are compared with the machine rotor speeds obtained by means of a real-time digital simulator (RTDS) using EMT models of the WSCC system, and with PMU bus frequency measurements as hardware in the loop (HiL). The contingency is a three-phase fault located at bus 7. The fault, with a reactance of 10^{-5} pu(Ω), is cleared after 70 ms. The interested reader can find a description of the three-stage RTDS-PMU setup used for this example in [164].

The rotor speed of the machines, and the frequencies of buses 1, 2, and 3 measured and estimated by means of the PMUs and the FDF respectively, are represented in Figure 5.9.

The amplitude of the oscillations of the frequencies estimated by means of the FDF and the PMU measurements are very similar. However, the oscillations of PMU signals show a delay with respect to those of the rotor speed of the machine (see Sections 3.3.2 and 3.3.3 for explanations of the causes of such a delay). Moreover, the PMU frequency signals show spikes that are due to the

Table 5.1 Matrices $\mathbf{B}_{BB}$, $\mathbf{B}^{\dagger}_{BG}$, and $\mathbf{D}^{\dagger}$ of the WSCC system.

	$\mathbf{B}_{BB}$								
Bus					Bus #				
#	1	2	3	4	5	6	7	8	9
1	−30.043	0	0	17.361	0	0	0	0	0
2	0	−22.315	0	0	0	0	16.000	0	0
3	0	0	−21.702	0	0	0	0	0	17.065
4	17.361	0	0	−39.309	11.604	10.511	0	0	0
5	0	0	0	11.604	−17.338	0	5.975	0	0
6	0	0	0	10.511	0	−15.841	0	0	5.588
7	0	16.000	0	0	5.975	0	−35.446	13.698	0
8	0	0	0	0	0	0	13.698	−23.303	9.784
9	0	0	17.065	0	0	5.588	0	9.784	−32.154

	$\mathbf{B}^{\dagger}_{BG}$								
Gen.					Bus #				
#	1	2	3	4	5	6	7	8	9
1	12.682	0	0	0	0	0	0	0	0
2	0	6.315	0	0	0	0	0	0	0
3	0	0	4.637	0	0	0	0	0	0

	$\mathbf{D}^{\dagger}$								
Gen.					Bus #				
#	1	2	3	4	5	6	7	8	9
1	0.823	0.251	0.285	0.693	0.584	0.587	0.350	0.358	0.362
2	0.125	0.650	0.233	0.216	0.321	0.248	0.512	0.425	0.296
3	0.104	0.171	0.567	0.180	0.203	0.278	0.238	0.329	0.449

computation of the numerical derivative of the fast time-varying quantity. On the other hand, the frequency swings of the FDF estimations show no spikes and no delays and are thus perfectly consistent with the machine rotor speeds. The difference of the oscillation amplitudes between the machine rotor speed and the frequency at the bus of connection is due to the internal impedance of the synchronous machine. The frequency measured and estimated at the load buses 6 and 8 and at the connection bus 9 are shown in Figure 5.10.

5.2.1.4. ENTSO-E Transmission System

For computer-based applications, it is relevant to discuss the computational burden of the FDF. With this aim, the FDF is tested through a real-world system, namely the ENTSO-E transmission system. The topology and the steady-state data of the system are based on the actual real-world systems

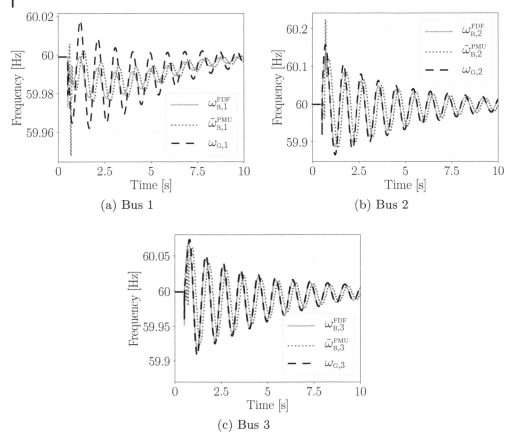

Figure 5.9 WSCC system – Synchronous machine rotor speeds and measured and estimated bus frequencies.

provided by ENTSO-E.[2] However, all dynamic data are guessed based on the knowledge of the technology of power plants.

The considered dynamic model of the ENTSO-E transmission system system includes only conventional power plants. Its large size enables the comparison of the computational burden of the conventional WF with the FDF, i.e. number of state and algebraic variables, size and sparsity of matrices and computing times.

The model of the ENTSO-E transmission system includes 21,177 buses (1,212 offline); 30,968 transmission lines and transformers (2,352 offline); 1,144 coupling devices, i.e. zero-impedance connections (420 offline); 15,756 loads

---

[2]The data of the ENTSO-E transmission system have been licensed to the first author by ENTSO-E. Data can be requested through an online application at www.entsoe.eu.

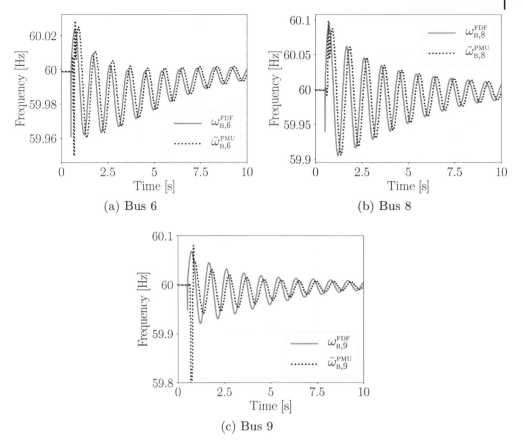

(a) Bus 6

(b) Bus 8

(c) Bus 9

Figure 5.10 WSCC system – Measured and estimated frequencies at load and connection buses.

(364 offline); and 4,828 power plants (1,160 offline). The system also includes 364 PSSs.

A comparison of the computational burden of the FDF and the WF when these are connected to all buses is considered. The case without any frequency estimator is also considered. Results are shown in Table 5.2.

The number of state and algebraic variables, and the size and sparsity of the state matrix in the three cases is first compared. Both the FDF and the WF add to the system a number of algebraic variables equal to the number of buses of the system. Then, each WF defines two state variables per bus, whereas the FDF does not include differential equations. This leads to an increase in the number of elements of the state matrix of 31% and 106% for the FDF and the WFs, respectively. The percentage of the nonzero elements with respect to the total number of elements, is reduced by about 13% by using the frequency divider, while the WFs decrease this number by about 40%.

Table 5.2 Computational burden of different washout filters and the frequency divider formula.

Feature	Base case	Frequency divider	Washout filter
State variables	49,396	49,396 (0.00%)	91,750 (85.74%)
Algebraic variables	96,768	117,945 (21.88%)	117,945 (21.88%)
Size of DAE system	146,164	167,341 (31.07%)	209,695 (105.82%)
NNZ % of Jacobian matrix	0.00256	0.00222 (−13.28%)	0.00153 (−40.23%)
DAEs initialization [s]	0.35087	0.40617 (15.76%)	0.40063 (14.18%)
Time domain analysis [s]	37.4006	41.2198 (10.21%)	44.3770 (18.65%)

A power flow analysis followed by the initialization of dynamic devices is then carried out. The computational time of the initialization is also reported in Table 5.2. This consists mainly in the setup of synchronous machines and primary regulators' state and algebraic variables, and the computation of the matrix $\mathbf{D}$ or of the initial values of the variables of the WFs, depending on the frequency estimator that is included in the model. Both the FDF and the WFs increase this value by about 16% and 14%, respectively.

Finally, a time domain simulation (TDS) is performed for each scenario. The simulation lasts 5 s, and the contingency considered is a three-phase fault, cleared after 200 ms. The time step of the TDS is 0.02 s. The implicit trapezoidal method is used for the time integration, and each integration step is solved by using the dishonest Newton method [127]. Table 5.2 shows that installing a WF at every bus increases the computational time of the TDS by about 19%, while this time is about 10% higher in the case of the FDF.

Table 5.3 shows the size and number of nonzero elements of the aforementioned matrices for the ENTSO-E transmission system. As opposed to matrices $\mathbf{B}_{\mathrm{BB}}$ and $\mathbf{B}_{\mathrm{BG}}$, which are extremely sparse, matrix $\mathbf{D}$ is almost dense and thus its computational burden is unacceptable for large systems. This fact justifies the implementation in a software tool of equation (5.13) rather than (5.14).

Table 5.3 Size and number of nonzero (NNZ) elements of matrices $\mathbf{B}_{\mathrm{BB}}$, $\mathbf{B}_{\mathrm{BG}}$, and $\mathbf{D}$ for the ENTSO-E transmission system.

Matrix	Size	NNZ	NNZ %
$\mathbf{B}_{\mathrm{BB}}$	$21{,}177 \times 21{,}177$	72,313	0.0161
$\mathbf{B}_{\mathrm{BG}}$	$21{,}177 \times 4{,}832$	4,832	0.0047
$\mathbf{D}$	$21{,}177 \times 4{,}832$	86,169,456	84.2

5.3. Equivalent Networks

Since the FDF is based on the network admittance matrix, all available techniques for network reduction and equivalencing can be used for the determination of bus frequencies. This comes particularly handy when only a reduced set of bus frequencies are required. Two well-known techniques are considered below, namely the Ward equivalent [223], and the Radial, Equivalent and Independent (REI) equivalent [49].

The Ward equivalent is a sort of generalized Thévenin equivalent, which has been widely utilized in the literature [11, 46, 194]. The network is split into two regions (see Figure 5.11), the internal region (I) and external one (E). With this notation and using adequate partitions of the extended admittance matrix $\bar{\mathbf{Y}}_{\text{ext}}$, the link between network currents and voltage and generator EMFs can be rewritten as follows:

$$
\begin{bmatrix} \bar{i}_G^E(t) \\ \bar{i}_B^E(t) \\ \bar{i}_G^I(t) \\ \bar{i}_B^I(t) \end{bmatrix} = \begin{bmatrix} \bar{\mathbf{Y}}_{GG}^{EE} & \bar{\mathbf{Y}}_{GB}^{EE} & \bar{\mathbf{Y}}_{GG}^{EI} & \bar{\mathbf{Y}}_{GB}^{EI} \\ \bar{\mathbf{Y}}_{BG}^{EE} & \bar{\mathbf{Y}}_{BB}^{EE} & \bar{\mathbf{Y}}_{BG}^{EI} & \bar{\mathbf{Y}}_{BB}^{EI} \\ \bar{\mathbf{Y}}_{GG}^{IE} & \bar{\mathbf{Y}}_{GB}^{IE} & \bar{\mathbf{Y}}_{GG}^{II} & \bar{\mathbf{Y}}_{GB}^{II} \\ \bar{\mathbf{Y}}_{BG}^{IE} & \bar{\mathbf{Y}}_{BB}^{IE} & \bar{\mathbf{Y}}_{BG}^{II} & \bar{\mathbf{Y}}_{BB}^{II} \end{bmatrix} \begin{bmatrix} \bar{e}_G^E(t) \\ \bar{v}_B^E(t) \\ \bar{e}_G^I(t) \\ \bar{v}_B^I(t) \end{bmatrix} , \tag{5.21}
$$

and, following the same assumptions and derivations discussed in Section 5.2 and deriving with respect to time, the FDF for the set of buses of the internal system becomes:

$$
\begin{aligned}
0 &= \mathbf{B}_{BB}^{II} \Delta\boldsymbol{\omega}_B^I(t) + \mathbf{B}_{BG}^{IE} \Delta\boldsymbol{\omega}_G^E(t) + \mathbf{B}_{BG}^{II} \Delta\boldsymbol{\omega}_G^I(t) \\
&= \mathbf{B}_{BB}^{II} \Delta\boldsymbol{\omega}_B^I(t) + \mathbf{B}_{BG}^{I} \Delta\boldsymbol{\omega}_G(t) ,
\end{aligned} \tag{5.22}
$$

where:

$$
\mathbf{B}_{BB}^{II} = \text{Im}\{ \bar{\mathbf{Y}}_{BB}^{II} - \bar{\mathbf{Y}}_{BB}^{IE} (\bar{\mathbf{Y}}_{BB}^{EE})^{-1} \bar{\mathbf{Y}}_{BB}^{EI} \} ,
$$

$$
\mathbf{B}_{BG}^{IE} = \text{Im}\{ \bar{\mathbf{Y}}_{BG}^{IE} - \bar{\mathbf{Y}}_{BB}^{IE} (\bar{\mathbf{Y}}_{BB}^{EE})^{-1} \bar{\mathbf{Y}}_{BG}^{EE} \} ,
$$

$$
\mathbf{B}_{BG}^{II} = \text{Im}\{ \bar{\mathbf{Y}}_{BG}^{II} - \bar{\mathbf{Y}}_{BB}^{IE} (\bar{\mathbf{Y}}_{BB}^{EE})^{-1} \bar{\mathbf{Y}}_{BG}^{EI} \} ,
$$

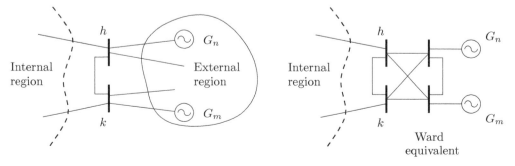

Figure 5.11 Ward equivalent.

and $\mathbf{B}_{BG}^I = [\mathbf{B}_{BG}^{IE} \quad \mathbf{B}_{BG}^{II}]$. If $\mathbf{B}_{BB}^{II}$ is full rank, the expression (5.22) can be also written in explicit form:

$$\Delta\boldsymbol{\omega}_B^I(t) = \mathbf{D}^{IE} \Delta\boldsymbol{\omega}_G^E(t) + \mathbf{D}^{II} \Delta\boldsymbol{\omega}_G^I(t)$$
$$= \mathbf{D}^I \Delta\boldsymbol{\omega}_G(t) ,$$

(5.23)

where:

$$\mathbf{D}^{IE} = -\left(\mathbf{B}_{BB}^{II}\right)^{-1} \mathbf{B}_{BG}^{IE} ,$$
$$\mathbf{D}^{II} = -\left(\mathbf{B}_{BB}^{II}\right)^{-1} \mathbf{B}_{BG}^{II} ,$$

and $\mathbf{D}^I = [\mathbf{D}^{IE} \ \mathbf{D}^{II}]$. Since they depend on the inverse of the admittance matrix of the external region, Ward equivalents tend to be characterized by dense matrices, i.e. everything connected with everything. A typical approach is to neglect the elements below a given threshold.

The REI equivalent is another well-known approach, with several applications [55, 153, 209]. It consists in the determination of a radial network at each retained bus of the internal region (see Figure 5.12). If bus h is a frontier bus of the internal region of the network, i.e. bus h is belongs to the internal region and is connected to some buses of the external region, and the subscript "R" indicates the union of the buses of the generators included in the external region and bus h, one has:

$$\begin{bmatrix} \bar{\imath}_R^E(t) \\ \bar{\imath}_B^E(t) \end{bmatrix} = \begin{bmatrix} \bar{\mathbf{Y}}_{RR}^E & \bar{\mathbf{Y}}_{RB}^E \\ \bar{\mathbf{Y}}_{BR}^E & \bar{\mathbf{Y}}_{BB}^E \end{bmatrix} \begin{bmatrix} \bar{v}_R^E(t) \\ \bar{v}_B^E(t) \end{bmatrix} ,$$

(5.24)

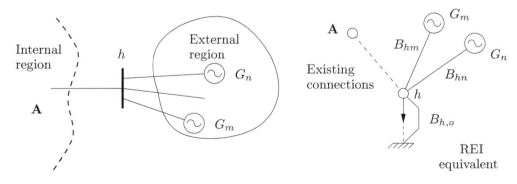

Figure 5.12 REI equivalent.

where $\bar{v}_R^E = [\bar{e}_G^{E\dagger} \ \bar{v}_h^\dagger]^\dagger$ and, with the usual assumptions and notation:

$$\mathbf{B}_R^E = \mathrm{Im}\{\bar{\mathbf{Y}}_{RR}^E - \bar{\mathbf{Y}}_{RB}^E (\bar{\mathbf{Y}}_{BB}^E)^{-1} \bar{\mathbf{Y}}_{BR}^E\} \ , \tag{5.25}$$

which, remembering the meaning of the subscript "R," can be rewritten as:

$$\mathbf{B}_R^E = \begin{bmatrix} \mathbf{B}_{GG}^E & \mathbf{B}_{Gh}^E \\ \mathbf{B}_{hG}^E & B_{hh}^E \end{bmatrix} \ . \tag{5.26}$$

The elements of the row vector $\mathbf{B}_{hG}^E$ are the series admittances of the fictitious lines that connect the considered frontier bus h with the retained generator buses of the external region of the grid. The shunt element at bus h that results from the REI equivalent is given by:

$$B_{h,o} = -B_{hh}^E - \mathbf{B}_{hG}^E \mathbf{1}_{n_{GE}} \ . \tag{5.27}$$

where n_{GE} is the number of generators that belong to the external region.

Finally, the row vector $\mathbf{B}_{hG}^E$ and the element B_{hh}^E are added to the row corresponding to bus h of the matrix $\mathbf{B}_{BG}^I$ and the diagonal element corresponding to bus h of matrix $\mathbf{B}_{BB}^I$ of the FDF of the retained grid.

The REI equivalent is symmetrical, i.e. one can deduce the REI equivalent also at bus h for the internal region and deduce the matrix:

$$\mathbf{B}_R^I = \begin{bmatrix} \mathbf{B}_{GG}^I & \mathbf{B}_{Gh}^I \\ \mathbf{B}_{hG}^I & B_{hh}^I \end{bmatrix} \ . \tag{5.28}$$

Finally, the frequency variation at bus h is given by:

$$\begin{aligned} 0 &= (B_{hh}^I + B_{hh}^E)\Delta\omega_h(t) + \mathbf{B}_{hG}^E \Delta\boldsymbol{\omega}_G^E(t) + \mathbf{B}_{hG}^I \Delta\boldsymbol{\omega}_G^I(t) \\ &= (B_{hh}^I + B_{hh}^E)\Delta\omega_h(t) + \mathbf{B}_{hG} \Delta\boldsymbol{\omega}_G(t) \ , \end{aligned} \tag{5.29}$$

where $\mathbf{B}_{hG} = [\mathbf{B}_{hG}^E \ \mathbf{B}_{hG}^I]$. In explicit form, (5.29) becomes:

$$\begin{aligned} \Delta\omega_h(t) &= \mathbf{D}_h^E \Delta\boldsymbol{\omega}_G^E(t) + \mathbf{D}_h^I \Delta\boldsymbol{\omega}_G^I(t) \\ &= \mathbf{D}_h \Delta\boldsymbol{\omega}_G(t) \ , \end{aligned} \tag{5.30}$$

where:

$$\begin{aligned} \mathbf{D}_h^E &= -(B_{hh}^I + B_{hh}^E)^{-1} \mathbf{B}_{hG}^E \ , \\ \mathbf{D}_h^I &= -(B_{hh}^I + B_{hh}^E)^{-1} \mathbf{B}_{hG}^I \ , \end{aligned}$$

and $\mathbf{D}_h = [\mathbf{D}_h^{\mathrm{E}} \ \mathbf{D}_h^{\mathrm{I}}]$. Thus the REI equivalent enables the calculation of a single bus frequency of the system as a function of the rotor speeds of the grid. This is convenient for large systems, for which the full FDF defined in (5.13) may require the storage of large matrices, as discussed in Section 5.2.1.4.

5.3.1. Example

This example illustrates the application of the Ward and REI equivalents to the FDF through the WSCC system.

Ward Equivalent

Let the internal and external regions consist of the following two bus sets $\mathbb{B}_{\mathrm{I}} = \{3, 6, 8, 9\}$ and $\mathbb{B}_{\mathrm{E}} = \{1, 2, 4, 5, 7\}$, respectively. Two external generators are connected to buses 1 and 2, whereas the internal generator is connected to bus 3. Then, one obtains:

$$
\mathbf{B}_{BB}^{\mathrm{II}} =
\begin{bmatrix}
-21.702 & 0 & 0 & 17.065 \\
0 & -10.550 & 1.259 & 5.588 \\
0 & 1.259 & -14.445 & 9.784 \\
17.065 & 5.588 & 9.784 & -32.154
\end{bmatrix},
\tag{5.31}
$$

$$
\mathbf{B}_{BG}^{\mathrm{IE}} =
\begin{bmatrix}
0 & 0 \\
3.734 & 0.423 \\
0.888 & 2.934 \\
0 & 0
\end{bmatrix}, \qquad
\mathbf{B}_{BG}^{\mathrm{II}} =
\begin{bmatrix}
4.637 \\
0 \\
0 \\
0
\end{bmatrix},
\tag{5.32}
$$

and

$$
\mathbf{D}^{\mathrm{IE}} =
\begin{bmatrix}
0.285 & 0.233 \\
0.589 & 0.247 \\
0.359 & 0.425 \\
0.363 & 0.296
\end{bmatrix}, \quad
\mathbf{D}^{\mathrm{II}} =
\begin{bmatrix}
0.566 \\
0.277 \\
0.328 \\
0.448
\end{bmatrix}, \quad
\mathbf{D}^{\mathrm{I}} =
\begin{bmatrix}
0.285 & 0.233 & 0.566 \\
0.589 & 0.247 & 0.277 \\
0.359 & 0.425 & 0.328 \\
0.363 & 0.296 & 0.448
\end{bmatrix}.
\tag{5.33}
$$

REI Equivalent at Bus 9

Assuming that the external and internal regions consist of the sets of buses $\mathbb{B}_{\mathrm{E}} = \{1, 2, 4, 5, 6, 7, 8\}$ and $\mathbb{B}_{\mathrm{I}} = \{3, 9\}$, respectively, then one has:

$$
\mathbf{B}_R^{\mathrm{E}} =
\begin{bmatrix}
-3.350 & 0.719 & 2.967 \\
0.719 & -2.934 & 2.410 \\
2.967 & 2.410 & -4.787
\end{bmatrix},
\tag{5.34}
$$

and, hence:

$$B_{99}^{E} = -4.787 , \qquad B_{9G}^{E} = [2.967 \quad 2.410] .$$

Similarly, for $\mathbf{B}_{R}^{I}$, one has:

$$\mathbf{B}_{R}^{I} = \begin{bmatrix} -3.646 & 3.646 \\ 3.646 & -3.646 \end{bmatrix} , \tag{5.35}$$

and, hence:

$$B_{99}^{I} = -3.646 , \qquad B_{9G}^{I} = [3.646] .$$

Finally, the FDF matrices at bus 9 are:

$$\mathbf{D}_{9}^{E} = [0.352 \ 0.286] , \quad \mathbf{D}_{9}^{I} = [0.432] , \quad \mathbf{D}_{9} = [0.352 \ 0.286 \ 0.432] . \tag{5.36}$$

As expected, the matrices $\mathbf{D}^{I}$ and $\mathbf{D}_{9}$ are slices of the matrix $\mathbf{D}$ shown in Table 5.1. The advantage of using Ward or REI equivalents is that one needs to factorize smaller matrices than the full $\mathbf{B}_{BB}$. For large networks, this can lead to a reduction of the computational burden.

5.4. Inclusion of Measurements

This section discusses how the FDF (5.13) can be modified to include frequency measurements as provided, for example, by PMU devices. The set of network buses can be split as $\boldsymbol{\omega}_{B} = [\tilde{\boldsymbol{\omega}}_{M}^{\dagger} \ \boldsymbol{\omega}_{U}^{\dagger}]^{\dagger}$, where $\tilde{\boldsymbol{\omega}}_{M}$ are the known measured bus frequencies and $\boldsymbol{\omega}_{U}$ are the remaining unknown bus frequencies. Using same notation as for (5.1), and neglecting the time dependence of the admittance matrix, one has:

$$\begin{bmatrix} \bar{\imath}_{G}(t) \\ \bar{\imath}_{M}(t) \\ \bar{\imath}_{U}(t) \end{bmatrix} = \begin{bmatrix} \bar{\mathbf{Y}}_{GG} & \bar{\mathbf{Y}}_{GM} & \bar{\mathbf{Y}}_{GU} \\ \bar{\mathbf{Y}}_{MG} & \bar{\mathbf{Y}}_{MM} & \bar{\mathbf{Y}}_{MU} \\ \bar{\mathbf{Y}}_{UG} & \bar{\mathbf{Y}}_{UM} & \bar{\mathbf{Y}}_{UU} \end{bmatrix} \begin{bmatrix} \bar{e}_{G}(t) \\ \bar{v}_{M}(t) \\ \bar{v}_{U}(t) \end{bmatrix} , \tag{5.37}$$

and, following the same derivations discussed in the previous section, the FDF (5.13) becomes:

$$\mathbf{0} = \mathbf{B}_{UU} \, \Delta \boldsymbol{\omega}_{U}(t) + \begin{bmatrix} \mathbf{B}_{UG} & \mathbf{B}_{UM} \end{bmatrix} \begin{bmatrix} \Delta \boldsymbol{\omega}_{G}(t) \\ \Delta \tilde{\boldsymbol{\omega}}_{M}(t) \end{bmatrix} . \tag{5.38}$$

A relevant application of (5.38) is the estimation of the frequency variations at the buses located in a given region of the grid, provided that the set $[\Delta\boldsymbol{\omega}_G^\dagger \; \Delta\tilde{\boldsymbol{\omega}}_M^\dagger]^\dagger$ contains the machine rotor speeds and/or the PMU measurements required to define all boundary conditions of such a region.

Consider an AC grid (Grid 1) connected synchronously with other neighboring AC grids (Grid 2, 3, ..., n). The system operator of Grid 1 will likely not know the topology nor the synchronous machine rotor speeds of the neighboring grids. Therefore, the FDF of Equation (5.13) cannot be applied. The system operator can, instead, measure the frequency at the border buses with the other grids, and use them as the boundary conditions of the problem.

To apply (5.38), one needs to properly define $\mathbf{B}_{UU}$ and $\mathbf{B}_{UM}$. It has been shown in this chapter that the rotor speed of a synchronous machine does not coincide with the frequency measured at the connection bus, due to the influence of the internal reactance of the machine (see Figure 5.9). This is collected in both the system extended admittance matrix, and in $\mathbf{B}_{UG}$.

In the same vein, the frequency measured at a border bus can be viewed as the rotor speed of a *fictitious synchronous machine* connected to that bus, and the reactance of the line connecting such a bus to the rest of the monitored region is the reactance of such a fictitious machine. With this assumption, the coefficients of matrix $\mathbf{B}_{UM}$ are given by the susceptances of the lines that belong to the subsystem under study and that are connected to each border bus. Note that, while $\mathbf{B}_{UG}$ has only one nonzero element per row because the EMF of the generator is always connected in antenna to the grid through its internal reactance, each row of $\mathbf{B}_{UM}$ has as many nonzero elements as the number of connections of the bus corresponding to that row with the buses of the internal region.

5.4.1. Examples

This concept of *measurement-based* FDF is illustrated by means of two examples based on the WSCC system simulated with the RTDS-PMU setup considered in Section 5.2.1.3.

In the first scenario, the WSCC system is split into three regions, as shown in Figure 5.13. The contingency is the same as that considered in Section 5.2.1.3, i.e. a three-phase fault at bus 7. The region where the FDF is applied is Grid 1, which comprises buses 3 (synchronous machine bus), 6 and 8 (load and interconnection buses), and 9 (internal connection bus). Two PMUs are placed in buses 6 and 8, and the rotor speed of the machine at bus 3 is known. Thus, the known frequencies are $\boldsymbol{\omega}_G = [\omega_{G,3}]$ and $\tilde{\boldsymbol{\omega}}_M^\dagger = [\tilde{\omega}_{B,6} \; \tilde{\omega}_{B,8}]$, and the vector of unknown frequencies is $\breve{\boldsymbol{\omega}}_U^\dagger = [\breve{\omega}_{B,3} \; \breve{\omega}_{B,9}]$. Particularizing equation

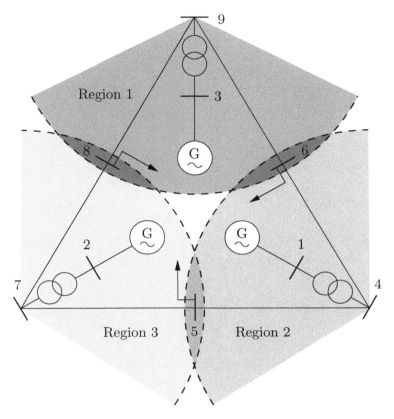

Figure 5.13 WSCC system divided into three regions.

(5.38) to the example described above, one has:

$$
\begin{bmatrix} \Delta \breve{\omega}_{\mathrm{B},3}(t) \\ \Delta \breve{\omega}_{\mathrm{B},9}(t) \end{bmatrix} = - \begin{bmatrix} -22.702 & 17.065 \\ 17.065 & -32.154 \end{bmatrix}^{-1} \begin{bmatrix} 4.637 & 0 & 0 \\ 0 & 5.588 & 9.784 \end{bmatrix} \begin{bmatrix} \Delta \omega_{\mathrm{G},3}(t) \\ \Delta \tilde{\omega}_{\mathrm{B},6}(t) \\ \Delta \tilde{\omega}_{\mathrm{B},8}(t) \end{bmatrix}
$$

$$
= \begin{bmatrix} 0.340 & 0.217 & 0.381 \\ 0.180 & 0.289 & 0.506 \end{bmatrix} \begin{bmatrix} \Delta \omega_{\mathrm{G},3}(t) \\ \Delta \tilde{\omega}_{\mathrm{B},6}(t) \\ \Delta \tilde{\omega}_{\mathrm{B},8}(t) \end{bmatrix} , \tag{5.39}
$$

where the elements of $\mathbf{B}_{\mathrm{UU}}$ are the elements $(3,3)$, $(3,9)$, $(9,3)$, and $(9,9)$ of the bottom matrix of Table 2.5; and the elements of $\mathbf{B}_{\mathrm{UG}}$ and $\mathbf{B}_{\mathrm{UM}}$ represent the internal susceptance of the synchronous machine at bus 3, and the susceptances of the lines 9-6 and 9-8, respectively. No information regarding buses 1, 2, 4, 5, nor 7 is required.

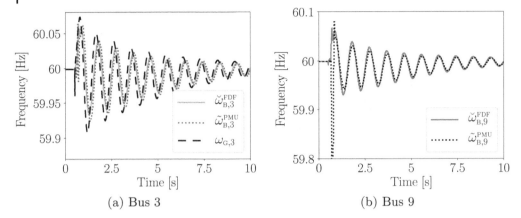

(a) Bus 3   (b) Bus 9

Figure 5.14 WSCC system – Frequency measured and estimated by means of the reduced FDF at buses 3 and 9.

Figure 5.14 shows the frequency measured and estimated by means of the FDF at the machine bus 3, and at the connection bus 9. Figure 5.14.a also includes the rotor speed of the machine at bus 3. The estimation provided by the FDF is highly accurate. Moreover, the inclusion of the machine rotor speed in the estimation naturally smooths the spikes of the PMU measurements, and reduces their delay.

In the second scenario, no rotor speed is measured. This is a relevant case for TSOs that do not have access, at least in a deregulated environment, to the measurements of the rotor speeds of the synchronous machines. If the TSO can measure the frequencies at the high-voltage winding of the three step-up transformers of the WSCC system, i.e. buses 4, 7, and 9, then $\boldsymbol{\omega}_{\mathrm{G}} = \varnothing$ and $\tilde{\boldsymbol{\omega}}_{\mathrm{M}}^{\dagger} = [\tilde{\omega}_{\mathrm{B},4} \quad \tilde{\omega}_{\mathrm{B},7} \quad \tilde{\omega}_{\mathrm{B},9}]$. With these measurements and the knowledge of the matrix $\mathbf{B}_{\mathrm{BB}}$,[3] it is possible to estimate the frequencies at the buses 5, 6, and 8, i.e. $\breve{\boldsymbol{\omega}}_{\mathrm{U}}^{\dagger} = [\breve{\omega}_{\mathrm{B},5} \quad \breve{\omega}_{\mathrm{B},6} \quad \breve{\omega}_{\mathrm{B},8}]$, as follows:

$$
\begin{bmatrix} \Delta\breve{\omega}_{\mathrm{B},5}(t) \\ \Delta\breve{\omega}_{\mathrm{B},6}(t) \\ \Delta\breve{\omega}_{\mathrm{B},8}(t) \end{bmatrix} = - \begin{bmatrix} -17.34 & 0 & 0 \\ 0 & -15.84 & 0 \\ 0 & 0 & -23.30 \end{bmatrix}^{-1} \begin{bmatrix} 11.60 & 5.98 & 0 \\ 10.51 & 0 & 5.59 \\ 0 & 13.70 & 9.78 \end{bmatrix} \begin{bmatrix} \Delta\tilde{\omega}_{\mathrm{B},4}(t) \\ \Delta\tilde{\omega}_{\mathrm{B},7}(t) \\ \Delta\tilde{\omega}_{\mathrm{B},9}(t) \end{bmatrix}
$$

$$
= \begin{bmatrix} 0.67 & 0.34 & 0 \\ 0.66 & 0 & 0.35 \\ 0 & 0.59 & 0.42 \end{bmatrix} \begin{bmatrix} \Delta\tilde{\omega}_{\mathrm{B},4}(t) \\ \Delta\tilde{\omega}_{\mathrm{B},7}(t) \\ \Delta\tilde{\omega}_{\mathrm{B},9}(t) \end{bmatrix} .
$$

---

[3] Since no rotor speed measurements are utilized, the knowledge of matrix $\mathbf{B}_{\mathrm{bus}}$ will suffice in this case.

The WSCC system is a small network and, hence, the number of estimated bus frequencies are only three in this example. It is relevant to note, however, that measuring the frequency at all generator buses enables the estimation of the frequencies at *all* other network buses, independently on their number. The only condition to be satisfied is $\mathbf{B}_{UU}$ being full rank.

The measurement-based FDF appears thus as a useful tool for system operators. Among its potential applications, the reduced FDF enables the monitoring of the frequency at every bus of the grid without requiring the knowledge of any information of neighboring networks or of generator rotor speeds.

5.5. Frequency Participation Factors

The problem of how to define the participation of generators in losses and transmission rights in transmission systems has been under intense research for more than two decades [17, 34, 40–43, 50, 70, 106, 166]. The solution of such a problem is particularly relevant in electricity markets where the costs of the utilization of the network should be shared among all participants.

In the literature, the concept of *participation factors* has been extensively applied to a variety of electrical quantities. In [34], two analytical expressions, referred to as *power divider laws*, are proposed to define the sensitivities of line current flows to nodal current injections, called *current injection sensitivity factors*. The aim of the power divider laws is to map nodal active- and reactive-power injections to active- and reactive-power flows on the lines of AC transmission networks.

In [167], the authors propose a model predictive control (MPC)-based coordinated control for multi-terminal HVDC grids to ensure power balance in the system while avoiding current and or voltage limit violations, in a time frame of seconds or a few minutes. To this aim, [167] considers the concept of power participation factors (PPFs) to update the reference powers of the converters of the HVDC systems, being the summation of all PPFs equal to unity.

An MPC-based AGC for secondary frequency control of multi-area power systems is proposed in [63]. In the optimization problem, participation factors for all synchronous machines are defined as optimization variables to determine the outputs of the AGC that are distributed to each generator.

Finally, in [73] and [144], the authors propose to evaluate the sensitivities of voltage and reactive power variations based on a partition of the Jacobian matrix of the power flow equations. The voltage participation factors (VPFs) defined in these references are computed based on the right and left eigenvectors of the eigenvalues of such a reduced Jacobian matrix.

An analogous emerging problem is the participation of synchronous machines or the border buses between regions to the frequency variations at the buses of an area of interest. To be able to evaluate such participation is relevant as the high penetration of nonsynchronous, often renewable generation leads to a drastic reduction of the inertia and frequency control of the system [207, 212] and, potentially, to a considerable impact on the variation and rate of change of the frequency [35, 56, 151, 202, 216, 222]. This section presents a procedure to evaluate the frequency participation factors (FPFs) based on the FDF, and shows quantitative tools to define how the inertia present in the system influences frequency variations at network buses.

It is relevant to note that matrix $\mathbf{D}$ defined in (5.15), from a formal point of view, is structurally similar to the matrix $\bar{\mathbf{F}}_{\mathrm{LG}}$ defined in [1, 103, 206],[4] which gives the percentage of participation that each generator has in establishing the no-load voltage at each bus. Both matrices, in fact, are obtained from the partitioning or augmentation of the network admittance matrix $\bar{\mathbf{Y}}_{\mathrm{bus}}$, as originally proposed in [103] in the context of voltage stability analysis. The observation that matrices $\mathbf{D}$ and $\bar{\mathbf{F}}_{\mathrm{LG}}$ are formally equivalent leads to the following relevant property: *the elements of each row of* $\mathbf{D}$ *are the FPFs of each synchronous machine rotor speed (or any other device and node at which the frequency is imposed) to the bus frequencies* [158].

It can be easily observed that all hypotheses that are assumed in Theorem 2.1 in [45] and that apply to $\bar{\mathbf{Y}}_{\mathrm{bus}}$ and $\bar{\mathbf{F}}_{\mathrm{LG}}$ also apply to $\bar{\mathbf{Y}}_{\mathrm{ext}}$ and $\mathbf{D}$, respectively. This leads to the conclusion that $\bar{\mathbf{F}}_{\mathrm{LG}}$ and $\mathbf{D}$ can be shown to have same properties and, hence, the rows of $\mathbf{D}$ sum close to 1. Note also that $\mathbf{D}$ is real by construction and, hence, the condition that R/X ratio has to be the same for every network branch does not apply (see Proposition 2.1 in [45]).

As stated in Section 5.4, the quantities forming the vector $\boldsymbol{\omega}_{\mathrm{G}}$ do not need to be obtained from synchronous machines. Boundary buses that define the interconnections with external grids or buses at which the frequency is controlled by large nonsynchronous generators can be used in (5.15), provided that accurate frequency measures are available at those buses, e.g. by means of PMU devices. In the remainder of this chapter, without loss of generality, it is assumed that $\boldsymbol{\omega}_{\mathrm{G}}$ consists of synchronous machine rotor speed measurements.

The FPFs show relevant differences both conceptually and/or in their applications with respect to the references above, as follows. First, the FPFs are not based on eigenvalue analysis as opposed to the VPFs presented in [73, 144]. Rather, the FPFs are the components of the linear expressions that relate each bus frequency to the rotor speeds of synchronous machines. Then, the proposed FPFs are a consequence of the topology of the system and machine parameters. Therefore, FPFs cannot be chosen/modified for/by a controller or any other agent such as system operators. This is in contrast with the participation

---

[4]In the notation of these papers, "L" stands for load buses and "G" for generator buses.

factors described in [63] and [167], as they are chosen based on the capability of each converter to reschedule their power flows, and on the result of an optimal control problem, respectively.

Despite all the aforementioned advantages of matrix **D**, it nevertheless shows a significant limitation from the practical implementation point of view. As discussed in Section 5.2.1.4, in fact, **D** is dense, i.e. each bus frequency depends on the rotor speeds of *all* machines. It is to be expected, however, that rotor speeds do not all weight in the same way when calculating the frequency at a given bus. One thus needs an efficient and robust criterion to define which elements of the rows of **D** are most relevant for the calculation of bus frequencies without loss of accuracy. The remainder of this section defines such a trade-off.

5.5.1. Calculation

This section presents two approaches to reduce the density of the frequency divider matrix **D** while retaining the accuracy of bus frequency estimation. The aim of this density reduction is to define the most relevant FPFs. The features of each approach are duly discussed.

Let ς_h be the summation of the elements of the h-th row of **D**. As stated in the previous section, the following applies:

$$\varsigma_h = \sum_{k=1}^{n_G} D_{hk} \approx 1, \quad \forall h = 1, \ldots, n_B . \tag{5.40}$$

Each element D_{hk} of the frequency divider matrix **D** thus represents the FPF – or normalized weight – of the synchronous machine rotor speed or the frequency measurement $\omega_{G,k}$ to the frequency of bus $\omega_{B,h}$.

Since matrix **D** is dense, the FPFs D_{hk} tend to be all nonzero.

Let ρ_{FD} be the density index of matrix **D**, such that:

$$\rho_{FD} = 100 \cdot \frac{n_{NNZ}}{n_G \, n_B} , \tag{5.41}$$

where n_{NNZ} is the number of nonzero elements of **D**. According to the discussion above, one has, in general, $\rho_{FD} \approx 100\%$. For large networks, however, one would expect that generators that are geographically (and electrically) far away from a given bus, do not significantly participate in the frequency of that bus. This intuition is confirmed by the observation that, at least in large networks, a large number of elements of **D** are *small*. The following sections describe two complementary approaches to quantify how small an element has to be such that it can be safely neglected without compromising the accuracy of the estimation of bus frequencies.

5.5.1.1. Approach 1 (A1)

The elements of each row of $\mathbf{D}$ are sorted in descending order according to their magnitudes. Then, the first, and thus the biggest m_h elements of each row of the sorted matrix $\mathbf{D}^>$ are summed such that:

$$\sum_{k=1}^{m_h} D_{hk}^> < \kappa_1 , \tag{5.42}$$

where $\kappa_1 \in [0, 1]$ is a given threshold. Finally, the reduced matrix $\mathbf{D}^{\mathrm{r}}$ is obtained by setting to zero all elements $D_{hk}^>$ with $k = m_h + 1, \ldots, n_{\mathrm{B}}$, and rearranging $D_{hk}^>$ according to their original positions before the sorting, i.e. D_{hk}. Therefore, if $\kappa_1 \to 0$, the sparsity of $\mathbf{D}^>$ and, thus, of $\mathbf{D}^{\mathrm{r}}$, increases and its accuracy decreases. Limit cases are:

$$\mathbf{D}^{\mathrm{r}} = \begin{cases} \mathbf{0} , & \text{if } \kappa_1 = 0 , \\ \mathbf{D} , & \text{if } \kappa_1 = 1 . \end{cases} \tag{5.43}$$

The main advantage of this approach is that it guarantees at least the specified level of accuracy of the frequency estimation at every bus. However, the sorting of $\mathbf{D}^>$, and the evaluation of (5.42) must be done for each row, thus leading to a high computational burden for large networks.

5.5.1.2. Approach 2 (A2)

The reduced matrix $\mathbf{D}^{\mathrm{r}}$ is obtained by neglecting all the elements of $\mathbf{D}$ that are below a threshold, as follows:

$$D_{hk}^{\mathrm{r}} = \begin{cases} 0 , & \text{if } D_{hk} < \kappa_2 \cdot \max(\mathbf{D}) , \\ D_{hk}, & \text{otherwise,} \end{cases} \tag{5.44}$$

where $\max(\mathbf{D})$ is the maximum value of the elements in $\mathbf{D}$, and $\kappa_2 \in [0, 1]$ is a given parameter and, generally, $\kappa_2 \ll 1$.

While this approach is considerably simpler and computationally more efficient than the previous one as it does not require sorting, it lacks the capability to control the desired accuracy of the estimated bus frequencies as provided by A1. If the FPFs of $\boldsymbol{\omega}_{\mathrm{G}}$ to the frequency of the h-th bus are similar, and if κ_2 is too high, there is the risk of neglecting several relevant measurements. On the other hand, if κ_2 is too low, all FPFs and thus also immaterial ones are taken into account. Therefore, a careful, network-based tuning of κ_2 is required.

5.5.2. Examples

The features, accuracy and computational efficiency of the approaches A1 and A2 are illustrated by means of the New England system, whose data are provided in Appendix A.4, as well as of two real-world systems, namely, a 1,479-bus model of the All-Island Irish Transmission System (AIITS), which is described in Appendix B, and the 21,177-bus model of the ENTSO-E transmission system considered in Section 5.2.1.4.

5.5.2.1. New England System

The results for A1 and A2 for the New England system are shown in Tables 5.4 and 5.5, respectively. Considering $\kappa_1 = 0.6$ and $\kappa_2 = 0.123$, the density of $\mathbf{D}^r$, ρ_{FD}^r, is the same for both approaches, and equal to 34.62%. Despite achieving the same ρ_{FD}^r in both cases, matrix $\mathbf{D}^r$ is substantially different. With A1, the values of the normalized summations of the rows of $\mathbf{D}^r$ over their respective rows of $\mathbf{D}$, i.e. $\varsigma_h^r/\varsigma_h$, are more uniformly distributed than those using A2. While the minimum $\varsigma_h^r/\varsigma_h$ obtained using A1 is 0.604 at bus 1 highlighted in light gray in Table 5.4, one can find 9 buses with equal or lower $\varsigma_h^r/\varsigma_h$ in Table 5.5, namely 2, 3, 4, 13, 14, 18, 26, 27, and 33. In fact, in buses 3 ($\varsigma_3^r/\varsigma_3 = 0.501$) and 18 ($\varsigma_{18}^r/\varsigma_{18} = 0.454$), these values are 17.1% and 24.8% lower than 0.604, respectively. On the other hand, while only 2 buses have a $\varsigma_h^r/\varsigma_h \geq 0.7$ using A1, highlighted in dark gray in Table 5.4, this number increases to 12 in the case of A2. This means that, using A1, one can estimate not only the frequency at a given bus, but also the accuracy of such a signal. On the other hand, A2 leads to a higher uncertainty of the accuracy of the estimated signal with respect to the full matrix $\mathbf{D}$.

5.5.2.2. Irish Transmission System

Matrix $\mathbf{D}$ indicates that, from the mathematical point of view, all $\boldsymbol{\omega}_G$ are *needed* for the estimation of the frequency variations at system buses, in the sense that all of them contribute to every $\boldsymbol{\omega}_{B,h}$ in some measure. However, the reduced matrix $\mathbf{D}^r$ enables the determination of the elements of $\boldsymbol{\omega}_G$ that are most significant for the frequency estimation. With this aim, for a given κ_1 or κ_2, one can determine the number of times a certain $\boldsymbol{\omega}_{G,k}$ contributes in matrix $\mathbf{D}^r$, and/or its average FPF. Table 5.6 illustrates this concept for the AIITS, where the number of FPFs, and their average values, are listed when using A1 ($\kappa_1 = 0.75$) and A2 ($\kappa_2 = 0.04$).

Table 5.4 Matrix $\mathbf{D}^{\mathrm{r}}$ of the New England system using A1. $\kappa_1 = 0.6$; $\rho^{\mathrm{r}}_{\mathrm{FD}} = 34.62\%$.

Bus #	1	2	3	4	5	6	7	8	9	10	ς_h	$\varsigma^{\mathrm{r}}_h/\varsigma_h$
1	0	0	0	0	0	0	0	0	0	0.67	1.109	0.604
2	0.3	0	0	0	0	0	0	0.093	0	0.358	1.152	0.652
3	0.236	0	0.085	0	0	0	0	0	0.09	0.35	1.171	0.65
4	0.182	0.098	0.118	0	0	0	0	0	0	0.378	1.171	0.662
5	0.159	0.117	0.126	0	0	0	0	0	0	0.41	1.165	0.697
6	0.157	0.122	0.129	0	0	0	0	0	0	0.407	1.164	0.7
7	0.152	0	0.122	0	0	0	0	0	0	0.436	1.162	0.612
8	0.15	0	0.118	0	0	0	0	0	0	0.451	1.16	0.62
9	0	0	0	0	0	0	0	0	0	0.693	1.115	0.622
10	0.158	0.106	0.162	0	0	0	0	0	0	0.373	1.167	0.685
11	0.158	0.111	0.151	0	0	0	0	0	0	0.385	1.167	0.689
12	0.161	0.108	0.152	0	0	0	0	0	0	0.38	1.175	0.682
13	0.162	0.103	0.152	0	0	0	0	0	0	0.371	1.17	0.674
14	0.172	0.095	0.129	0	0	0	0	0	0	0.362	1.175	0.645
15	0.174	0	0	0.111	0	0.118	0	0	0	0.313	1.181	0.606
16	0.174	0	0	0.123	0	0.13	0	0	0	0.291	1.181	0.608
17	0.197	0	0	0	0	0.111	0	0	0.109	0.308	1.186	0.612
18	0.212	0	0	0	0	0.101	0	0	0.102	0.325	1.182	0.626
19	0.152	0	0	0.205	0	0.113	0	0	0	0.255	1.17	0.62
20	0.139	0	0	0.187	0.109	0	0	0	0	0.233	1.101	0.607
21	0.162	0	0	0	0	0.167	0.13	0	0	0.271	1.176	0.621
22	0.149	0	0	0	0	0.205	0.153	0	0	0.249	1.167	0.649
23	0.148	0	0	0	0	0.184	0.177	0	0	0.248	1.166	0.65
24	0.171	0	0	0.12	0	0.138	0	0	0	0.286	1.18	0.605
25	0.263	0	0	0	0	0	0	0.14	0	0.326	1.166	0.625
26	0.219	0	0	0	0	0	0	0.102	0.194	0.302	1.208	0.676
27	0.209	0	0	0	0	0.097	0	0	0.155	0.306	1.201	0.639
28	0.196	0	0	0	0	0	0	0	0.309	0.27	1.216	0.637
29	0.187	0	0	0	0	0	0	0	0.344	0.258	1.209	0.652
30	0.481	0	0	0	0	0	0	0	0	0.256	1.091	0.675
31	0.128	0.224	0	0	0	0	0	0	0	0.333	1.077	0.636
32	0.13	0	0.254	0	0	0	0	0	0	0.307	1.08	0.639
33	0.13	0	0	0.261	0	0.097	0	0	0	0.218	1.086	0.65
34	0.132	0	0	0.177	0.149	0	0	0	0	0.22	1.087	0.623
35	0.133	0	0	0	0	0.272	0.136	0	0	0.222	1.126	0.677
36	0.128	0	0	0	0	0.159	0.29	0	0	0.214	1.143	0.692
37	0.226	0	0	0	0	0	0	0.241	0	0.279	1.121	0.666
38	0.163	0	0	0	0	0	0	0	0.406	0.225	1.161	0.684
39	0	0	0	0	0	0	0	0	0	0.846	1.062	0.796

Table 5.5 Matrix $\mathbf{D}^r$ of the New England system using A2. $\kappa_2 = 0.123$; $\max(\mathbf{D}) = 0.846$; $\rho_{\mathrm{FD}}^r = 34.62\%$.

Bus #	Generator # 1	2	3	4	5	6	7	8	9	10	ς_h	$\varsigma_h^r/\varsigma_h$
1	0.155	0	0	0	0	0	0	0	0	0.67	1.109	0.743
2	0.3	0	0	0	0	0	0	0	0	0.358	1.152	0.571
3	0.236	0	0	0	0	0	0	0	0	0.35	1.171	0.501
4	0.182	0	0.118	0	0	0	0	0	0	0.378	1.171	0.579
5	0.159	0.117	0.126	0	0	0	0	0	0	0.41	1.165	0.697
6	0.157	0.122	0.129	0	0	0	0	0	0	0.407	1.164	0.7
7	0.152	0.115	0.122	0	0	0	0	0	0	0.436	1.162	0.711
8	0.15	0.111	0.118	0	0	0	0	0	0	0.451	1.16	0.716
9	0	0	0	0	0	0	0	0	0	0.693	1.115	0.622
10	0.158	0.106	0.162	0	0	0	0	0	0	0.373	1.167	0.685
11	0.158	0.111	0.151	0	0	0	0	0	0	0.385	1.167	0.689
12	0.161	0.108	0.152	0	0	0	0	0	0	0.38	1.175	0.682
13	0.162	0	0.152	0	0	0	0	0	0	0.371	1.17	0.585
14	0.172	0	0.129	0	0	0	0	0	0	0.362	1.175	0.564
15	0.174	0	0	0.111	0	0.118	0	0	0	0.313	1.181	0.606
16	0.174	0	0	0.123	0	0.13	0.107	0	0	0.291	1.181	0.699
17	0.197	0	0	0.105	0	0.111	0	0	0.109	0.308	1.186	0.701
18	0.212	0	0	0	0	0	0	0	0	0.325	1.182	0.454
19	0.152	0	0	0.205	0	0.113	0	0	0	0.255	1.17	0.62
20	0.139	0	0	0.187	0.109	0	0	0	0	0.233	1.101	0.607
21	0.162	0	0	0.114	0	0.167	0.13	0	0	0.271	1.176	0.718
22	0.149	0	0	0.105	0	0.205	0.153	0	0	0.249	1.167	0.739
23	0.148	0	0	0.105	0	0.184	0.177	0	0	0.248	1.166	0.74
24	0.171	0	0	0.12	0	0.138	0.117	0	0	0.286	1.18	0.705
25	0.263	0	0	0	0	0	0	0.14	0.114	0.326	1.166	0.723
26	0.219	0	0	0	0	0	0	0	0.194	0.302	1.208	0.592
27	0.209	0	0	0	0	0	0	0	0.155	0.306	1.201	0.558
28	0.196	0	0	0	0	0	0	0	0.309	0.27	1.216	0.637
29	0.187	0	0	0	0	0	0	0	0.344	0.258	1.209	0.652
30	0.481	0	0	0	0	0	0	0	0	0.256	1.091	0.675
31	0.128	0.224	0.106	0	0	0	0	0	0	0.333	1.077	0.735
32	0.13	0	0.254	0	0	0	0	0	0	0.307	1.08	0.639
33	0.13	0	0	0.261	0	0	0	0	0	0.218	1.086	0.561
34	0.132	0	0	0.177	0.149	0	0	0	0	0.22	1.087	0.623
35	0.133	0	0	0	0	0.272	0.136	0	0	0.222	1.126	0.677
36	0.128	0	0	0	0	0.159	0.29	0	0	0.214	1.143	0.692
37	0.226	0	0	0	0	0	0	0.241	0	0.279	1.121	0.666
38	0.163	0	0	0	0	0	0	0	0.406	0.225	1.161	0.684
39	0	0	0	0	0	0	0	0	0	0.846	1.062	0.796

Table 5.6 FPFs of the generators of the AIITS to the bus frequencies. Left: A1 ($\kappa_1 = 0.75$, $\rho^r_{FD} = 40.54\%$). Right: A2 ($\kappa_2 = 0.04$, $\rho^r_{FD} = 40.39\%$).

Generator bus #	Number of participations	Average FPF	Generator bus #	Number of participations	Average FPF
715	1,468	0.086	715	1,234	0.099
699	1,459	0.066	682	1,231	0.178
682	1,432	0.155	699	1,228	0.075
857	1,256	0.085	857	1,177	0.089
989	1,224	0.057	989	1,170	0.064
953	1,214	0.061	953	1,154	0.063
987	888	0.050	987	1,140	0.049
988	746	0.054	988	1,104	0.049
1,283	534	0.038	756	717	0.050
756	513	0.055	992	548	0.059
992	457	0.064	505	533	0.046
1,221	455	0.034	993	526	0.057
505	422	0.049	1,283	442	0.041
1,220	342	0.033	1,221	288	0.038
993	316	0.070	1,220	181	0.049
1,353	218	0.044	1,353	102	0.068
1,174	160	0.024	1,010	82	0.038
1,011	29	0.052	1,011	82	0.038
1,013	29	0.052	1,012	82	0.038
1,143	18	0.163	1,013	82	0.038
1,010	5	0.140	1,174	21	0.079
1,012	5	0.140	1,143	18	0.163

The columns of Table 5.6 have been sorted in descending order according to the total number of nonnull FPFs of each generator in $\mathbf{D}^r$. The six generators with the highest average value have been highlighted in gray. The most relevant generators are those located in buses 682, 699, 715, 857, 953, and 989, regardless of the approach used. This indicates that, in practical applications, it is important to have an accurate and reliable measurement of at least the rotor speeds of these machines, or of the frequency variations at these generator buses recorded with PMUs.

Next the accuracy of matrix $\mathbf{D}^r$ as obtained using A1 and A2 for different values of κ_1 and κ_2 is studied by means of TDSs solved for the AIITS. With this aim, a three-phase fault is simulated at $t = 0.5$ s, and cleared after 250 ms by disconnecting the line where the fault has occurred.

Figure 5.15.a shows the frequency estimated at a bus where a wind power plant is installed, for different values of κ_1, and their respective densities ρ^r_{FD}. The absolute errors ϵ_ω between the trajectories and the ideal case, i.e. using the full matrix $\mathbf{D}$, are depicted in Figure 5.15.b. The frequency of the center

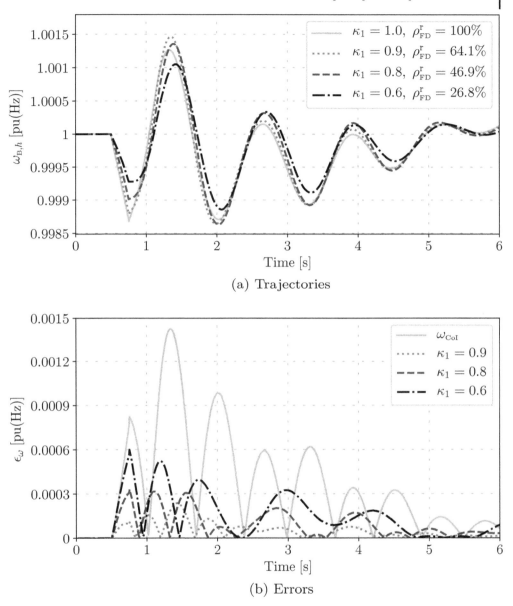

(a) Trajectories

(b) Errors

Figure 5.15 AIITS – Frequency estimated at a nonsynchronous wind generation bus facing a three-phase fault using A1.

of inertia, ω_{CoI}, is also included as a limit case for the desired accuracy of $\mathbf{D}^{\text{r}}$. While the density of matrix $\mathbf{D}^{\text{r}}$ can be reduced considerably with the A1, it nevertheless captures the local frequency oscillations with a high level of accuracy. In all considered cases, the reduced FDF outperforms the accuracy of the ω_{CoI} widely used in these type of studies.

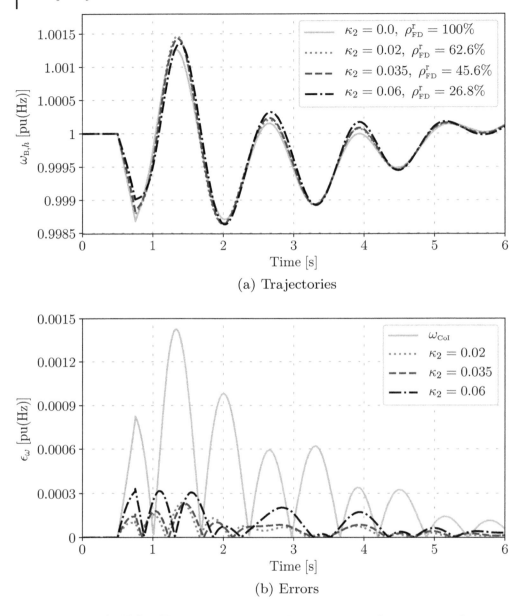

Figure 5.16 AIITS – Frequency estimated at a nonsynchronous wind generation bus facing a three-phase fault using A2.

A similar analysis is solved using A2, and results are shown in Figure 5.16. The values of κ_2 have been chosen such that the respective densities $\rho_{\mathrm{FD}}^{\mathrm{r}}$ are similar to the A1 scenario. In this case, a better accuracy of the estimation of the frequency at the bus is obtained for equivalent $\rho_{\mathrm{FD}}^{\mathrm{r}}$. However, if other buses are to be analyzed, one must take into account the uncertainty that characterizes the accuracy of A2 discussed in Section 5.5.2.1.

5.5.2.3. ENTSO-E Transmission System

This example considers the same ENTSO-E transmission system model utilized in Section 5.2.1.4. The size of matrix $\mathbf{D}$, including the off-line buses and power plants, is thus $21{,}177 \times 4{,}832$, with a density $\rho_{\mathrm{FD}} = 84.21\%$ (see Table 5.3). This means that, on average, the estimation of the frequency variations of each bus depends on more than 4,000 power plants. From the practical point of view, it is clearly not realistic to assume such a dependency, and the need of a reduction of the density of matrix $\mathbf{D}$ becomes apparent.

To demonstrate that one does not require to retrieve the information of such a large number of generator rotor speeds to estimate the frequency variations at a certain bus, a sensitivity analysis of the ENTSO-E transmission system is carried out, and results are shown in Figure 5.17. The intervals $\kappa_1 \in [0.8, 1]$ and $\kappa_2 \in [0, 0.01]$ have been split into 50 segments with logarithmic increments. The curves show very stiff saturation, confirming the intuition that only a small number of generator rotor speeds have significant weights (from Figure 5.17.a), and that a very high percentage of these weights are extremely small (from Figure 5.17.b).

To ensure a minimum $\varsigma_h^r / \varsigma_h = 0.75$, using A1 (i.e. $\kappa_1 = 0.75$), it is required a density $\rho_{\mathrm{FD}}^r = 1.48\%$. However, the time needed to initialize the set of DAEs (including the computation of matrix $\mathbf{D}^r$) using A1 is about 37 s. On the other hand, this time is reduced to about 8 s when using A2 to obtain the same ρ_{FD}^r ($\kappa_2 = 0.0042$). As opposed to using A1, the minimum $\varsigma_h^r / \varsigma_h = 0.75$ can not be guaranteed using A2, and a significant number of buses are below this threshold, as shown in Figure 5.18.

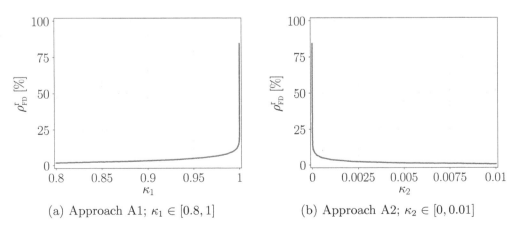

(a) Approach A1; $\kappa_1 \in [0.8, 1]$ (b) Approach A2; $\kappa_2 \in [0, 0.01]$

Figure 5.17 ENTSO-E transmission system – Density of matrix $\mathbf{D}^r$.

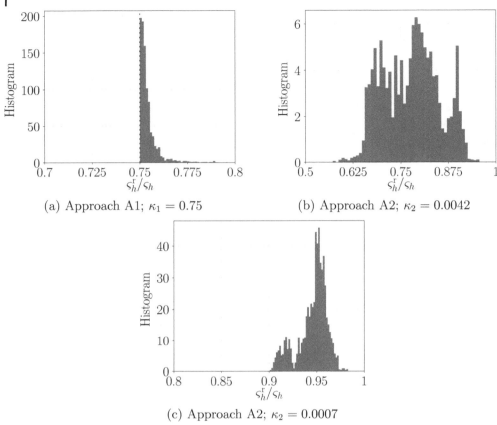

(a) Approach A1; $\kappa_1 = 0.75$ (b) Approach A2; $\kappa_2 = 0.0042$

(c) Approach A2; $\kappa_2 = 0.0007$

Figure 5.18 ENTSO-E transmission system – Histograms of the normalized summations of the FPFs. (a) and (b) $\rho_{\mathrm{FD}}^{\mathrm{r}} = 1.48\%$. (c) $\rho_{\mathrm{FD}}^{\mathrm{r}} = 4.45\%$.

A trade-off between accuracy and speed of computation can be achieved by combining A1 and A2, as follows. A2 is firstly applied with a relatively low κ_2. The aim is to reduce considerably $\rho_{\mathrm{FD}}^{\mathrm{r}}$ without implying a relevant impact on the accuracy of the estimations, as observed from Figure 5.17. In this way, the computational burden of the sorting and calculation of (5.42) can be considerably reduced. Finally, A1 can be then applied with the desired κ_1.

The effectiveness of the combined approach is illustrated in Figure 5.18. A2 is applied first with $\kappa_2 = 0.0007$ and $\rho_{\mathrm{FD}}^{\mathrm{r}}$ is reduced from the initial 84% to 4.5%. As expected, the accuracy has not been deteriorated significantly, as shown in Figure 5.18.c, where all $\varsigma_h^{\mathrm{r}}/\varsigma_h$ are above 0.9. A1 is then applied with $\kappa_1 = 0.75$, with a resulting $\rho_{\mathrm{FD}}^{\mathrm{r}} = 1.48\%$ and with a similar distribution of $\varsigma_h^{\mathrm{r}}/\varsigma_h$ to the one shown in Figure 5.18.a. However, the computational time required to initialize the set of DAEs has been reduced from 37 to 19 s.

Chapter 6
Frequency Makers and Frequency Takers

6.1. Introduction

A current challenge for the secure operation of the grid is the ability of TSOs to determine through simple measurements whether nonconventional devices provide frequency control at a given time. Some TSOs have resolved the problem by measuring the active power output to estimate their mileage. Other TSOs "trust" the operators of nonconventional devices, which clearly exposes the system to potential security issues if the control is not provided or not available when needed. A third approach consists in allocating conventional frequency reserve, which guarantees a secure operation but leads to higher energy costs.

This is a major concern for system operators and prevents relying on and properly rewarding the devices that provide such frequency support. Smart metering is already a reality but it is mostly utilized on the device side to implement the frequency control itself, e.g. [183], rather than on the system operator side.

The need for metrics to define the frequency response and control in a transmission system has been recognized since a decade ago. The report [65], for example, defines three obvious metrics, namely frequency nadir, nadir-based frequency response, and primary frequency response. These are, however, "global" metrics and are adequate only for off-line adequacy and reliability studies. Existing techniques to evaluate the primary frequency and inertial responses are qualitative and based on statistical analysis of time series [52, 53] or on the Kalman filter [16].

This chapter presents an alternative derivation of the FDF, which leads to determine, at least in transient conditions and under certain hypothesis, whether a device provides inertial response and/or frequency control (*frequency*

Frequency Variations in Power Systems: Modeling, State Estimation, and Control, First Edition.
Federico Milano and Álvaro Ortega Manjavacas.

maker) or not (*frequency taker*).[1] The formulation is a generalization of the FDF presented in Chapter 5 as it is based exclusively on frequency and power measurements at network buses [136]. This formulation leads to the concept of "rate of change of power," which refers to the "regulated" active power injected at a bus.

6.2. Derivation

As introduced in Section 2.4.3.5, the DC power flow formulation is a simplification of the conventional power flow equations, which leads to a set of linear equations linking total active power injections, p_B, with nodes and bus voltage phase angles θ_B, as follows [199]:

$$p_B(t) = \mathbf{B}_{\text{bus}}\, \theta_B(t) \ , \tag{6.1}$$

where matrix $\mathbf{B}_{\text{bus}}$ is the conventional susceptance matrix of the network as it is well-known from power flow analysis (see Section 2.4.3.3). The hypotheses and approximations that lead to the final expression of the DC power flow are also well-known and are not repeated here. The interested reader can refer to [199] for an up-to-date review of the DC power flow problem.

The total active power injections can be split into two components:

$$p_B(t) = p_B'(t) + p_B''(t) \ , \tag{6.2}$$

where the component p_B' is the quota of power that results from a *regulation* put in place by the device, whereas p_B'' is the quota of power that the device consumes or generates passively as a consequence of the voltage imposed by the grid at its point of connection. Of these two components, only p_B' is of interest in the context of HV transmission systems. The fact that the variations of p_B'' are actually immaterial (and *invisible*) for the frequency maker devices defined in Section 6.3 is a relevant property of the theory discussed in this chapter. This point is further elaborated in Section 6.3.2.3 and illustrated in the example of Section 6.4.1.3.

According to the assumption above, differentiating (6.1) with respect to time leads to:

$$\boxed{\dot{p}_B'(t) = \omega_o\, \mathbf{B}_{\text{bus}}\, \Delta\omega_B(t) = \hat{\mathbf{B}}_{\text{bus}}\, \Delta\omega_B(t)} \ , \tag{6.3}$$

---

[1]The notation *frequency maker* and *frequency taker* is preferred to the notation *grid forming* (or *grid feeding*) and *grid following*, as these refer exclusively to the control setup of power electronics converters. The concepts discussed in this chapter are more general and do not refer to a specific device technology.

where ω_o is the synchronous reference angular speed in rad/s. Multiplying by ω_o is necessary to take into account the fact that $\boldsymbol{\theta}_{\mathrm{B}}$ in (6.1) are in radians, while $\boldsymbol{\omega}_{\mathrm{B}}$ in (6.3) and in all other equations in the remainder of this chapter are expressed in per unit. The vector $\dot{\boldsymbol{p}}'_{\mathrm{B}}$ represents the vector of *rates of change of powers (RoCoPs)* at the network buses. It is important to note that, in general and according to (6.2), the RoCoP is not trivially the time derivative of the total active power $\boldsymbol{p}_{\mathrm{B}}$.

An additional step is needed to link (6.3) and (5.13). Matrix $\mathbf{B}_{\mathrm{BB}}$ can be obtained from $\mathbf{B}_{\mathrm{bus}}$ as follows:

$$\mathbf{B}_{\mathrm{BB}} = \mathbf{B}_{\mathrm{bus}} + \mathbf{B}_{\mathrm{G}} , \tag{6.4}$$

where $\mathbf{B}_{\mathrm{G}}$ is a diagonal matrix where the i-th element is $-1/X_{\mathrm{G},i}$, with $X_{\mathrm{G},i}$ the internal impedance of the generator connected to bus i; and 0 if no generator is connected to bus i. Merging together (5.13), (6.3), and (6.4), one obtains:

$$\dot{\boldsymbol{p}}'_{\mathrm{B}}(t) = -\omega_o \left[\mathbf{B}_{\mathrm{BG}} \, \Delta\boldsymbol{\omega}_{\mathrm{G}}(t) + \mathbf{B}_{\mathrm{G}} \, \Delta\boldsymbol{\omega}_{\mathrm{B}}(t) \right] , \tag{6.5}$$

or, equivalently:[2]

$$\dot{\boldsymbol{p}}'_{\mathrm{B}}(t) = -\hat{\mathbf{B}}_{\mathrm{BG}} \left[\Delta\boldsymbol{\omega}_{\mathrm{G}}(t) - \Delta\boldsymbol{\omega}_{\mathrm{BG}}(t) \right] , \tag{6.6}$$

or, equivalently:

$$\mathbf{B}_{\mathrm{BG}} \left[\Delta\boldsymbol{\omega}_{\mathrm{G}}(t) - \Delta\boldsymbol{\omega}_{\mathrm{BG}}(t) \right] = -\mathbf{B}_{\mathrm{bus}} \, \Delta\boldsymbol{\omega}_{\mathrm{B}}(t) , \tag{6.7}$$

where $\Delta\boldsymbol{\omega}_{\mathrm{BG}} \subset \Delta\boldsymbol{\omega}_{\mathrm{B}}$ is the subset of frequency variations at the terminal buses of the synchronous machines; and $\hat{\mathbf{B}}_{\mathrm{BG}} = \omega_o \mathbf{B}_{\mathrm{BG}}$.

The expression (6.7) is the sought alternative formulation of the FDF, which will be used in the next section. Equations (6.5) and (6.6) indicate that, whenever $\dot{p}'_{\mathrm{B},h} \neq 0$ at a bus h, the frequency at that bus is modified. Any device that shows $\dot{p}'_{\mathrm{B},h} \neq 0$ is thus, according to the definitions proposed in this book, a frequency maker. Similarly, a device characterized by $\dot{p}'_{\mathrm{B},h} = 0$ is a frequency taker. It is also relevant to note that in steady-state there is no meaningful way to distinguish between frequency makers and frequency takers. The concept itself lacks of meaning, as in stationary conditions, the frequency is the same at every bus of the network and every device, with its power injection or consumption, equally contributes to keep the power balance and, hence, the synchronous operation.

From (6.3), it also descends that any device, not only synchronous machines, can be frequency makers. Equation (6.5) as well as the FDF in (5.13)

---

[2]The absolute values of the nonzero elements of each row of $\mathbf{B}_{\mathrm{BG}}$ and $\mathbf{B}_{\mathrm{G}}$ are the same, but have opposite signs.

are thus accurate only as long as the active powers of loads and other nonsynchronous generators connected to the grid are constant. But, as soon as loads or other devices show quick and externally imposed active power variations, they become frequency makers. Of course, the impact on the frequency depends on the amount and velocity of power variations. This point is thoroughly elaborated in the next section.

6.3. Taxonomy

The observation that motivates this section originates from the structure of the revisited FDF given in (6.7). The simple example of Figure 6.1 is discussed first.

Particularizing (6.7) for bus 1 of the scheme of Figure 6.1 leads to the following scalar expression:

$$B_G \left[\Delta\omega_G(t) - \Delta\omega_{B,1}(t) \right] = (B_{12} + B_{10}) \Delta\omega_{B,1}(t) - B_{12} \Delta\omega_{B,2}(t) , \qquad (6.8)$$

where $B_G = 1/\langle X \rangle_{dq}$ is the inverse of the internal impedance the synchronous machine;[3] B_{10} is the shunt susceptance at bus 1, e.g. the capacitive charging of the π-model of transmission lines, and B_{12} is the susceptance of the branch that connects buses 1 and 2. In (6.8), the signs are a consequence of (6.7) and of assuming B_G, B_{10}, and B_{12} to be positive if inductive.

Equation (6.8) is written with the knowledge that the device connected to bus 1 is a synchronous machine, which is certainly a frequency maker as it imposes the frequency at the EMF behind the susceptance B_G.

A more interesting case is when the model of the device connected to bus 1 is not known. It is important to keep in mind that *no information* about this

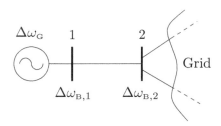

Figure 6.1 Synchronous machine connected in antenna to the grid.

---

[3]The important point is that bus 1 is *accessible* by the TSO. Hence, it can also represent the high-voltage winding of the step-up transformer that connects the generator to the grid. In that case, $B_G = 1/(\langle X \rangle_{dq} + X_T)$, i.e. it will also take into account the impedance of such a transformer.

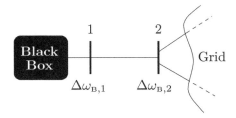

Figure 6.2 Black-box device connected in antenna to the grid.

device is available but some measurements at its terminal bus. This device is effectively a *black box*, as shown in Figure 6.2.

Regardless of the actual behavior of the black-box device, (6.8) can be rewritten by assuming that the device has an internal frequency, $\Delta\omega_\blacksquare$, and an equivalent, possibly time-variant susceptance, $B_\blacksquare$, both of which are unknown:

$$B_\blacksquare(t)\left[\Delta\omega_\blacksquare(t) - \Delta\omega_{B,1}(t)\right] = B_{12}\left[\Delta\omega_{B,1}(t) - \Delta\omega_{B,2}(t)\right] , \tag{6.9}$$

where, consistently with the assumptions that lead to the FDF and without loss of generality, $B_{12} \gg B_{10}$ is assumed. The term $B_\blacksquare[\Delta\omega_\blacksquare - \Delta\omega_{B,1}]$ in (6.9) is not known. According to (6.6), (6.9) can be also written as:

$$\dot{p}'_\blacksquare(t) = \hat{B}_{12}\left[\Delta\omega_{B,1}(t) - \Delta\omega_{B,2}(t)\right] , \tag{6.10}$$

where $\hat{B}_{12} = \omega_o B_{12}$ and $\dot{p}'_\blacksquare$ is the RoCoP at the point of connection of the black-box device with the grid.

Expression (6.10) can be conveniently generalized assuming that there is more than one branch connected to the monitored bus. Hence, assuming to monitor the h-th bus, the proposed general formula to discriminate between frequency makers and frequency takers is:

$$\boxed{\dot{p}'_{B,h}(t) = \sum_{k\in\mathbb{B}_h} \hat{B}_{hk}\left[\Delta\omega_{B,h}(t) - \Delta\omega_{B,k}(t)\right] ,} \tag{6.11}$$

where $\mathbb{B}_h$ is the set of buses connected to bus h and B_{hk} is the susceptance of the branch connecting bus h to bus k. Equation (6.11) only requires the knowledge of the system admittance matrix and the estimation of the frequencies at the neighboring buses of the device to be monitored. This information is easily available to the TSOs. No confidential information has to be provided by the device itself.

In (6.11), the term $\sum_{k\in\mathbb{B}_h} \hat{B}_{hk}\Delta\omega_{B,h}$ represents the combined effect on the frequency of both the device connected to bus h and the rest of the network; whereas the term $\sum_{k\in\mathbb{B}_h} \hat{B}_{hk}\Delta\omega_{B,k}$ represents the effect of the whole network on the frequency at bus h. Subtracting the latter to the former, what remains is the effect of the black-box device on the frequency variation at bus h. Equation (6.11) is a direct consequence of the rationale of the FDF, which expresses the variation of the frequencies at any node of the system as an algebraic boundary value problem.

The quantity $\dot{p}'_{B,h}$, i.e. the RoCoP at bus h, is more sensitive the higher are the differences between the frequency variations $\Delta\omega_{B,h}$ and $\Delta\omega_{B,k}$ that appear in (6.11). It is interesting to note that the inverse of the impedance of the branches (lines or transformers) that connect the monitored bus to the grid "amplifies" such differences. A consequence of the FDF is, in fact, that the smaller the branch impedance, the smaller is the difference of the frequencies at its ends, but also the higher the inverse value of such impedance. Note also that, if a branch impedance is too small and, hence, the accuracy with which frequency variations have to be measured is too high, one can utilize measurements at further buses as described in the discussion of (6.29).

6.3.1. Frequency Takers

As consequence of the FDF, there cannot be any variation of frequency within a passive circuit because the boundary conditions on the frequency are imposed externally from the device. For a pure frequency taker, thus, the relationship:

$$\Delta\omega_{\blacksquare,h}(t) \equiv \Delta\omega_{B,h}(t) \tag{6.12}$$

holds. Merging (6.12) into (6.11) leads to define a frequency taker as a device for which $\ddot{p}'_{B,h} = 0$. Note that for frequency takers, the value of $B_{\blacksquare,h}$ is not important.

While analytically irreproachable, the condition $\ddot{p}'_{B,h} = 0$ is rarely satisfied in practice. It is thus convenient to relax it and assume that a frequency taker satisfies the following condition:

$$\left|\dot{p}'_{B,h}(t)\right| \approx \left|\frac{\Delta p'_{B,h}(t)}{\Delta t}\right| < \epsilon, \tag{6.13}$$

where $\epsilon > 0$ is a given empirical threshold that, once agreed upon by all parties, can be used by the TSO to define network codes and ancillary services. The numerical evaluation of this threshold is discussed in Section 6.4.2.

The condition (6.13) can be satisfied in two cases, both relevant in this context and that are discussed next.

6.3.1.1. Slow Power Variations

The device does vary its power consumption/production, but the RoCoP is small in the considered time frame, i.e. the time scale of the primary frequency control (PFC). According to the notation of (6.13), this situation is characterized by a nonnegligible $\Delta p'_{\mathrm{B},h}$ and a *large* Δt. Thus, the secondary frequency control (SFC) and the daily ramp-up of loads are expected to not vary the frequency at the buses.

Note that the variation of the total power injection at the bus, $\Delta p_{\mathrm{B},h}$, does not need to be slow. Since the only component of the bus power injections that is relevant for the frequency variations is $\dot{p}'_{\mathrm{B},h}$, the condition $\dot{p}_{\mathrm{B},h} \approx \dot{p}''_{\mathrm{B},h} \neq 0$ still qualifies as "slow" power variation.

6.3.1.2. Small Power Variations

No load consumption or generation is ever perfectly constant. Stochastic white noise creates, at least, local tiny fluctuations. If the noise is small enough, however, such variations are unable to impact on the frequency. According to the notation of (6.13), this situation is characterized by *small* $\Delta p'_{\mathrm{B},h}$ per unit of time.

For example, assuming that a load can be modeled as two components, one time-varying ramp, p_{ramp}, and a stochastic component, p_{stoch} (see, for example, the model described in [134]), one has:

$$p_{\mathrm{D},h}(t) = p_{\mathrm{ramp},h}(t) + p_{\mathrm{stoch},h}(t) \;, \tag{6.14}$$

then $\dot{p}_{\mathrm{D},h}$ will be negligible, even though the load consumption is not perfectly constant, because $p_{\mathrm{ramp},h}$ varies slowly and $p_{\mathrm{stoch},h}$ varies little. The simulation results discussed in Section 6.4.1 support this conclusion. Note also that *sudden* load variations, due for example to the connection or disconnection of part of it, will cause an instantaneous variation of local frequency, which can be identified as "spikes" in the frequency estimated by the PMU, as it has already been discussed in the literature [161, 163].

Finally, the variation of the total power injection at the bus, $\Delta p_{\mathrm{B},h}$, does not need to be small. Since the only component of the bus power injections that is relevant for the frequency variations is $\Delta p'_{\mathrm{B},h}$, the condition $\Delta p_{\mathrm{B},h} \approx \Delta p''_{\mathrm{B},h} \neq 0$ still qualifies as "small" power variation.

6.3.2. Frequency Makers

A device becomes a frequency maker whenever the condition $\dot{p}'_{\mathrm{B},h} \neq 0$ holds. Following the discussion above on frequency takers, however, the only variations of power of interest are those that are sufficiently big to be able to actually vary the local frequency above a certain threshold and sufficiently fast to be comparable to the time scale of the inertial response and PFC of synchronous machines. Relevant cases are discussed below.

6.3.2.1. Synchronous Machines

The synchronous machines is without doubt the most relevant frequency maker. The dynamics of the machine rotor speed can be approximated by the following well-known equation:[4]

$$M_G \dot{\omega}_G(t) \approx p_m(t) - p_G(t) , \qquad (6.15)$$

where M_G is the inertia constant and p_m is the mechanical power provided by the turbine. The mechanical power can be decomposed into three terms:

$$p_m(t) = p_{UC}(t) + p_{PFC}(t) + p_{AGC}(t) , \qquad (6.16)$$

where p_{UC} is the power set point as defined by the solution of the unit commitment problem; p_{PFC} is the regulating power due to PFC (turbine governor) of the machine; and p_{AGC} is the regulating power due to the SFC (AGC) if any and if the machine participates in it. Hence, the active power injected by a synchronous machine into its terminal bus can be written as:

$$p_G(t) = p_{UC}(t) + p_{PFC}(t) + p_{AGC}(t) - M_G \dot{\omega}_G(t) . \qquad (6.17)$$

p_{UC} is piece-wise constant, while p_{AGC} varies slowly. Of the four components above, the ones that actually contribute to modify the frequency at the machine bus and thus make the machine a frequency maker are p_{PFC} and the machine inertial response. Therefore:

$$\dot{p}_G(t) \approx \dot{p}_{PFC}(t) - M_G \ddot{\omega}_G(t) . \qquad (6.18)$$

In the very first instants after a contingency, the dominant effect is due to the inertial response but, in general, the two terms are deeply intertwined and their effect cannot be separated. However, if a machine does not provide PFC, then the lack of regulation can be inferred by observing the transient behavior of $\dot{p}_G$. This point is illustrated in the example of Section 6.4.1.1.

6.3.2.2. Controlled Nonsynchronous Devices

This category of devices includes frequency-controlled power electronic converters of nonsynchronous generation (such as wind turbines) or energy storage systems, and thermostatically controlled loads. Thus such devices can be assumed to be equivalent to a frequency control loop, with a given reference frequency, ω^{ref}, as illustrated in Figure 6.3.

The actual implementation of the controller, which is accounted for with a time-dependent susceptance, $B_\blacksquare$, is unknown. However, whatever is the

---

[4]In the equations of the remainder of this section, the index h is dropped for simplicity.

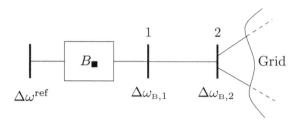

Figure 6.3 Generic device that regulates the frequency.

transfer function of the controller, it tracks a reference frequency and, in turn, $\Delta\omega_{\blacksquare,h} = \Delta\omega^{\text{ref}}$. In practice, ω^{ref} is constant and hence $\Delta\omega^{\text{ref}} = 0$. While not known in detail, it can be assumed that $B_{\blacksquare} \neq 0$ and, hence, $\dot{p}'_{\text{B},h} \neq 0$ for any transient condition for which $\Delta\omega_{\text{B},h} \neq \Delta\omega^{\text{ref}}$.

Note that synchronous machines also fall in the category of the frequency controlled devices shown in Figure 6.3, as conventional generators generally include a turbine governor and provide PFC. However, synchronous machines are a special case and deserve to be treated separately. Synchronous machines are further discussed in Section 8.3.3.

The following two cases are relevant: (i) wind turbines with frequency control, and (ii) energy storage devices that provide frequency support.

Wind Turbines

For a wind turbine, one has:

$$p_{\text{WT}}(t) = p_{\text{stoch}}(t) + p_{\text{PFC}}(t) , \tag{6.19}$$

where the stochastic term p_{stoch} depends on the uncertainty and volatility of the wind speed and the second term p_{PFC} is given by the PFC of the wind turbine. It is important to note that large and fast stochastic variations such as wind gusts are indistinguishable, in principle, from power variations aimed at regulating the frequency. The only difference is statistical. Wind gusts, in fact, will show, for about 50% of the times, a variation that further increases the actual frequency variation with respect to the synchronous reference. Instead, a power variation imposed by a frequency controller always aims at tracking the synchronous frequency. Wind gusts, however, are relatively uncommon and, very often, wind power plants are not operated at their maximum capacity.[5] Moreover, typical values of the autocorrelation coefficients of wind speeds (see, for example, [229]) lead to conclude that wind turbulence has a small local effect on active power fluctuations. On the other hand, the average value of

---

[5]See, for example, the so-called "wind dispatch down" periods defined by EirGrid Group and SONI, that effectively make constant the power production of a wind power plant [58].

the wind speed varies relatively slowly in time. Except for rare cases of strong wind gusts, the RoCoP of a wind turbine can be assumed to be:

$$\dot{p}_{\text{WT}}(t) = \dot{p}_{\text{stoch}}(t) + \dot{p}_{\text{PFC}}(t) \approx \dot{p}_{\text{PFC}}(t) \ , \tag{6.20}$$

at least for all variations such that $|\dot{p}_{\text{WT}}| > \epsilon_{\text{WT}}$, where the threshold ϵ_{WT} can be chosen based on statistical properties (shape factor and autocorrelation) of the wind at the location of the wind power plant. This point is further discussed through the examples presented in Section 6.4.2.

Energy Storage Systems

The case of energy storage systems (ESSs) is probably the simplest one. These devices are based on deterministic controllers and, when used to provide PFC, they generate/absorb power only if the frequency is outside a band around the synchronous reference speed:

$$p_{\text{ESS}}(t) = p_{\text{PFC}}(t) \quad \Rightarrow \quad \dot{p}_{\text{ESS}}(t) = \dot{p}_{\text{PFC}}(t) \ . \tag{6.21}$$

The power production/consumption of an ESS, and hence, also its RoCoP, is a direct (obvious) indication that the storage device is regulating the frequency and this modifies the frequency at the point of connection of the storage itself.

6.3.2.3. Voltage Dependent Loads

The examples discussed in Section 5.2.1.2 have shown that the impact of load models on bus frequency variations are negligible. Using the notation proposed in this chapter, loads are frequency takers. However, a relevant and probably unexpected consequence of equation (6.2) is that voltage dependent loads (VDLs) can behave as frequency makers, frequency takers or both, depending on the exponent of the voltage.

This result is not inconsistent with the simulations shown in Section 5.2.1.2. Simply, even when behaving as frequency makers, the impact of VDLs on bus frequencies is very weak compared to devices that actually regulate the frequency. In practice, thus, VDLs can be safely approximated as frequency takers. The purpose of this section is to show that equation (6.2) allows precisely identifying the behavior of VDLs.

The model of VDLs is described in Section 2.4.2 and, for clarity, the expression of the active power is recalled below:

$$p_{\text{D}}(t) = p_{\text{D},o} \left(\frac{v_h(t)}{v_{h,o}} \right)^{\gamma_p} \ , \tag{6.22}$$

where the explicit dependency on the frequency is neglected, i.e. $\beta_p = 0$.

The total power consumption of the VDL, p_D, can be split into the two components as in (6.2):

$$-p_\mathrm{D}(t) = p_{\mathrm{B},h}(t) = p'_{\mathrm{B},h}(t) + p''_{\mathrm{B},h}(t) , \tag{6.23}$$

where the minus sign indicates that p_D is a negative power injection at the node h. Relevant cases are discussed below.

Constant Power

In this case $\gamma_p = 0$ and $p_\mathrm{D} = p_{\mathrm{D},o}$ is constant. This does not mean, however, that $p'_{\mathrm{B},h}$ is constant. The ability of a load to maintain a constant active power consumption, in fact, implies that for any variation of the voltage, the current has to be modified to compensate such a variation and keep constant value of the product $v_h i_h \cos \phi_h$. Since varying the current injection of the load, ultimately requires to vary the voltage phase angle, for an ideal constant power load, the following conditions hold:

$$\dot{p}_{\mathrm{B},h}(t) = 0 , \qquad \dot{p}'_{\mathrm{B},h}(t) = -\dot{p}''_{\mathrm{B},h}(t) . \tag{6.24}$$

Even if the total power injection at the load is perfectly constant, the VDL is a frequency maker and shows $\dot{p}'_{\mathrm{B},h} \neq 0$ for any variations of the voltage at its bus.

The rationale behind (6.24) can be deduced for a purely constant active power load connected to a bus with constant voltage $\bar{v}_k = v_k \angle 0$ through a lossless line with admittance B_{hk}:

$$p_{\mathrm{D},o} = -B_{hk}\, v_h(t)\, v_k\, \sin\big(\theta_h(t)\big)$$
$$\Rightarrow \quad 0 = -B_{hk}\, v_k\, \sin\big(\theta_h(t)\big)\, dv_h(t) - B_{hk}\, v_h(t)\, v_k\, \cos\big(\theta_h(t)\big)\, d\theta_h(t) ,$$
$$0 \approx -B_{hk}\, dv_h(t)\, v_k\, \big(\theta_h(t)\big) - B_{hk}\, v_h(t)\, v_k\, d\theta_h(t)$$
$$\Rightarrow \quad \theta_h(t)\, dv_h(t) \approx -v_h(t)\, d\theta_h(t) ,$$

where it is assumed that $\sin(\theta_h) \approx \theta_h$ and $\cos(\theta_h) \approx 1$. The development above indicates that a variation of the voltage magnitude v_h is compensated by a variation of the phase angle θ_h, and *vice versa*, which confirms the conditions given in (6.24).

Constant Admittance

In this case $\gamma_p = 2$ and $p_\mathrm{D} \propto v_h^2$. The active power depends exclusively on the voltage magnitude at the bus. Its current consumption follows instantaneously the bus voltage variations. For an ideal constant admittance load, thus, the following conditions hold:

$$\dot{p}_{\mathrm{B},h}(t) = \dot{p}''_{\mathrm{B},h}(t) , \qquad \dot{p}'_{\mathrm{B},h}(t) = 0 . \tag{6.25}$$

The rationale behind (6.25) can be deduced for a purely resistive load connected to a bus with nonconstant voltage $\bar{v}_k = v_k \angle \theta_k$ through a lossless line with admittance B_{hk}:

$$G_{\mathrm{D},o}\, v_h^2(t) = -B_{hk}\, v_h(t)\, v_k(t)\, \sin\left(\theta_h(t) - \theta_k(t)\right)$$
$$\Rightarrow \quad G_{\mathrm{D},o}\, v_h(t) = -B_{hk}\, v_k(t)\, \sin\left(\theta_h(t) - \theta_k(t)\right) ,$$

where $G_{\mathrm{D},o} = p_{\mathrm{D},o}/v_{h,o}^2$. The results of the examples discussed in Chapter 5 indicate that, for a constant admittance load, the phase angle difference $\theta_h - \theta_k$ is constant because $\Delta\omega_h = \Delta\omega_k$. This result implies that any variation of v_k has to be compensated by a variation of v_h, which confirms the conditions given in (6.25).

Constant Current

In this case $\gamma_p = 1$ and $p_{\mathrm{D}} \propto v_h$. This is the dual case with respect to the constant admittance load. In fact, for an ideal constant current load, the following conditions hold:

$$\dot{p}_{\mathrm{B},h}(t) = \dot{p}'_{\mathrm{B},h}(t) , \qquad \dot{p}''_{\mathrm{B},h}(t) = 0 . \tag{6.26}$$

The rationale behind (6.26) can be deduced for a purely constant current load connected to a bus with nonconstant voltage $\bar{v}_k = v_k \angle \theta_k$ through a lossless line with admittance B_{hk}:

$$i_{\mathrm{D},o}\, v_h(t) = -B_{hk}\, v_h(t)\, v_k(t)\, \sin\left(\theta_h(t) - \theta_k(t)\right)$$
$$\Rightarrow \quad i_{\mathrm{D},o} = -B_{hk}\, v_k(t)\, \sin\left(\theta_h(t) - \theta_k(t)\right) ,$$

where $i_{\mathrm{D},o}$ is the constant current of the load and which indicates that any variations of v_k or θ_k have to be compensated by a variation of θ_h, which confirms the conditions given in (6.26).

General Voltage Dependent Load

In general, a VDL shows $\dot{p}'_{\mathrm{B},h} \neq 0$ if the voltage phase angle varies in time and $\dot{p}''_{\mathrm{B},h} \neq 0$ if the voltage magnitude varies in time. Since $p_{\mathrm{B},h}$ can be measured directly at the load bus and $\dot{p}_{\mathrm{B},h}$ can be calculated with (6.3), then $\dot{p}''_{\mathrm{B},h}$ is obtained as:

$$\dot{p}''_{\mathrm{B},h}(t) = \dot{p}_{\mathrm{B},h}(t) - \dot{p}'_{\mathrm{B},h}(t) , \tag{6.27}$$

and, hence, the dependency of the load on the voltage magnitude and phase angle at its point of connection can be always characterized, at least for small variations, as:

$$\Delta p'_{\mathrm{B},h}(t) \approx \dot{p}'_{\mathrm{B},h}(t)\, \Delta\theta_h(t) ,$$
$$\Delta p''_{\mathrm{B},h}(t) \approx \dot{p}''_{\mathrm{B},h}(t)\, \Delta v_h(t) . \tag{6.28}$$

6.4. Examples

This section presents several examples based on the WSCC system and a case study based on the All-Island Irish Transmission System (AIITS). The simulations carried out with the WSCC system illustrate the behavior of relevant frequency maker and frequency taker devices, whereas the AIITS serves to discuss the effect of stochastic generation and noise in the information provided by the RoCoP.

6.4.1. WSCC System

The WSCC system is described in Appendix A.2. Bus frequencies are estimated with SRF-PLLs whose model is described in Section 3.3.2. For its relevance in the following examples, Table 6.1 shows the matrix $\mathbf{B}_{bus}$ of the WSCC system. As discussed in Section 6.3.2.3, shunt admittances are "transparent" and not needed for the calculation of $p'_{B,h}$. Thus, shunt charging of transmission lines are not included in $\mathbf{B}_{bus}$ and, hence, the sum of the elements of each row of $\mathbf{B}_{bus}$ is zero.[6]

6.4.1.1. Synchronous Machines

The features of the RoCoP, i.e. $p'_{B,h}$, to evaluate frequency maker and frequency taker devices are first tested considering the base-case scenario with inclusion of an AGC implemented as perfect tracking integral controller. In all simulations presented in this example, loads are modeled as constant admittances.

Table 6.1 Matrix $\mathbf{B}_{bus}$ for the WSCC system.

Bus #	Bus # 1	2	3	4	5	6	7	8	9
1	−17.36	0	0	17.36	0	0	0	0	0
2	0	−16.00	0	0	0	0	16.00	0	0
3	0	0	−17.06	0	0	0	0	0	17.06
4	17.36	0	0	−39.47	11.60	10.51	0	0	0
5	0	0	0	11.60	−17.575	0	5.975	0	0
6	0	0	0	10.51	0	−16.98	0	0	5.588
7	0	16.00	0	0	5.975	0	−35.675	13.70	0
8	0	0	0	0	0	0	13.70	−23.484	9.784
9	0	0	17.06	0	0	5.588	0	9.784	−32.432

---

[6]For this reason the diagonal elements of the matrix shown in Table 6.1 are different from those shown in Table 2.4.

Example 1: Frequency Maker versus Frequency Taker

Figure 6.4 shows the RoCoP for the generator connected to bus 3 and the load connected to bus 8 following the outage of 20% of the load connected to bus 5. The RoCoP indices of buses 3 and 8 are given by the following expressions, based on (6.11):

$$\ddot{p}'_{\text{B},3}(t) = 17.06 \, \Delta \tilde{\omega}_{\text{B},3}(t) - 17.06 \, \Delta \tilde{\omega}_{\text{B},9}(t) \ ,$$

$$\ddot{p}'_{\text{B},8}(t) = 23.484 \, \Delta \tilde{\omega}_{\text{B},8}(t) - \left[13.7 \, \Delta \tilde{\omega}_{\text{B},7}(t) + 9.784 \, \Delta \tilde{\omega}_{\text{B},9}(t) \right] \ , \tag{6.29}$$

where the coefficients are obtained from Table 6.1, and the symbol "∼" on top of a bus frequency represents indirect measurements from the SRF-PLL. As thoroughly discussed later in Section 8.1, for practical applications, if a measurement is not available, this can be calculated using measurements from other buses. For example, if there is no PMU at bus 3, $\Delta \tilde{\omega}_{\text{B},3}$ can be replaced in (6.29) with:

$$\Delta \tilde{\omega}_{\text{B},3}(t) = \frac{32.432}{17.06} \, \Delta \tilde{\omega}_{\text{B},9}(t) - \left[\frac{5.588}{17.06} \, \Delta \tilde{\omega}_{\text{B},6}(t) + \frac{9.784}{17.06} \, \Delta \tilde{\omega}_{\text{B},8}(t) \right] \ , \tag{6.30}$$

which can be readily deduced from the last row of Table 6.1.

Simulation results show that, as expected, at the generator bus, $\ddot{p}'_{\text{B},3} \neq 0$ after the load outage, whereas at the load bus, $\ddot{p}'_{\text{B},8} \approx 0$, $\forall t$. The spike at beginning of the simulation of the load bus in Figure 6.4.a is due to the numerical integration of the SRF-PLL, and thus it does not represent any physical behavior of the system.

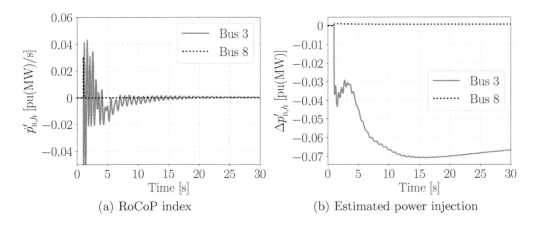

(a) RoCoP index　　　　　　　(b) Estimated power injection

Figure 6.4 WSCC system – Response at buses 3 (generator) and 8 (passive load).

In steady state, the quantity $p'_{B,h}$ goes to zero, which also indicates that, in stationary conditions, all machines rotate at the same speed. Note that it is immaterial the value of such a stationary speed. In fact, even if it is different from the synchronous reference, e.g. if no AGC is installed, (6.11) compensates the frequency steady-state error. This result is expected as, in steady state, all frequency controllers are inactive, as discussed in the next example.

Example 2: Layers of System Frequency Control

Figure 6.5 shows the effect of removing the AGC from the system as well as the PFC from the generator at bus 3. The effect of the AGC on the RoCoP and the estimated power injection is negligible. This result is consistent with the transient nature of the RoCoP. On the other hand, if a machine disconnects its turbine governor, the effect on the RoCoP and on the estimated power injection is evident a few seconds after the contingency. It is also interesting to note that, in the first instants, the response of the RoCoP is driven exclusively by the inertia of the machine, as the three trajectories are fairly similar up to about two seconds after the loss of the load.

Example 3: Actual versus "Inertial" Active Power Injection

The estimated power injection at bus 3 shown in Figures 6.4.b and 6.5.b indicates that there is a jump in the trajectory of the active power generated by the synchronous machines. However, the turbine governors of the machines require several seconds to vary their mechanical power generation to match the power unbalance that, in this case, is due to the loss of part of the load at bus 5. This is graphically represented in Figure 6.6, where the estimated $\Delta p'_{B,3}$

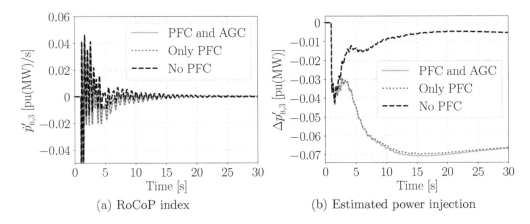

(a) RoCoP index

(b) Estimated power injection

Figure 6.5 WSCC system – Response at generator bus 3 for different frequency regulation strategies.

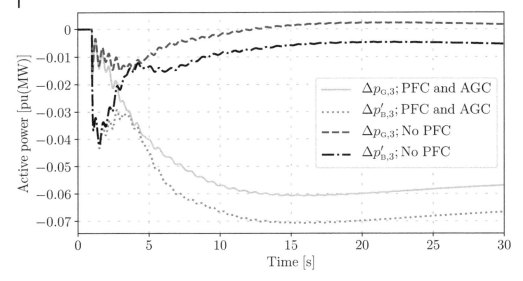

Figure 6.6 WSCC system – Variations of the active power injected at bus 3 and equivalent "inertial" power for different frequency regulation strategies.

is compared with the variations of the actual active power generated by the machine at bus 3 for the cases without any frequency control, and with both PFC and SFC.

The results shown in Figure 6.6 indicate that the estimated $\Delta p'_{B,3}$ includes the effect of the (fast) inertial response of the machine, $M_G \dot{\omega}_G$, and the turbine governor, Δp_{PFC}. In Figure 6.6, $\Delta p'_{B,3}$ is calculated through the integral of (6.3) with respect to time, whereas $\Delta p_{G,3}$ is calculated as follows:

$$\Delta p_{G,3}(t) = p_{G,3}(t) - p_{G,3}(0) \ . \tag{6.31}$$

The actual variations of the active power injected by the synchronous machine are smooth even in the first instants after the contingency, and following later a similar behavior than that estimated by the "inertial" power $\Delta p'_{B,3}$, i.e. after the inertial response of the machine gives way to the PFC. Note also that the differences in the steady-state offset between $\Delta p'_{B,3}$ and $\Delta p_{G,3}$ are of little relevance, as they mainly depend on the quota of power $\Delta p''_{B,3}$ which is not captured by the RoCoP.

6.4.1.2. ESS and TCL

This example compares the performance of the RoCoP of two families of devices, namely converter-interfaced ESSs and TCLs, whose FFR and PFC schemes are outlined in Section 4.4. These devices show significantly different response times of their primary frequency regulators. While some ESSs show a response time against frequency variations of tens of milliseconds, e.g. flywheels and superconducting magnetic coils [132], TCLs generally require several seconds to reach the active power consumption set by the controller [127].

Figure 6.7 shows the RoCoP, the regulating power estimation and the frequency of the CoI when an ESS or a TCL is connected to bus 8. The ESS models a battery storage system [162], while the TCL represents 25% of the total load of bus 8. In this scenario, PFC is included in all machines. The difference between power nadirs is less than 50% and the difference between the nadirs of the RoCoP $p'_{B,8}$ is about 70%. The ESS has a higher RoCoP than the TCL, and this leads to a significant improvement of the transient response of the system as shown by the trajectories of the frequency of the CoI.

The accuracy of the inertial power estimation is validated next. If the control loop that regulates the active power output of the ESSs, p_{ESS}, is designed

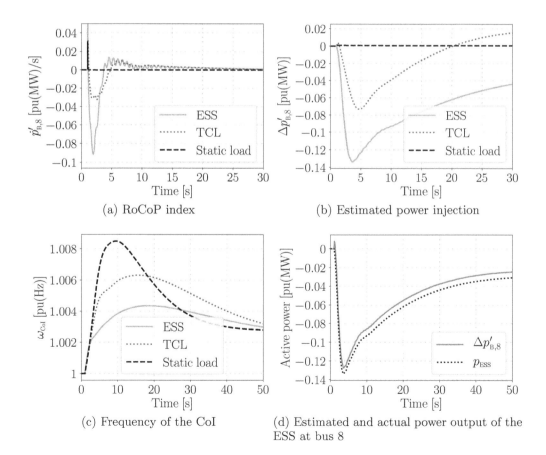

(a) RoCoP index

(b) Estimated power injection

(c) Frequency of the CoI

(d) Estimated and actual power output of the ESS at bus 8

Figure 6.7 WSCC system – Response with and without an ESS or a TCL regulating the local frequency.

to provide FFR then, in steady-state conditions, $p_{\text{ESS}} = 0$, therefore:

$$\Delta p_{\text{ESS}}(t) = p_{\text{ESS}}(t) - p_{\text{ESS}}(0) = p_{\text{ESS}}(t) . \tag{6.32}$$

Figure 6.7.d also shows the trajectories of Δp_{ESS} and $p'_{\text{B,8}} = \int_0^t \dot{p}'_{\text{B,8}}(r)dr$, which are fairly similar. The small offset (about 4%) that can be observed for $t > 5$ s is due to the fact that (6.11) does not take into account transmission line losses, nor voltage-driven power variations.

6.4.1.3. Voltage-Dependent Load

The examples presented so far consider exclusively frequency makers such as generators and frequency controlled devices for which $p_{\text{B},h} \simeq p'_{\text{B},h}$. This section elaborates on VDLs and their ability to behave as frequency makers, frequency takers or both, depending on their voltage exponent γ_p.

Figure 6.8 shows the total power variations $\Delta p_{\text{B,8}} = p_{\text{B,8}}(t) - p_{\text{B,8}}(0)$ and $\Delta p'_{\text{B,8}} = \int_0^t \dot{p}'_{\text{B,8}}(r)dr$ when the load connected to bus 8 is modeled as a VDL for $\gamma_p \in \{0, 0.5, 1, 1.5, 2, 2.5\}$. The results confirm the qualitative discussion given in Section 6.3.2.3. In particular, $\Delta p_{\text{B,8}} = 0$ for the constant load model ($\gamma_p = 0$); $\Delta p'_{\text{B,8}} \approx 0$ for the constant admittance model ($\gamma_p = 2$); and $\Delta p_{\text{B,8}} \approx \Delta p'_{\text{B,8}}$ for the constant current model ($\gamma_p = 1$). For all other values of γ_p, $\Delta p_{\text{B,8}} \neq \Delta p'_{\text{B,8}} \neq 0$ and the two quantities show a consistent "dual" transient behavior.

Thus, during a transient, VDLs show $\dot{p}'_{\text{B},h} \neq 0$ only if $\gamma_p \neq 2$. This, in turn, means that only constant admittances are frequency takers. For all other conditions, the behavior of the load shows some level of power or current control, either intrinsic, e.g. the self-regulating active power of asynchronous motors, or artificial, e.g. TCLs. Note that these results are not in contrast with those of the example discussed in Section 5.2.1.2. The FDF can be implemented neglecting VDLs even if they are frequency makers. Their impact on the frequency is marginal with respect of that of generators and other frequency controlled devices. Then, the effect of VDLs on the grid is taken into account as they affect the overall dynamic of the system and, hence, also of synchronous machines and their rotor speeds that, ultimately, are used in the FDF.

The results shown in Figure 6.8 also suggest an empirical formula for the estimation of the parameter γ_p of VDLs in the WSCC system, as follows:

$$\breve{\gamma}_p(t) = 2 \frac{\Delta p_{\text{B},h}(t)}{\Delta p'_{\text{B},h}(t) + \Delta p_{\text{B},h}(t)} . \tag{6.33}$$

The transient behavior of $\breve{\gamma}_p$ for the same VDLs and contingency discussed above are shown in Figure 6.9, which confirms the overall good accuracy of the expression (6.33).

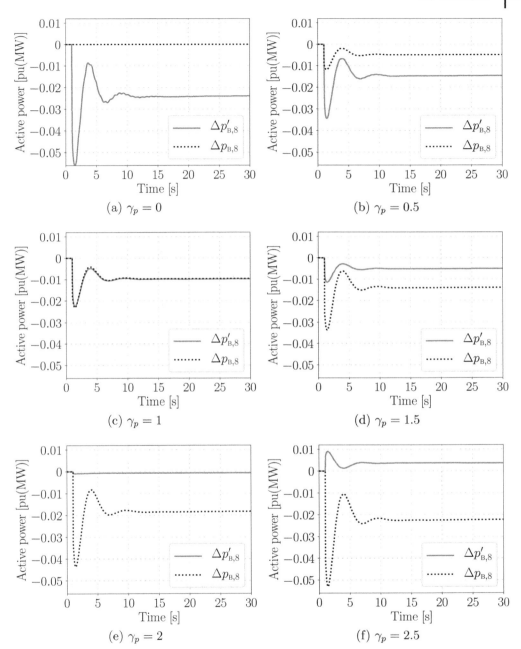

Figure 6.8 WSCC system – Dynamic behavior of $\Delta p'_{\mathrm{B},8} = \int_t \dot{p}'_{\mathrm{B},8}(r)dr$ and $\Delta p_{\mathrm{B},8} = p_{\mathrm{B},8}(t) - p_{\mathrm{B},8}(0)$ for the VDL connected at bus 8 for various voltage exponents.

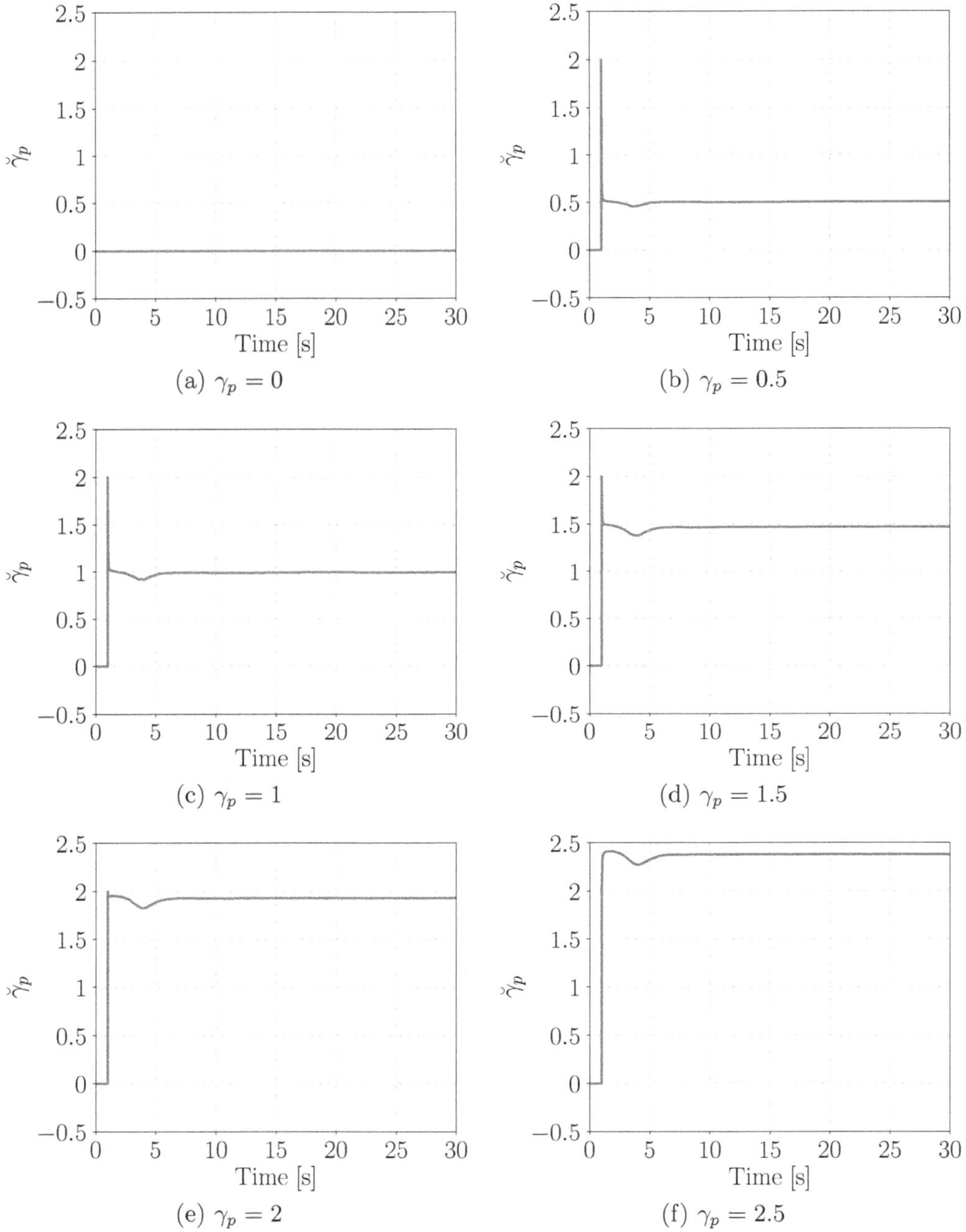

Figure 6.9 WSCC system – Estimated voltage exponent $\breve{\gamma}_p$ in transient conditions for the VDL connected at bus 8.

The most noteworthy and unexpected result of this analysis is that, from the measurements of the frequencies at the point of connection of the load and at neighboring buses, and the measurement of the active power consumption of the load, one is able to deduce the voltage dependency of the load itself.

6.4.2. All-Island Irish Transmission System

This example consists of two parts. Section 6.4.2.1 discusses the ability of the RoCoP to detect the frequency regulation of nonsynchronous generation, in particular, wind power plants. This section also serves to illustrate the capability of the RoCoP to discriminate between wind energy conversion systems (WECSs) that do or do not provide FFR. This is done by studying the statistical variations of the power injections that are due to variations of the wind speed and those that are a result of the FFR provided by the WECSs [59, 125]. The nonlinear behavior of converter-interfaced generation operating near its rated power is studied in Section 6.4.2.2. Finally, Section 6.4.2.3 discusses the impact of measurement noise on the accuracy and reliability of the RoCoP.

The study considers a small area of the AIITS described in Appendix B and shown in Figure 6.10. The area includes a synchronous machine, four wind power plants, four loads, and five interties with the rest of the system.

The devices connected to buses A to E are as follows.

- Bus A – 55 MVA synchronous machine and starting time $M_{\mathrm{G}} = 16.3$ s.

- Bus B – 60 MVA DFIG-based wind power plant.

- Bus C – 60 MVA DFIG-based wind power plant with FFR.

- Bus D – 8.35 MVA DFIG-based wind power plant.

- Bus E – 1.2 MVA CSWT.

All wind power plants are subject to stochastic wind speed fluctuations around an average value. Due to the short term of the simulations, these fluctuations are modelled using the Ornstein-Uhlenbeck process given in equation (4.31).

6.4.2.1. Stochastic Generation

A proper setup of the threshold ϵ_{WT} defined in equation (6.13) allows discriminating between WECSs that do and do not regulate the frequency. The value of ϵ_{WT} is system dependent and can be determined using the standard deviation of $\dot{p}'_{\mathrm{B},h}$ of a device or an subsystem connected to a particular bus over an

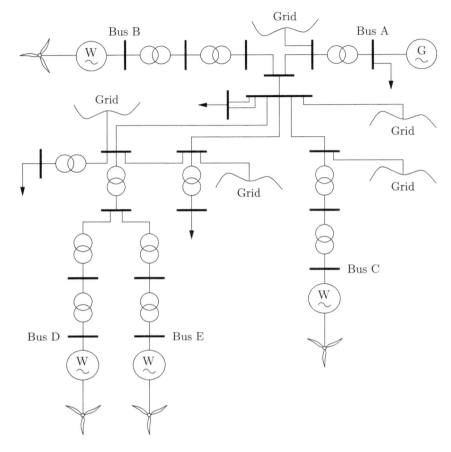

Figure 6.10 Single-line diagram of an area of the AIITS.

extended time window. Note that the specific nature of the device/subsystem connected to bus h to define ϵ_{WT} does not need to be known because (6.11) is calculated exclusively with grid-side measurements.

Figure 6.11 shows the estimation of the RoCoP for the WECSs connected to buses B, C, and D. 1,000 trajectories of 120 s of duration have been generated with different wind profiles, and no FFR is implemented in any WECSs. Figure 6.11.a shows the histograms of the trajectories at the end of the simulations, whereas Figure 6.11.b shows the probability density function (PDF)-fit of each histogram to a normal distribution with zero mean and standard deviations:

$$\sigma_{\dot{p}',\mathrm{B}} = 2.18 \cdot 10^{-3} \quad \mathrm{pu(MW)/s} \;,$$
$$\sigma_{\dot{p}',\mathrm{C}} = 3.14 \cdot 10^{-3} \quad \mathrm{pu(MW)/s} \;,$$
$$\sigma_{\dot{p}',\mathrm{D}} = 4.24 \cdot 10^{-4} \quad \mathrm{pu(MW)/s} \;.$$

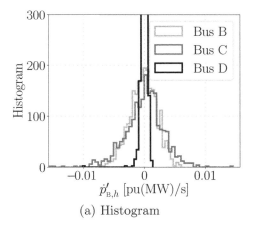

(a) Histogram

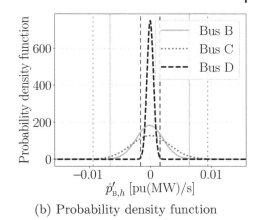

(b) Probability density function

Figure 6.11 AIITS – 1,000 trajectories. FFR is not included in the WECSs. The vertical lines indicate the value of $\pm 3\,\sigma_{\dot{p}',h}$.

As expected, WECSs with larger capacities lead to higher values of $\sigma_{\dot{p}',h}$. The WECS at bus D shows a $\sigma_{\dot{p}',h}$ that is about an order of magnitude smaller than that of buses B and C. For this reason, the CSWT connected to bus E has been omitted in this comparison. Less expected is the fact that $\sigma_{\dot{p}',h}$ shows nonnegligible differences (about 30%) between WECSs of same technology and capacity such as those of buses B and C.

The different behavior of the WECSs is due to various factors. These include: (i) how the WECS is connected to the main grid (topological reasons); (ii) the location of the WECS and the typical wind profile that is characteristic of that location (geographical reasons); and (iii) the number of wind turbines in operation within the WECS in the considered time period (technical reasons). Therefore, the effectiveness of the FFR provided by a WECS does not only depend on its power capacity. The RoCoP can thus be a valuable tool for system operators as it allows to take into account all topological, geographical, and technical aspects discussed above.

Figure 6.12 shows the PDF-fit of the RoCoP at buses B, C, and D when an FFR is implemented in the WECS at bus C. The wind variations are not large enough to trigger the FFR, and thus, $\sigma_{\dot{p}',C}$ shows a fairly similar value as for the case without FFR: $\sigma_{\dot{p}',C} = 3.29 \cdot 10^{-3}$ pu(MW)/s. This is an expected result from the discussion presented in Sections 6.3.1.2 and 6.3.2.2.

Figure 6.13 shows the trajectories of the RoCoP at buses B, C, and D, and the variations of the active power injected and equivalent inertial power at buses B and C for a single trajectory with stochastic wind perturbations and FFR implemented at bus C.

The thresholds $\pm\epsilon_{\mathrm{WT}} = \pm 3\,\sigma_{\dot{p}',B} \approx \pm 6.5 \cdot 10^{-3}$ pu(MW)/s are marked with horizontal dashed lines in Figure 6.13.a. Since its RoCoP is consistently below

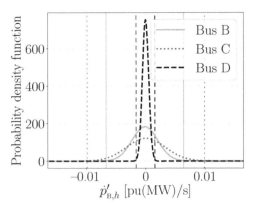

Figure 6.12 AIITS – PDF of the RoCoP at buses B, C, and D for 1,000 trajectories. FFR is included in the WECS at bus C. The vertical lines indicate the value of $\pm 3\,\sigma_{\dot{p}',h}$.

such thresholds, it can be safely concluded that the WECS at bus D is a *frequency taker*. For this reason, the WECS at bus D (and consequently, also the WECS at bus E) is not considered for study in the remainder of this section.

The quantities $\dot{p}'_{B,B}$ and $\dot{p}'_{B,C}$ are, for most of the time, within the thresholds $\pm\epsilon_{WT}$. On the other hand, if only the measurements of the variations of the active power injected by the WECS (or the estimation of its equivalent inertial power as shown in Figure 6.13.b) are available, intuitively one can assume that the WECS at bus B without FFR will have, *a priori*, a greater impact on its local frequency than that at bus C with FFR. This is shown in Figure 6.14, where the loss of a 44 MW, 13.2 MVAr load is simulated at $t = 50$ s. The RoCoP clearly indicates that the WECS at bus C is providing FFR as observed from the values of $\dot{p}'_{B,C}$ in the few seconds after the contingency. This is not so clear from the active power injection shown in Figure 6.14.b. After $t = 50$ s, $\Delta p'_{B,C}$ shows a ramp that, in magnitude, is even smaller than other ramps observed in $\Delta p'_{B,B}$, e.g. the variations in the interval $t \in [160, 220]$ s.

These results indicate that measuring the active power injection is, in general, not a sufficient criterion to remunerate the owners of the WECSs for the provision of FFR. With this regard, the RoCoP appears as a valuable tool for system operator.

As a final remark, different ϵ_{WT} can be defined for different purposes. For instance, the system operator can define a threshold $\epsilon_{WT,o}$, common for all devices, above which specific devices or systems are required to provide FFR. Then, individual, possibly multiple thresholds $\epsilon_{WT,h}$ can be defined for each device providing FFR to quantify the *amount of control* that is provided by the device.

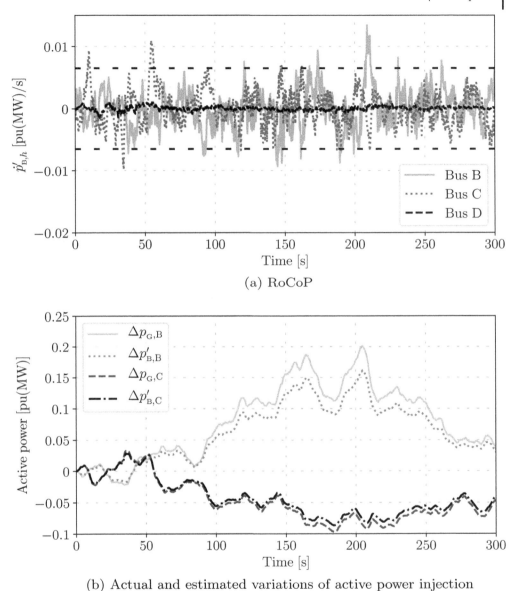

(a) RoCoP

(b) Actual and estimated variations of active power injection

Figure 6.13 AIITS – Response following stochastic wind variations. FFR is included in the WECS at bus C.

6.4.2.2. Power Saturation of the WECSs

This section studies the behavior of nonsynchronous devices that provide FFR when operating near their rated power output. With this aim, a comparison of the trajectories of the RoCoP and of the estimated "inertial" power variations when the WECS at bus C operates *close* and *far* from its rated power $p_{B,C,n}$

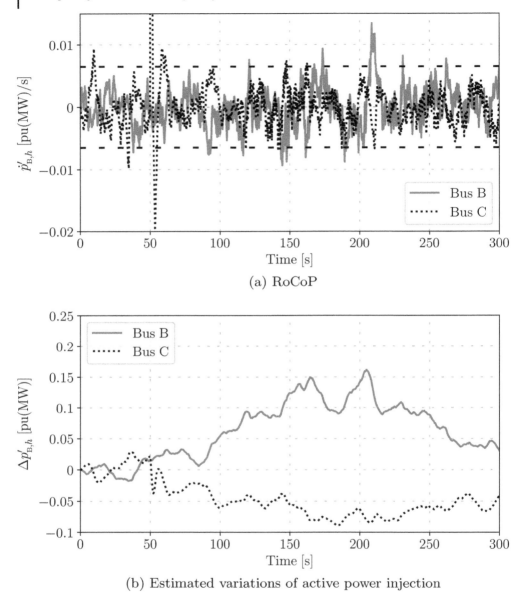

(a) RoCoP

(b) Estimated variations of active power injection

Figure 6.14 AIITS – Response following the loss of a load. FFR is included in the WECS at bus C.

is shown in Figure 6.15. The contingency is the loss of a 44 MW, 13.2 MVAr load at $t = 100$ s.

If the WECS operates close to its rated power, its output power is almost constant and the WECS is basically a frequency taker. According to the discussion on constant power loads given in Section 6.3.2.3, a constant power

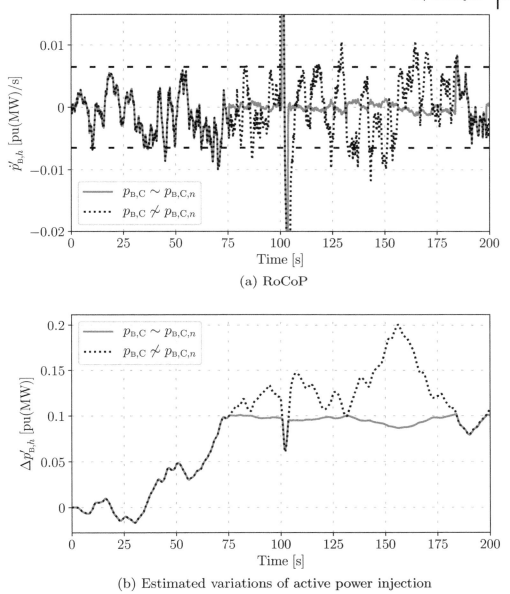

(a) RoCoP

(b) Estimated variations of active power injection

Figure 6.15 AIITS – Response following the loss of a load. FFR is included in the WECS at bus C, which operates near its rated power.

injection is actually a sort of weak frequency maker but its effect on frequency is very small and, consequently, a constant power generator can be assumed to be a frequency taker.

Following the loss of load, the control of the WECS reduces its active power output to regulate the frequency and, thus, during the transient, the WECS clearly behaves as a frequency maker. A few seconds after the occurrence of the disturbance, the WECS operates again at its rated power, thus recovering the behavior as frequency taker.

The main conclusion that can be drawn from Figure 6.15 is that being a frequency maker or taker is not an immutable property of the device itself but, rather, a consequence of its operation and control.

6.4.2.3. Measurement Noise

The robustness of the RoCoP against the presence of noise in the frequency measurements retrieved by SRF-PLLs or PMUs is discussed next. With this aim, an Ornstein-Uhlenbeck process as in (4.31) is applied to all bus voltage phase angles of the AIITS.

The noise added to the bus angles creates a distorted frequency measurement such as the one shown in Figure 6.16 for bus B. All PLLs are assumed to have same technology and thus show similar noise profiles. The resulting frequency signal shows maximum variations of the order of 10^{-5} pu(Hz), i.e. 5 mHz. This value is 5 times higher than the expected maximum error at the fundamental frequency from PMUs discussed in [47].

The top panel of Figure 6.17 shows the trajectories of $\dot{p}'_{B,h}$ at buses B and C under the presence of noise in the SRF-PLL signals for the same scenario as that of Figure 6.14. As expected, the RoCoP is highly sensitive to fast variations of the frequency signal, thus limiting the utility of the RoCoP unless the signals are properly filtered. With this aim, a simple yet effective approach consists

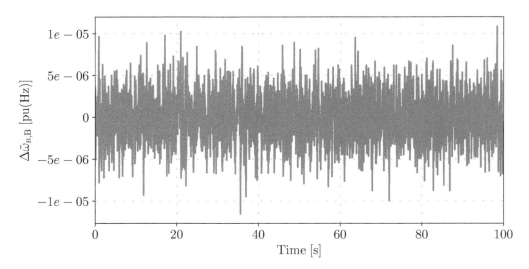

Figure 6.16 AIITS – White noise in the frequency signal measured at bus B with an SRF-PLL.

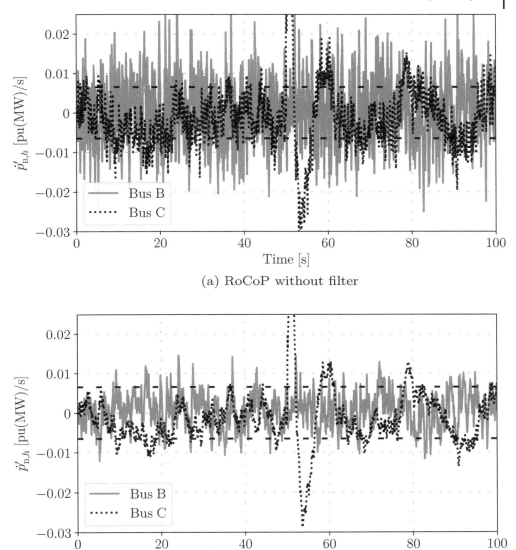

(a) RoCoP without filter

(b) RoCoP with filter ($T_{\mathrm{f},p} = 1$ ms)

Figure 6.17 AIITS – RoCoP at buses B and C following a loss of load with inclusion of white noise in the frequency signal of the SRF-PLLs. FFR is included in the WECS at bus C.

in applying an LPF to $\dot{p}'_{\mathrm{B},h}$, with time constant $T_{\mathrm{f},p} = 1$ ms. The effect of the LPF is shown in Figure 6.17.b.

The inclusion of the LPF helps prevent the issues associated to the deterioration of the RoCoP due to the presence of noise in the measurements. The features of the resulting filtered RoCoP are fairly similar to those discussed in the previous cases that do not consider noise. The RoCoP also shows relatively high sensitivity to the time constant of the LPF, thus this parameter must be tuned carefully. Too small values of $T_{f,p}$ may not remove the effect of the noise on $\dot{p}'_{B,h}$, whereas too high values of $T_{f,p}$ lead to a reduction of the overall values of $\dot{p}'_{B,h}$, thus reducing its reliability.

Ultimately, the ability of the RoCoP to determine whether a device is providing frequency control or not depends on the time scale of the frequency control itself. If such a time scale does not overlap with that of noise and other stochastic variations of the power of the device itself, the RoCoP can be a very effective index.

Part III
Applications

Chapter 7
Frequency Control

7.1. Impact of Frequency Signals

The move from conventional synchronous generation to CIG leads to the reduction of the total inertia of the system, which is one of the most limiting constraints to the integration of renewable resources based on wind and solar energy [131]. Section 4.4 addresses this issue and discusses the potential role of DERs (CIG, ESSs and TCLs) in the RoCoF, FFR, and PFC of low-inertia power systems in their path to first complement, and later substitute the inertial response and PFC of synchronous machines described in Chapter 4.

While the provision of frequency control through DERs is considered to be inevitable, there is still no clear understanding on what the best approach to control such devices is. In particular, there are several concerns for DERs with "small" capacity, which are typically connected at the distribution (medium voltage) level. The main issues that are anticipated for the control of these DERs are: (i) large number of small devices; and (ii) relatively high noise in the distribution network due to the proximity to loads.

Having to deal with several devices would suggest to adopt some kind of centralized control, to make sure that the response of all devices is consistent and contributes to the stable operation of the system. A sort of centralized control has been proposed in the so-called "virtual power plants" [64, 108]. A centralized control, however, is expected to introduce delays and other communication issues that can significantly reduce the effectiveness of the control itself.

For this reason, conventional PFC (see Section 4.2.1) has been implemented in a decentralized way, also because typical power plants have a "good" local estimation of the frequency, i.e. the rotor speed of the synchronous machines. Since DERs are typically nonsynchronous, the frequency has to be estimated based on voltage/current phasors at the bus of connection of the DER.

In practice, the frequency regulated by DERs is thus measured locally through PLLs. Such electronic devices are crucial for the proper synchronization and regulation of VSCs and several implementations exist (see Section 3.3.2). It has to be expected that, depending on the implementation, different

Frequency Variations in Power Systems: Modeling, State Estimation, and Control, First Edition.
Federico Milano and Álvaro Ortega Manjavacas.
© 2020 John Wiley & Sons Ltd. Published 2020 by John Wiley & Sons Ltd.

signals are originated and these lead to different responses of the frequency controllers and, hence, of the overall system.

As discussed in Section 3.3.2, PLLs introduce errors, e.g. due to the calculation of the numerical derivative of phasor components. Reference [163] shows that the impact of PLLs on the frequency regulation of nonsynchronous generation at the HV transmission system level can create instabilities. Similar issues have to be anticipated for the frequency estimation of PLLs at the MV distribution system level.

From a simulation point of view, the frequency control through converter-interfaced or nonsynchronous devices poses the problem of properly defining the frequency signal to be used as input of the regulators. Chapter 2 argues that fully-fledged electromagnetic models are still impractical to simulate large power systems, while conventional electromechanical models for transient stability analysis neglect *a priori* frequency variations in transmission lines and loads.

The following examples, which constitute the first part of the chapter, consider exclusively a conventional decentralized frequency control approach, where each DER utilizes a local frequency measurement, and discuss the impact of different approaches to estimate the frequency, from the *ideal* signals obtained with the FDF and the CoI, to numerical techniques based on the WF and practical implementations based on PLLs devices. A comparison of different strategies to estimate the frequency when frequency control approaches other than the decentralized one are implemented is then provided in Section 7.2.

7.1.1. Examples

This section collects a set of examples that illustrate the contribution of DERs to the frequency control of transmission and distribution grids. The impact of using different strategies to synthesize the input signal of the frequency regulators of DERs is comprehensively analyzed. With this aim, Sections 7.1.1.1 and 7.1.1.2 discuss the performance of the frequency control of WECSs and TCLs, respectively, by comparing different approaches to estimate the frequency at the bus of connection, namely the PLL, the CoI, the WF, and the FDF.

7.1.1.1. Wind Power Plant

The dynamic response of the PFC provided by a WECS (see Section 4.4) is studied by comparing two scenarios: (i) PFC where the input signal is obtained through a PLL; and (ii) the same control fed by the frequency estimation approaches based on the CoI and the FDF. This example utilizes the WSCC system described in Appendix A.2 where the generator at bus 3 is substituted with a WECS with DFIGs with same capacity as that of the original synchronous machine (see also Section 4.4.3).

In order to provide a realistic evaluation of the accuracy and dynamic performance of the different frequency estimation techniques, noise modeled as an Ornstein-Uhlenbeck process (see equation (4.31)) is applied to the magnitudes and angles of all bus voltage phasors of the grid.

The input signals of the wind plant controller and the subsequent active power outputs are represented in Figures 7.1 and 7.2, respectively, for the different frequency estimation approaches. The difference between the signals is relatively small, leading to similar active powers supplied by the WECS.

The response of the frequency of the CoI is shown in Figure 7.3 for each of the frequency estimation approaches, and compared with that without the controller of the WECS. The inclusion of such a controller reduces the frequency peak caused by the line outage by about 30–40%, being the lowest when the FDF is used as input signal of the controller. Note that the spikes of the frequency signals that can be observed in Figure 7.1 when using the PLL are filtered by the WECS controller.

To study the sensitivity of the three frequency estimation techniques to fast system dynamics, the flux dynamics of the DFIGs are included in the model. Results are shown in Figure 7.4. In this case, the behavior of the CoI, the FDF and the response of the controller based on the PLL is similar to the one when no flux dynamics were included.

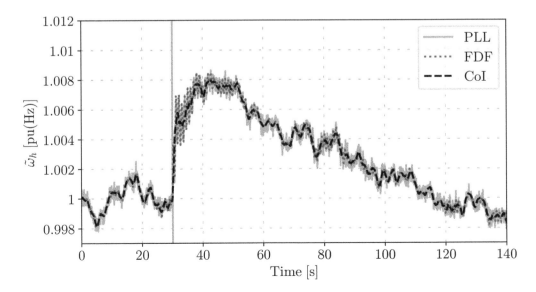

Figure 7.1 WSCC system – Input frequency signal of the WECS control estimated by different approaches.

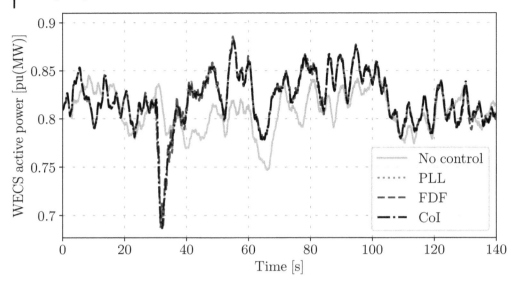

Figure 7.2 WSCC system – Active power supplied by the WECS.

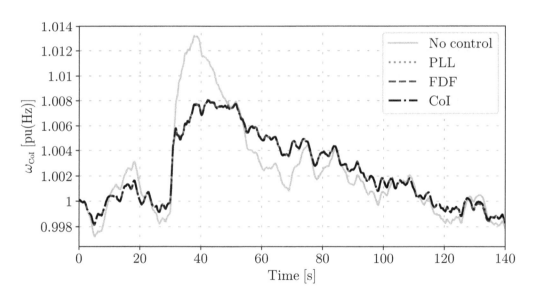

Figure 7.3 WSCC system – Frequency of the CoI.

These results suggest that a proper tuning of the PLL can remove the effect of fast dynamics on the controllers based on local frequency measurements. This result, however, is not general as discussed in Example 2 on TCLs below, which is based on the New England system. Note also that the filters of the PLLs should not introduce a delay in the frequency measurement to prevent a potential deterioration in the dynamic response.

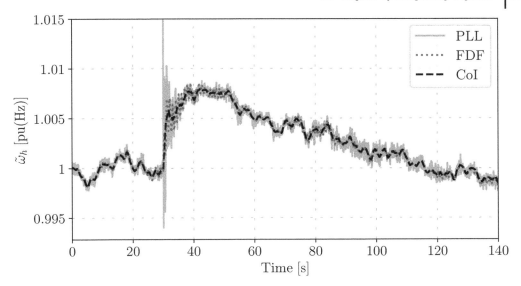

Figure 7.4 WSCC system – Input frequency signal of the WECS control estimated by different approaches with inclusion of DFIG flux dynamics.

7.1.1.2. Thermostatically Controlled Loads

The performance of TCLs in a power system when their input signal is provided by the CoI, WF, and FDF is studied in this section. Two benchmark networks are considered: the IEEE 14-bus system and the New England system, which are described in Sections A.3 and A.4, respectively.

Example 1: IEEE 14-Bus Test System

Primary and secondary frequency regulations of synchronous machines are not included in order to study the effect of the frequency regulation of the TCLs solely. The contingency is the outage of the line connecting buses 2 and 4 in base loading conditions, as well as with 20% of overload. The amount of TCLs in the system is 30% of the total load.

Figure 7.5 shows the rotor speed of the synchronous machine at bus 2, the frequency of the bus estimated by the WF and the FDF, and the frequency of the CoI, for the base case loading conditions and without TCLs. While both WF and FDF estimators show a trend similar to the rotor speed of the machine, the signal provided by the FDF is more accurate, since it includes oscillations of same period and similar amplitude than those of the machine rotor speed. On the other hand, the CoI "filters" such oscillations, providing only information of the average frequency variation.

The differences in the estimated frequencies appear to be negligible when the TCLs are included, as shown in Figure 7.6. The rotor speed of the machine

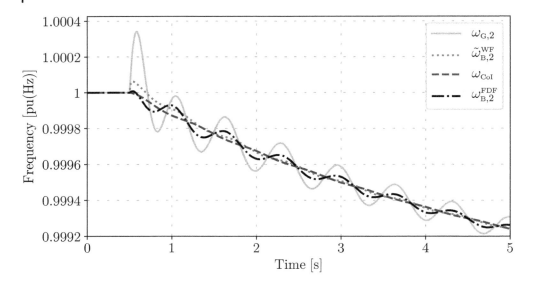

Figure 7.5 IEEE 14-bus system – Dynamic response following line 2-4 outage without TCLs.

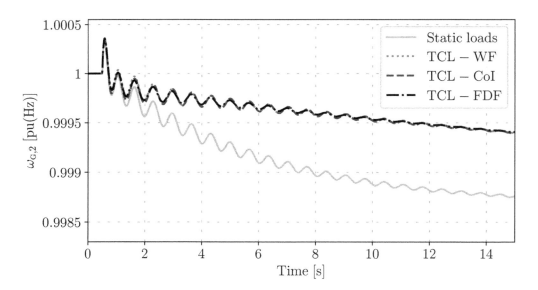

Figure 7.6 IEEE 14-bus system – Dynamic response following line 2-4 outage with TCLs.

at bus 2 is depicted for the cases without and with TCLs. In the latter case, the dynamic response of TCLs is compared in three scenarios, corresponding to using as control input signal the frequency estimation provided by the WF, the CoI and the FDF, respectively.

The inclusion of the TCLs reduces the steady-state frequency drop by about 50% for the three cases. While in this case the transient response of the system does not appear to be affected by the model of the frequency estimation, one cannot conclude that this is always the case. The following example demonstrates, in fact, that depending on the loading level the choice of the input signal of the TCLs can provide a considerably different performance of these devices, and therefore, a different behavior of the whole system.

Figure 7.7 is obtained for a 20% increase of the load with respect to the base case without TCLs and shows the trajectories of the rotor speed of the machine at bus 2, as well as the estimations of the frequency at buses 2 and 14 with the WF, the CoI, and the FDF.

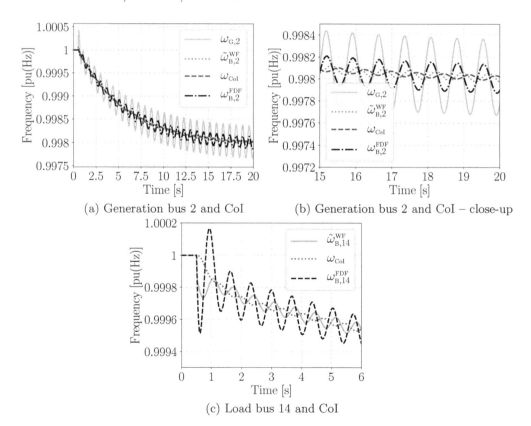

(a) Generation bus 2 and CoI

(b) Generation bus 2 and CoI – close-up

(c) Load bus 14 and CoI

Figure 7.7 IEEE 14-bus system – Dynamic response with 20% overload following line 2-4 outage without TCLs.

Undamped oscillations can be observed due to the presence of a limit cycle. Figures 7.7.a and 7.7.b show that the FDF is again the most accurate, while the WF includes a delay in the signal, and the CoI contains counter-phase oscillations due to the large size of the synchronous machine at bus 1. Moreover, depending of the chosen frequency estimation technique, considerably different input signals are introduced into the TCLs controllers. This is shown in Figure 7.7.c, where the estimated frequencies at the load bus 14 are compared.

Finally, the rotor speed of the machine at bus 2 is shown in Figure 7.8, which is obtained including TCLs in the system. The frequency drop is again reduced by about 50%. However, while the WF and the CoI signals lead to a stationary limit cycle, the usage of the FDF indicates that oscillations are actually damped. It is worth noticing that the only difference in the model is the method to estimate the frequency signal sent to the TCL. This difference in the frequency signal affects the response of frequency control devices such as the TCLs studied in this section, thus leading to a different behavior of the overall system. In this case, the WF and the CoI appear to be more conservative from the control point of view. This is a consequence of the fact that these models do not capture the variations of local bus frequencies as accurately as the FDF.

Example 2: New England System

The contingency is a three-phase fault at bus 21, cleared by the opening of the line connecting buses 16 and 21 after 160 ms. TCLs represent 20% of the total load.

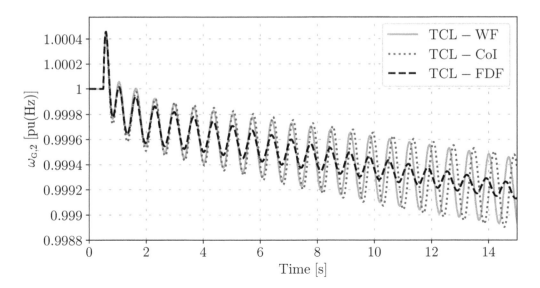

Figure 7.8 IEEE 14-bus system – Dynamic response with 20% overload following line 2-4 outage with TCLs.

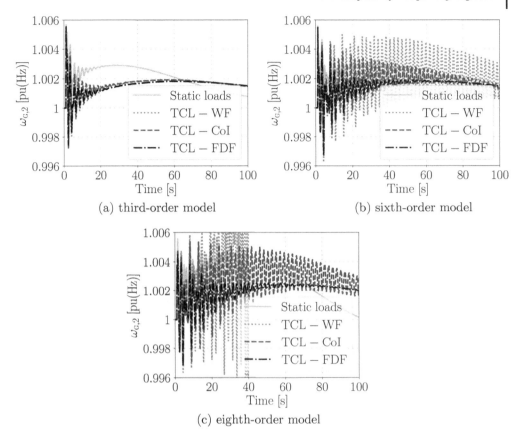

(a) third-order model

(b) sixth-order model

(c) eighth-order model

Figure 7.9 New England system – Dynamic response with TCLs following a three-phase fault for different synchronous machine models.

Figure 7.9 shows the rotor speed of the machine at bus 31 (G2) without TCLs, and with TCLs considering the three control input signals, namely FDF, WF and CoI. Three different models of the synchronous machines of the system are compared (see Section 2.4.1), namely: the one-axis third-order model (Figure 7.9.a), the Sauer and Pai's sixth-order model (Figure 7.9.b), and the fully-fledged eighth-order model (Figure 7.9.c).

Figure 7.9.a shows that including TCLs into the system enables the reduction of the frequency variations due to the fault. The TCLs reduce the damping of dominant modes. Such modes are slightly better damped if the model includes the FDF signal.

Figures 7.9.b and 7.9.c show the response of the system for more detailed and accurate models of the synchronous machines. In these cases, the dynamic interaction of the machine transient and sub-transient dynamics with the TCLs causes poorly damped frequency oscillations. It is interesting to note that, when considering stator flux dynamics and the WF signal (see Figure 7.9.c) such oscillations become unstable and lead the system to collapse. On the other hand, the TCLs controller coupled to the FDF shows that oscillations are properly damped.

7.1.1.3. Remarks

The following remarks on the frequency control through DERs are relevant.

- The numerical derivative of the bus voltage phase angle of the WFs and PLLs can lead to nonphysical oscillations and, possibly to numerical instabilities.

- The average rotor speed provided by the CoI filters local frequency variations. This fact may cause poorly-damped frequency oscillations, especially if coupled to devices with a slow response, such as TCLs.

- Controllers using signals obtained with the FDF are less prone to introduce numerically-driven undamped oscillations.

From the results above, it appears that a proper modeling of the control signals can make a significant difference in the transient stability analysis of a power system with inclusion of frequency controllers other than primary frequency regulators of synchronous machines.

7.2. Synthesis of Frequency Signals

It is relevant to study the impact of the frequency control of transmission- and distribution-level DERs on the overall system transient behavior when strategies other than the decentralized one are utilized to synthesize the frequency signals used by the DERs. With this aim, two strategies to synthesize the signal used as input of the DER frequency regulators, namely, *centralized* and *averaged* are discussed [165]. For completeness, also the conventional decentralized approach discussed in Section 7.1 is considered. Figure 7.10 illustrates these strategies.

It is important to note that the term *centralized* is not used here with the conventional meaning that it has in control theory. The frequency controllers of DERs are *always* local in all approaches considered in this section. *Centralized* as well as *decentralized* refers only to the strategy used to define the input signals of such controllers. The proposed strategies are detailed below.

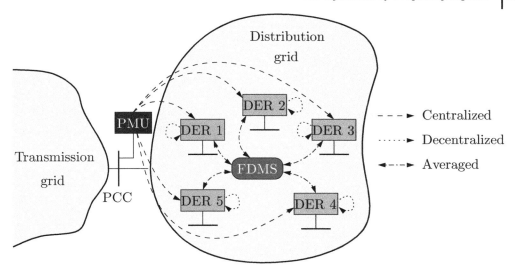

Figure 7.10 Illustration of three strategies to retrieve the frequency control input signal.

Centralized

The frequency at the point of common coupling (PCC) between the transmission and distribution grids is measured by means of, e.g. a PMU, and then the signal is sent to every DER installed. As all DERs use the same frequency signal in their regulators, a good overall control performance can be expected from this strategy. However, it is also characterized by a communication delay related to the measurement and dispatch of the frequency signal that can deteriorate such a performance. As the overall frequency control relies on only one measurement, it is desirable to have a redundancy by means of, e.g. a second PMU connected to the PCC, to avoid the loss of all regulation capability in case of possible PMU malfunctions.[1]

Decentralized

This strategy considers that each DER measures, fully independently from any other DER, the frequency at its own bus of connection, by means of the PLLs

---

[1]The interested reader can refer to [61] for a report on the largest absolute frequency deviation since 2006, recorded on 10 January 2019, which stretched across 26 countries of the Continental Europe Power System. The event was, in part, the result of a frozen measurement on four interconnection lines between Germany and Austria.

included in its power electronic converter. The main advantage of this strategy is that it does not include any communication delay in the process, as both measurement and control are done locally. On the other hand, this strategy does not provide any form of coordination between DERs. Moreover, while the frequency variations estimated at every DER bus of a distribution grid should be the same, this estimation can significantly differ from bus to bus during transients due mainly to the numerical issues that derive from the numerical derivation of the bus voltage phase angle.

Averaged

The PLL frequency estimations from every DER bus are sent and collected by a frequency data management system (FDMS) located generally within the distribution grid. The FDMS then computes the average value of all signals, and then this average signal is sent back to every DER. This strategy shows the same main advantage of the centralized approach, as the DERs regulate the frequency using the same signal. The averaging process of such a signal help reduce the impact of the spikes and other numerical issues present in the measured signals, as well as that of losing one or more of the measurements. However, as this strategy is based on a bidirectional communication channel, latency must be carefully taken into account.

7.2.1. Examples

A modified version of the WSCC system described in Appendix A.2 serves to compare the performance of the three strategies described above. The impact of measurement issues, including the calculation of the numerical derivative of the bus voltage phase angle, as well as noise, delays, and loss of information of the frequency measurements are considered in the comparison.

The modification of the WSCC system consists in the inclusion of a distribution grid, that is connected to bus 6 and replaces the original load as shown in Figure 7.11. The distribution system resembles a small Irish MV grid that includes both radial and meshed configurations, and is composed of eight buses and lines, six loads, two WECSs, one SPVG, and one ESS [148]. The distribution grid is operated at 38 kV and is connected through an ULTC type step-down transformer to the transmission grid. The total active and reactive power consumed by the loads of the distribution grid is 0.578 MW and 0.117 MVAr, respectively. The active power generation at the initial operating point of the WECSs at buses D2 and D5, and of the SPVG at bus D8 are of 15 MW each. The power rate of the ESS at bus D2 is 10 MW. Given the short time scales considered in this example, energy limits of the ESS are neglected.

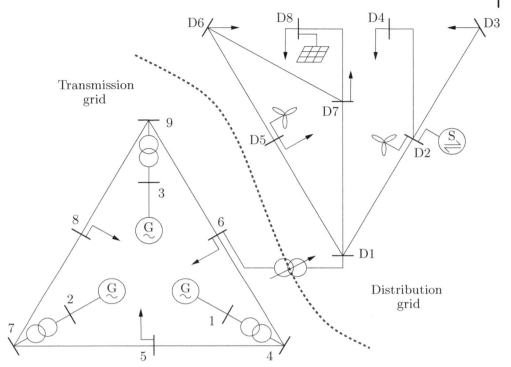

Figure 7.11 Modified WSCC system – the load at bus 6 has been substituted with an 8-bus 38 kV distribution system.

To accommodate DERs, the net power consumption of the distribution grid is lower than the original load connected to bus 6 of the WSCC. This has been taken into account by reducing accordingly the active power generation of the synchronous machines at the initial operating point. Each of the DERs connected to buses D2, D5, and D8 include an SRF-PLL that retrieves the bus frequency signal.[2] Identical gains $K_P = 0.2$ and $T_I = 20$ s of the LF of the PLL scheme described in Section 3.3.2.1 have been used for all DERs. The frequency signals of the regulators of the wind power plant and the ESS at bus D2 are generated with a single PLL.

Two scenarios are considered in this example, as follows. In Section 7.2.1.1, the impact of noise and latency in the frequency signals for the three strategies described in Section 7.2 is compared. Then, the robustness of each strategy against the loss of one of the PLL measurements is discussed in Section 7.2.1.2. The contingency considered for both scenarios is a three-phase fault at bus 7 at $t = 1$ s, cleared after 150 ms by tripping the line that connects buses 5 and 7.

---

[2] In the remainder of this section, the acronym PLL will be used to refer to the SRF-PLL.

7.2.1.1. Noise and Latency

Noise is introduced at the bus voltage angle of every bus of the distribution grid. This noise accounts for possible unbalances, proximity of the loads, harmonics of the power electronic devices, etc. The Ornstein-Uhlenbeck process described in equation (4.31) is utilized to model the noise. Same parameters to generate the stochastic processes that define the noise are used for all buses of the distribution grid.

The delays associated to the processing and communication of the frequency control signals are also taken into account. To this aim, depending on the strategy used to retrieve such a signal, different values of the time delay τ_θ are assigned accordingly. The default τ_θ used for every PLL is 5 ms, which accounts for the time needed to perform the signal measurement and variable transformation. This value is thus utilized for the decentralized strategy in the remainder of this section.

The scenario where communication delays are neglected is first studied, i.e. $\tau_\theta = 5$ ms for all PLLs for the three control signal retrieval strategies. The response of the system for the different strategies is compared by observing the frequency at bus 6, i.e. the PCC of the transmission and the distribution grids, and results are shown in Figure 7.12. Such trajectories are obtained by applying the FDF, which, as discussed in Chapter 5, provides a highly accurate frequency estimation that is free from noise, delays, and numerical issues. Figure 7.12 also includes, as a reference for the comparison, the *ideal* response obtained by using the FDF to generate the input signal of the frequency regulators.

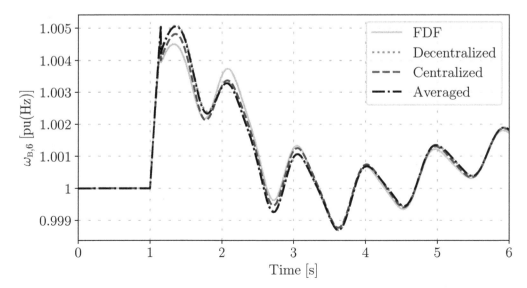

Figure 7.12 WSCC system – Frequency of bus 6 after a three-phase fault. Communication latency is not modeled.

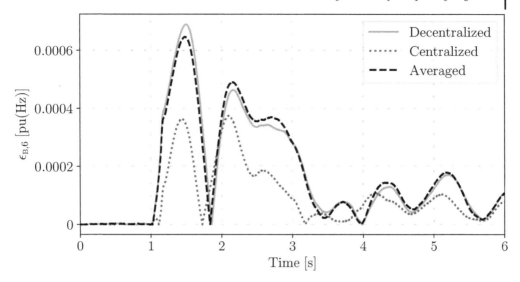

Figure 7.13 WSCC system – Absolute error of the frequency at bus 6 after a three-phase fault. Communication latency is not modeled.

As expected, the PLL-based responses are worse than the one based on the FDF, as they show larger frequency variations at the first instants after the contingency, even for the scenario where no communication delays are considered. This can be better appreciated in Figure 7.13 which shows the absolute errors with respect to the FDF-based trajectory, $\epsilon_{B,6}$.

The centralized strategy (which takes the measurement from bus D1) shows a better response than the other two, which show a fairly similar profile. This is because the signal is measured at the bus that is closest to the transmission grid, thus the least affected by voltage fluctuations and, in turn, shows lower "spikes" derived from the numerical derivative of the bus voltage phasor during the fault, and few instants after the line outage.

This observation is confirmed in Figure 7.14, where the PLL measurements of the DER buses D2, D5, and D8, as well as their average signal, are represented. The farther is the PLL device from the PCC, the more sensitive to numerical issues is the generated signal. The results shown in Figure 7.14, also justify why the decentralized and the averaged strategies are similar. In fact, by averaging the signals one can reduce the impact of the worst signal (bus D8), but also the DERs at buses D2 and D5 cannot benefit from the better quality of the signals generated by their PLLs.

The impact of noise on the performance of the frequency controllers is negligible for all cases, as the noise is filtered out by the LPFs of the regulators.

The latency that results from communication delays of the frequency signals is another relevant parameter that affects the performance of the controller.

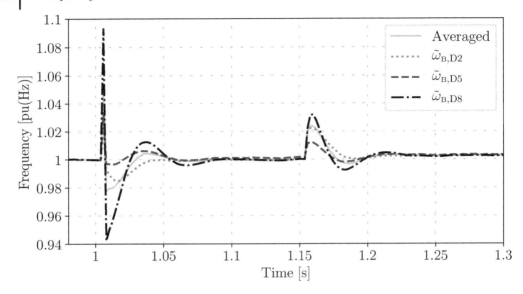

Figure 7.14 WSCC system – Frequency signals measured with different PLLs, and their average value. Communication latency is not modeled.

Assume that the time needed to send the measured signal from bus D1 and to retrieve the three PLL signals from buses D2, D5, and D8, compute and send the average signal back to the regulators has a conservative value of 20 ms for both cases. Therefore, $\tau_\theta = 5$ ms for the decentralized strategy, and $\tau_\theta = 25$ ms for both the centralized and the averaged strategy. The resulting $\epsilon_{B,6}$ of trajectories of the frequency at bus 6 are shown in Figure 7.15.

The performance of the averaged strategy deteriorates due to the inclusion of the communication delays. A 25 ms delay is also enough to lead to frequency oscillations of small amplitude and with a period of about 0.1 s. The centralized strategy, on the other hand, does not appear to be affected by such a delay. However, if the time required to send the signal from bus D1 to the DERs increases, similar "high-frequency" oscillations are observed, as shown in Figure 7.16, where a communication delay of 55 ms for the centralized strategy is included, i.e. $\tau_\theta = 60$ ms. Figure 7.16 also shows the unstable response of the averaged strategy for a communication delay of 35 ms.

7.2.1.2. Loss of Information

This scenario studies the robustness of the three strategies against the loss of a frequency measurement signal. With this aim, the communication system of the DER connected to bus D8 is assumed to malfunction and does not properly send the frequency measurements of the PLL to the decentralized and the

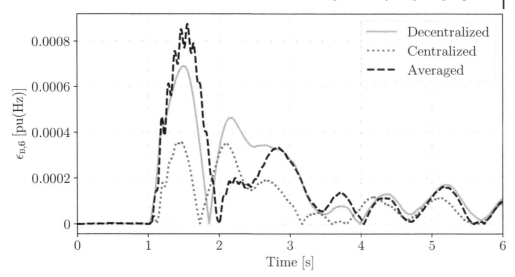

Figure 7.15 WSCC system – Absolute error of the frequency at bus 6 after a three-phase fault. Communication delay is 20 ms.

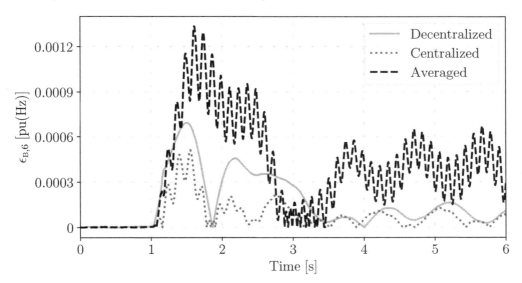

Figure 7.16 WSCC system – Absolute error of the frequency at bus 6 after a three-phase fault. Communication delays are 35 and 55 ms for the averaged and centralized strategies, respectively.

averaged strategies. As for the centralized strategy, the PLL is assumed to work well and thus the local frequency control of the DERs is not affected. The performance of the centralized strategy is thus the same as that shown in Section 7.2.1.1 above.

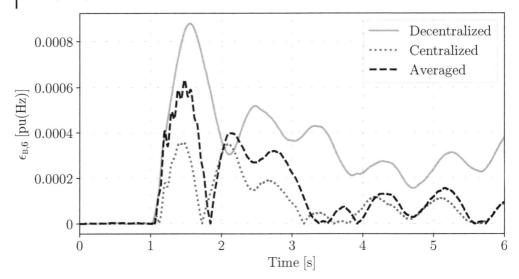

Figure 7.17 WSCC system – Absolute error of the frequency at bus 6 after a three-phase fault. The frequency signal of bus D8 is lost for the averaged and decentralized strategies.

Figure 7.17 shows the $\epsilon_{B,6}$ of the three strategies when the PLL at bus 8 fails to retrieve/send its frequency measurement. The total delays τ_θ are 5 ms for the decentralized strategy and 25 ms for the centralized and averaged ones.

The error $\epsilon_{B,6}$ of the decentralized strategy increases by about 20% with respect to that of Figure 7.15 in the first swing after the contingency. The error of the decentralized strategy is also considerably higher than that of the other strategies throughout the rest of the simulation. On the other hand, the loss of the measurement at bus D8 improves the performance of the averaged strategy by reducing $\epsilon_{B,6}$ during the first instants after the contingency by about 30%. This interesting result is justified by the fact that the signal from the PLL of bus D8 is also the most affected by the spikes as shown previously in Figure 7.14. Therefore, the resulting average signal shows a better accuracy, thus leading to a better performance of the DER frequency regulator.

7.2.1.3. Remarks

The following remarks are relevant.

- While the centralized strategy shows a better overall performance, it highly depends on the associated signal communication delays. To avoid the loss of all regulation capability in case of malfunctions of the measurement device, a redundancy of such a measurement is desirable.

- The decentralized strategy works reasonably well, and it does not include any form of communication delay. However, its overall performance can be highly deteriorated in case of loss of any of the frequency measurements.

- The averaged strategy shows a good robustness against the loss of measurement signals without the need of redundant measures. It naturally filters out the largest spikes and other numerical issues of the measurements during transients. Similarly to the centralized strategy, its performance highly depends on the communication delays.

The averaged strategy appears interesting for practical applications, provided that communication issues are properly addressed. For instance, one way of improvement of the averaged strategy is by means of the identification of the areas within the distribution grid with high density of DERs installed, with the aim of minimizing the related communication delays.

Chapter 8

Dynamic State Estimation

8.1. Machine Rotor Speeds

In simulations, the vector of rotor speeds is accessible at any time and thus the FDF is a consistent and reliable tool. Transmission system operators, however, can install PMUs at network buses, and have thus access to the bus frequencies estimated by such PMUs, while the rotor speeds of the synchronous machines are not accessible to them. This section considers a practical application of the FDF based on available real-world measurements [133]. As it turns out, this is a consequence of the *acausal* form of (5.13) and the fact that inputs and outputs are not set *a priori* in such an expression.

The synchronous machines rotor speed estimation method discussed in this section is intended to be a tool for TSOs, which generally have full knowledge of the transmission network, but do not have the detailed dynamic models and measurements of internal quantities of the power plants and the loads connected to the grid. The assumption, implied in the FDF, of knowing generator internal impedances is acceptable as the values of such impedances can be easily obtained from the literature and manufacturer data sheets once the technology and the capacity of the power plant is known. Apart from that, the estimation of the rotor speeds discussed below is model-agnostic, as it does not requires any knowledge of the model of the machines and of the controllers included in the power plants.

The remainder of this section is organized as follows. The proposed rotor-speed estimation techniques, namely the convex quadratic programming problem and the weighted least squares (WLS) are presented in Section 8.1.1. Finally, a variety of examples based on the WSCC system as well as the AIITS are presented in Section 8.1.2.

Frequency Variations in Power Systems: Modeling, State Estimation, and Control, First Edition.
Federico Milano and Álvaro Ortega Manjavacas.

8.1.1. DSE Problem Formulation

The FDF-based DSE problem to estimate the rotor angular speed of synchronous machines is formulated as the following optimization problem:

$$\min_{\left(\varepsilon_{\mathrm{B}}(t),\, \Delta\omega_{\mathrm{G}}(t)\right)} \quad J(\varepsilon_{\mathrm{B}}(t)) = \frac{1}{2}\varepsilon_{\mathrm{B}}^{\dagger}(t)\, \mathbf{U}^{-1}\, \varepsilon_{\mathrm{B}}(t) \tag{8.1}$$

$$\text{s.t.} \quad \mathbf{0}_{n_{\mathrm{B}},1} = \mathbf{B}_{\mathrm{BG}}\Delta\omega_{\mathrm{G}}(t) + \mathbf{B}_{\mathrm{BB}}\big(\Delta\tilde{\omega}_{\mathrm{B}}(t) + \varepsilon_{\mathrm{B}}(t)\big) \quad : \boldsymbol{\mu}_{\mathrm{B}}(t)\,, \tag{8.2}$$

where $\Delta\tilde{\omega}_{\mathrm{B}}$ is the input vector of measured bus frequency variations as provided by the PMUs; ε_{B} is the vector of measurement errors; $\Delta\omega_{\mathrm{G}}$ is the vector of estimated rotor speed variations of the synchronous machines; $\boldsymbol{\mu}_{\mathrm{B}}$ are the dual variables associated with the equality constraints; and $\mathbf{U}$ is the covariance matrix of the measurements as defined in Section 3.1.

A relevant property of the constraints (8.2) is that they are linear. Linear formulations are not common in DSE but do exist, e.g. [179] and [220]. However, in existing works, the models of the synchronous machines and their controllers as well as of dynamic loads have to be defined *a priori*. The same requirement affects the approaches discussed in [76, 192, 231], which are aimed at estimating the states of the machines using local PMU measurements.

Matrix $\mathbf{B}_{\mathrm{BG}}$ is maximum rank, i.e. n_{G}, and $\mathbf{B}_{\mathrm{BB}}$ is full rank, i.e. n_{B}, hence the gradient vectors of the constraints (8.2) at the solution are linearly independent. The problem (8.1)–(8.2) has thus the required regularity conditions. Moreover, since $\mathbf{U}$ and hence its inverse are positive definite, problem (8.1)–(8.2) is a convex quadratic programming problem, whose solution is unique [22]. Note that, since the solution is the global optimum, the problem (8.1)–(8.2) also coincides with the Kalman filter formulation.

Problem (8.1)–(8.2) assumes a continuous dependence on time. Discrete time has to be utilized if the measurements $\Delta\tilde{\omega}_{\mathrm{B}}$ are available as digital signals, as it is usual, but since the DSE problem does not involve differential equations, all variables have to be available at the same given time, which, for simplicity of notation, is indicated with t.

The Lagrangian function $\mathcal{L}(\Delta\omega_{\mathrm{G}}, \varepsilon_{\mathrm{B}}, \boldsymbol{\mu}_{\mathrm{B}})$ associated with the optimization problem (8.1)–(8.2) is:[1]

$$\mathcal{L}(t) = J\big(\varepsilon_{\mathrm{B}}(t)\big) - \boldsymbol{\mu}_{\mathrm{B}}^{\dagger}(t)\Big[\mathbf{B}_{\mathrm{BG}}\Delta\omega_{\mathrm{G}}(t) + \mathbf{B}_{\mathrm{BB}}\big(\Delta\tilde{\omega}_{\mathrm{B}}(t) + \varepsilon_{\mathrm{B}}(t)\big)\Big]. \tag{8.3}$$

Assuming an input vector of measurements $\tilde{\omega}_{\mathrm{B}}$ and the optimal solution $(\Delta\breve{\omega}_{\mathrm{G}}, \breve{\varepsilon}_{\mathrm{B}}, \breve{\boldsymbol{\mu}}_{\mathrm{B}})$ of problem (8.1)–(8.2), the first order KKT optimality conditions

---

[1] In (8.3) as well as in the remainder of this chapter, the dependency of $\mathcal{L}$ on $(\Delta\omega_{\mathrm{G}}, \varepsilon_{\mathrm{B}}, \boldsymbol{\mu}_{\mathrm{B}})$ is omitted for compactness.

are:

$$\mathbf{0}_{n_G,1} = \frac{\partial \mathcal{L}}{\partial \Delta \omega_G}\bigg|_{(\Delta \check{\omega}_G, \check{\varepsilon}_B, \check{\mu}_B)} = -\mathbf{B}_{BG}^\dagger \check{\mu}_B(t) \ , \tag{8.4}$$

$$\mathbf{0}_{n_B,1} = \frac{\partial \mathcal{L}}{\partial \varepsilon_B}\bigg|_{(\Delta \check{\omega}_G, \check{\varepsilon}_B, \check{\mu}_B)} = \mathbf{U}^{-1}\check{\varepsilon}_B(t) - \mathbf{B}_{BB}^\dagger \check{\mu}_B(t) \ , \tag{8.5}$$

$$\mathbf{0}_{n_B,1} = \frac{\partial \mathcal{L}}{\partial \mu_B}\bigg|_{(\Delta \check{\omega}_G, \check{\varepsilon}_B, \check{\mu}_B)} = -\mathbf{B}_{BG}\Delta \check{\omega}_G(t) - \mathbf{B}_{BB}\big(\Delta \tilde{\omega}_B(t) + \check{\varepsilon}_B(t)\big) \ . \tag{8.6}$$

The equations (8.4)–(8.6) enable the computation of a wide set of sensitivities and their solution is robust with respect to noise and bad data. These features are not provided by the WLS problem discussed in Section 8.1.1.1 below. Equations (8.4)–(8.6) can also be readily included in a time-domain integration algorithm and solved together with the set of DAEs that describes the dynamics of the system. This is the implementation utilized in the examples presented in Section 8.1.2.

The low computational burden of solving the linear set of equations (8.4)–(8.6) enables its implementation in practical real-world DSEs, as illustrated in Figure 8.1. As the power system evolves in time, PMUs measure the voltages at the network buses, $v_{B,abc}$, and return the estimated values of the bus frequencies $\tilde{\omega}_B$. These are then collected by the state estimator, which estimates the synchronous machine rotor speeds $\Delta \check{\omega}_G$ through the solution of (8.4)–(8.6). Finally, the estimated $\Delta \check{\omega}_G$ can be used by the system operator to take control actions u on the system.

The added value of (8.4)–(8.6) and, in general, of rotor speed estimation is the ability to monitor online the transient behavior of synchronous machines.

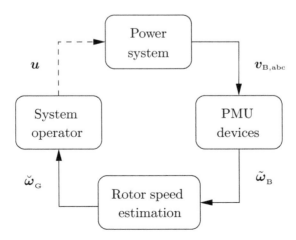

Figure 8.1 Scheme of an on-line frequency state estimator.

This is a valuable information for TSOs and can be utilized, for example, for online transient stability corrective control, wide area monitoring and control, and model validation in offline stability analysis.

8.1.1.1. Weighted Least Square Problem

For completeness and for the relevant topological properties that can be deduced from it, this section presents a linear measurement problem based on (5.14):

$$\mathbf{D}\Delta\boldsymbol{\omega}_{\mathrm{G}}(t) = \Delta\tilde{\boldsymbol{\omega}}_{\mathrm{B}}(t) + \boldsymbol{\varepsilon}_{\mathrm{B}}(t) \ , \tag{8.7}$$

whose optimal solution is:

$$\Delta\breve{\boldsymbol{\omega}}_{\mathrm{G}}(t) = \left(\mathbf{D}^{\dagger}\mathbf{U}^{-1}\mathbf{D}\right)^{-1}\mathbf{D}^{\dagger}\mathbf{U}^{-1}\Delta\tilde{\boldsymbol{\omega}}_{\mathrm{B}}(t) \ . \tag{8.8}$$

This problem is the conventional WLS state estimation problem formulation of (8.1)–(8.2).

Assuming that the diagonal elements of $\mathbf{U}$ are equal, i.e. all frequency measurements are based on the same technology, the solution of (8.8) reduces to that of a linear least square problem:

$$\Delta\breve{\boldsymbol{\omega}}_{\mathrm{G}}(t) = \left(\mathbf{D}^{\dagger}\mathbf{D}\right)^{-1}\mathbf{D}^{\dagger}\Delta\tilde{\boldsymbol{\omega}}_{\mathrm{B}}(t) = \mathbf{D}^{+}\Delta\tilde{\boldsymbol{\omega}}_{\mathrm{B}}(t) \ , \tag{8.9}$$

where $\mathbf{D}^{+}$ is the Moore-Penrose pseudo-inverse, or *left inverse*, of $\mathbf{D}$, which is unique as $\mathbf{D}$ has rank m_{G}. From (5.14), one has:

$$\mathbf{D}^{+} = -\left(\mathbf{B}_{\mathrm{BB}}^{-1}\mathbf{B}_{\mathrm{BG}}\right)^{+} \ . \tag{8.10}$$

Then, utilizing the matrix equivalence $(\mathbf{AB})^{+} = \mathbf{B}^{+}\mathbf{A}^{-1}$, with $\mathbf{A}$ square and invertible and $\mathbf{B}$ rectangular, the right-hand side of (8.10) can be rewritten as:

$$\mathbf{D}^{+} = -\mathbf{B}_{\mathrm{BG}}^{+}\left(\mathbf{B}_{\mathrm{BB}}^{-1}\right)^{-1} = -\mathbf{B}_{\mathrm{BG}}^{+}\mathbf{B}_{\mathrm{BB}} \ . \tag{8.11}$$

Note that equation (8.9) can be also obtained straightforwardly from (5.13), as follows:

$$\mathbf{B}_{\mathrm{BG}}\,\Delta\boldsymbol{\omega}_{\mathrm{G}}(t) = -\mathbf{B}_{\mathrm{BB}}\,\Delta\tilde{\boldsymbol{\omega}}_{\mathrm{B}}(t)$$
$$\Rightarrow \ \mathbf{B}_{\mathrm{BG}}^{+}\mathbf{B}_{\mathrm{BG}}\,\Delta\boldsymbol{\omega}_{\mathrm{G}}(t) = -\mathbf{B}_{\mathrm{BG}}^{+}\mathbf{B}_{\mathrm{BB}}\,\Delta\tilde{\boldsymbol{\omega}}_{\mathrm{B}}(t) \ , \tag{8.12}$$
$$\Delta\boldsymbol{\omega}_{\mathrm{G}}(t) = -\mathbf{B}_{\mathrm{BG}}^{+}\mathbf{B}_{\mathrm{BB}}\,\Delta\tilde{\boldsymbol{\omega}}_{\mathrm{B}}(t) \ .$$

From the discussion of (5.14), it is known that $\mathbf{D}$ is dense as it is obtained from the inverse of $\mathbf{B}_{\mathrm{BB}}$. The pseudo-inverse of $\mathbf{D}$, however, is extremely sparse. This can be readily deduced from the observation of the structure of $\mathbf{B}_{\mathrm{BG}}$. The elements of $\mathbf{B}_{\mathrm{BG}}$, say $B_{\mathrm{G},k}$, are nonzero only in the rows corresponding to the buses at which each synchronous machine is connected. Then, it is straightforward to show that the pseudo-inverse of $\mathbf{B}_{BG}$ is:

$$\mathbf{B}_{BG}^{+} = \left(\mathbf{B}_{BG}^{\dagger}\mathbf{B}_{BG}\right)^{-1}\mathbf{B}_{BG}^{\dagger}$$

$$\Rightarrow \quad B_{G,k}^{+} = \frac{1}{B_{G,k}^2} \cdot B_{G,k} = \frac{1}{B_{G,k}}, \quad \forall B_{G,k} \neq 0 . \tag{8.13}$$

For example, let:

$$\mathbf{B}_{BG} = \begin{bmatrix} B_{G,1} & 0 \\ 0 & B_{G,2} \\ 0 & 0 \end{bmatrix} ,$$

which represents the three-bus grid depicted in Figure 5.5, with the machines connected to buses 1 and 2, respectively. Then, from (8.13), one has:

$$\mathbf{B}_{BG}^{+} = \left(\begin{bmatrix} B_{G,1} & 0 & 0 \\ 0 & B_{G,2} & 0 \end{bmatrix} \begin{bmatrix} B_{G,1} & 0 \\ 0 & B_{G,2} \\ 0 & 0 \end{bmatrix}\right)^{-1} \begin{bmatrix} B_{G,1} & 0 & 0 \\ 0 & B_{G,2} & 0 \end{bmatrix}$$

$$= \begin{bmatrix} 1/B_{G,1}^2 & 0 \\ 0 & 1/B_{G,2}^2 \end{bmatrix} \begin{bmatrix} B_{G,1} & 0 & 0 \\ 0 & B_{G,2} & 0 \end{bmatrix}$$

$$= \begin{bmatrix} 1/B_{G,1} & 0 & 0 \\ 0 & 1/B_{G,2} & 0 \end{bmatrix} .$$

Hence, $\mathbf{B}_{BG}^{+}$ is as sparse as $\mathbf{B}_{BG}$ and is effectively the element-wise inverse of $\mathbf{B}_{BG}^{\dagger}$. Nonzero elements of $\mathbf{B}_{BG}^{+}$ are in the columns corresponding to the buses at which each synchronous machine is connected. The product $\mathbf{B}_{BG}^{+}\mathbf{B}_{BB}$ and, hence, $\mathbf{D}^{+}$, have also the property that the number of nonzero elements of each row are equal to the first order connectivity degree of each synchronous machine terminal bus. This property has several relevant consequences, as follows.

- The nonzero elements of $\mathbf{D}^{+}$ indicate the minimum number of PMUs that have to be installed and at which buses.

- Each rotor speed can be estimated independently from the others, i.e. each row of $\mathbf{D}^{+}$ enables the estimation of one rotor speed.

- The minimum set of measurements required to estimate a rotor speed consists of only the measurements at the machine bus and its neighboring buses (see Figure 8.2.b).

- A corollary of the point above is that no less than two bus frequency measurements are needed to estimate the rotor speed of a machine connected to the system (see Figure 8.2.a).

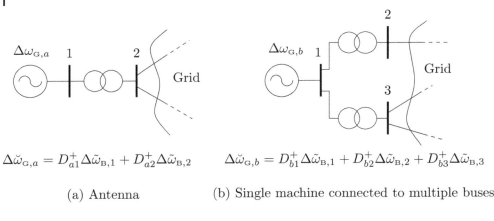

$$\Delta\breve{\omega}_{G,a} = D_{a1}^+ \Delta\tilde{\omega}_{B,1} + D_{a2}^+ \Delta\tilde{\omega}_{B,2} \qquad \Delta\breve{\omega}_{G,b} = D_{b1}^+ \Delta\tilde{\omega}_{B,1} + D_{b2}^+ \Delta\tilde{\omega}_{B,2} + D_{b3}^+ \Delta\tilde{\omega}_{B,3}$$

(a) Antenna   (b) Single machine connected to multiple buses

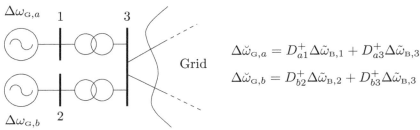

$$\Delta\breve{\omega}_{G,a} = D_{a1}^+ \Delta\tilde{\omega}_{B,1} + D_{a3}^+ \Delta\tilde{\omega}_{B,3}$$

$$\Delta\breve{\omega}_{G,b} = D_{b2}^+ \Delta\tilde{\omega}_{B,2} + D_{b3}^+ \Delta\tilde{\omega}_{B,3}$$

(c) Multiple machines connected to the same bus

Figure 8.2 Examples of synchronous machines connectivity. D_{kh}^+ are the nonzero elements of $\mathbf{D}^+$. The time dependency of $\Delta\tilde{\omega}_{B,h}$ and $\Delta\breve{\omega}_{G,k}$ is omitted.

- It is also possible to actually need less measurements than twice the number of machines of the system (see Figure 8.2.c).

The latter three properties are relevant for large networks, as only a few local measurements are needed per machine. Moreover, given that the vast majority of power plants of real-world networks are connected in antenna to the rest of the system, the minimum number of bus frequency measurements required to estimate all rotor speeds is about twice the number of power plants.

The nonzero pattern of $\mathbf{D}^+$ shows that each rotor speed estimation in (8.9) is fully decoupled. It is thus possible to reformulate (8.1) and (8.2) as a set of n_G decoupled (decentralized) problems, each of which returns a single rotor speed. The main issue of (8.9), however, is that it does not provide redundancy. $\mathbf{D}^+$ indicates the minimum number of required bus frequency measurements. All other measurements are redundant, improving the robustness of the solution of (8.1) and (8.2), but increasing the "centralization" and the computational burden of the estimation problem. Also, remote measurements can be affected by delays. In practical applications, thus, one should find a trade-off between the simplicity of (8.9) and the robustness of (8.1) and (8.2). This point is duly discussed in Section 8.1.2.2.

8.1.1.2. Sensitivities

A relevant byproduct of solving an optimization problem is the ability to calculate, with no extra computational effort, the sensitivities of *everything with respect to everything*. This property relies on the dual variables and has already had a variety of applications in mathematical programming [25, 26, 27] as well as in the analysis of power system operation [39, 130] and state estimation [138].

The sensitivities of the convex problem (8.1) and (8.2) are obtained following the procedure given in [138]. In this case, however, the procedure is quite straightforward as the starting optimization problem is convex quadratic. First, let differentiate the objective function (8.1) and the optimality conditions (8.4)–(8.6) with respect to the optimal values of the objective function and the solution (primal and dual), and the input bus frequency vector, namely $(\breve{J}, \Delta\breve{\omega}_{\mathrm{G}}, \breve{\varepsilon}_{\mathrm{B}}, \breve{\mu}_{\mathrm{B}}, \Delta\tilde{\omega}_{\mathrm{B}})$:

$$0 = -dJ(t) + \breve{\varepsilon}_{\mathrm{B}}^{\dagger}(t)\,\mathbf{U}^{-1}\,d\varepsilon_{\mathrm{B}}(t)\,, \tag{8.14}$$

$$\mathbf{0}_{n_{\mathrm{G}},1} = -\mathbf{B}_{\mathrm{BG}}^{\dagger}\,d\boldsymbol{\mu}_{\mathrm{B}}(t)\,, \tag{8.15}$$

$$\mathbf{0}_{n_{\mathrm{B}},1} = \mathbf{U}^{-1}\,d\varepsilon_{\mathrm{B}}(t) - \mathbf{B}_{\mathrm{BB}}^{\dagger}\,d\boldsymbol{\mu}_{\mathrm{B}}(t)\,, \tag{8.16}$$

$$\mathbf{0}_{n_{\mathrm{B}},1} = -\mathbf{B}_{\mathrm{BG}}\,d\boldsymbol{\omega}_{\mathrm{G}}(t) - \mathbf{B}_{\mathrm{BB}}\,d\varepsilon_{\mathrm{B}}(t) - \mathbf{B}_{\mathrm{BB}}\,d\tilde{\omega}_{\mathrm{B}}(t)\,, \tag{8.17}$$

or, equivalently:

$$\mathbf{0}_{n_{\mathrm{tot}},1} = \begin{bmatrix} -1 & \mathbf{0}_{1,n_{\mathrm{G}}} & \breve{\varepsilon}_{\mathrm{B}}^{\dagger}(t)\,\mathbf{U}^{-1} & \mathbf{0}_{1,n_{\mathrm{B}}} & \mathbf{0}_{1,n_{\mathrm{B}}} \\ \mathbf{0}_{n_{\mathrm{G}},1} & \mathbf{0}_{n_{\mathrm{G}},n_{\mathrm{G}}} & \mathbf{0}_{n_{\mathrm{G}},n_{\mathrm{B}}} & -\mathbf{B}_{\mathrm{BG}}^{\dagger} & \mathbf{0}_{n_{\mathrm{G}},n_{\mathrm{B}}} \\ \mathbf{0}_{n_{\mathrm{B}},1} & \mathbf{0}_{n_{\mathrm{B}},n_{\mathrm{G}}} & \mathbf{U}^{-1} & -\mathbf{B}_{\mathrm{BB}}^{\dagger} & \mathbf{0}_{n_{\mathrm{B}},n_{\mathrm{B}}} \\ \mathbf{0}_{n_{\mathrm{B}},1} & -\mathbf{B}_{\mathrm{BG}} & -\mathbf{B}_{\mathrm{BB}} & \mathbf{0}_{n_{\mathrm{B}},n_{\mathrm{B}}} & -\mathbf{B}_{\mathrm{BB}} \end{bmatrix} \begin{bmatrix} dJ(t) \\ d\boldsymbol{\omega}_{\mathrm{G}}(t) \\ d\varepsilon_{\mathrm{B}}(t) \\ d\boldsymbol{\mu}_{\mathrm{B}}(t) \\ d\tilde{\omega}_{\mathrm{B}}(t) \end{bmatrix}\,, \tag{8.18}$$

where $n_{\mathrm{tot}} = n_{\mathrm{G}} + 2n_{\mathrm{B}} + 1$. The sensitivities with respect to the vector of measured bus frequencies $\tilde{\omega}_{\mathrm{B}}$ can be written as:

$$\mathbf{J}\begin{bmatrix} dJ(t) & d\boldsymbol{\omega}_{\mathrm{G}}(t) & d\varepsilon_{\mathrm{B}}(t) & d\boldsymbol{\mu}_{\mathrm{B}}(t) \end{bmatrix}^{\dagger} = \mathbf{S}\,d\tilde{\omega}_{\mathrm{B}}(t)\,, \tag{8.19}$$

where

$$
\mathbf{J} =
\left[
\begin{array}{cc|c|c}
-1 & \mathbf{0}_{1,n_G} & \breve{\boldsymbol{\varepsilon}}_B^\dagger(t)\,\mathbf{U}^{-1} & \mathbf{0}_{1,n_B} \\
\hline
\mathbf{0}_{n_G,1} & \mathbf{0}_{n_G,n_G} & \mathbf{0}_{n_G,n_B} & -\mathbf{B}_{BG}^\dagger \\
\hline
\mathbf{0}_{n_B,1} & \mathbf{0}_{n_B,n_G} & \mathbf{U}^{-1} & -\mathbf{B}_{BB}^\dagger \\
\hline
\mathbf{0}_{n_B,1} & -\mathbf{B}_{BG} & -\mathbf{B}_{BB} & \mathbf{0}_{n_B,n_B}
\end{array}
\right],
\tag{8.20}
$$

and

$$
\mathbf{S} = -\left[\mathbf{0}_{1,n_B} \quad \mathbf{0}_{n_G,n_B} \quad \mathbf{0}_{n_B,n_B} \quad -\mathbf{B}_{BB}\right]^\dagger .
\tag{8.21}
$$

Then one has:

$$
\left[dJ(t) \quad d\boldsymbol{\omega}_G(t) \quad d\boldsymbol{\varepsilon}_B(t) \quad d\boldsymbol{\mu}_B(t)\right]^\dagger = \mathbf{J}^{-1}\,\mathbf{S}\,d\tilde{\boldsymbol{\omega}}_B(t) ,
\tag{8.22}
$$

from which one can obtain the sensitivities of all primal and dual variables with respect to $\Delta\tilde{\boldsymbol{\omega}}_B$:

$$
\left[\frac{\partial J(t)}{\partial\tilde{\boldsymbol{\omega}}_B(t)} \quad \frac{\partial\boldsymbol{\omega}_G(t)}{\partial\tilde{\boldsymbol{\omega}}_B(t)} \quad \frac{\partial\boldsymbol{\varepsilon}_B(t)}{\partial\tilde{\boldsymbol{\omega}}_B(t)} \quad \frac{\partial\boldsymbol{\mu}_B}{\partial\tilde{\boldsymbol{\omega}}_B(t)}\right]^\dagger = \mathbf{J}^{-1}\,\mathbf{S} .
\tag{8.23}
$$

For example, from (8.23), one can obtain:

$$
\frac{\partial\boldsymbol{\omega}_G(t)}{\partial\tilde{\boldsymbol{\omega}}_B(t)} = -[\mathbf{B}_{BG}^\dagger\,\mathbf{H}\,\mathbf{B}_{BG}]^{-1}\mathbf{B}_{BG}^\dagger\mathbf{H}\,\mathbf{B}_{BB} ,
\tag{8.24}
$$

where $\mathbf{H} = (\mathbf{B}_{BB}\mathbf{U}\mathbf{B}_{BB}^\dagger)^{-1}$. Equation (8.24) is consistent with (8.8). This can be shown by exploiting the following matrix properties:

- $(\mathbf{A}\,\mathbf{B}\,\mathbf{C})^{-1} = \mathbf{C}^{-1}\mathbf{B}^{-1}\mathbf{A}^{-1}$;
- $(\mathbf{A}^\dagger)^{-1} = (\mathbf{A}^{-1})^\dagger$;
- $(\mathbf{A}^{-1})^{-1} = \mathbf{A}$; and
- $(\mathbf{A}\,\mathbf{B})^\dagger = \mathbf{B}^\dagger\mathbf{A}^\dagger$.

Then, matrix $\mathbf{H}$ can be rewritten as:

$$
\begin{aligned}
\mathbf{H} &= (\mathbf{B}_{BB}\mathbf{U}\mathbf{B}_{BB}^\dagger)^{-1} = (\mathbf{B}_{BB}^\dagger)^{-1}\mathbf{U}^{-1}\mathbf{B}_{BB}^{-1} \\
&= (\mathbf{B}_{BB}^{-1})^\dagger\,\mathbf{U}^{-1}\,\mathbf{B}_{BB}^{-1} ,
\end{aligned}
\tag{8.25}
$$

and, substituting the obtained expression into (8.24), one has:

$$
\begin{aligned}
\frac{\partial\boldsymbol{\omega}_G(t)}{\partial\tilde{\boldsymbol{\omega}}_B(t)} &= -\left[\mathbf{B}_{BG}^\dagger(\mathbf{B}_{BB}^{-1})^\dagger\,\mathbf{U}^{-1}\,\mathbf{B}_{BB}^{-1}\,\mathbf{B}_{BG}\right]^{-1}\mathbf{B}_{BG}^\dagger(\mathbf{B}_{BB}^{-1})^\dagger\,\mathbf{U}^{-1}\,\mathbf{B}_{BB}^{-1}\,\mathbf{B}_{BB} \\
&= -\left[(\mathbf{B}_{BB}^{-1}\,\mathbf{B}_{BG})^\dagger\,\mathbf{U}^{-1}\,\mathbf{B}_{BB}^{-1}\,\mathbf{B}_{BG}\right]^{-1}(\mathbf{B}_{BB}^{-1}\,\mathbf{B}_{BG})^\dagger\,\mathbf{U}^{-1},
\end{aligned}
\tag{8.26}
$$

which, recalling the definition of $\mathbf{D}$ in (5.14), namely $\mathbf{D} = -\mathbf{B}_{BB}^{-1}\mathbf{B}_{BG}$, is the same expression that can be obtained by differentiating (8.8) with respect to $\Delta\tilde{\omega}_B$ and $\Delta\breve{\omega}_G$.

Another relevant sensitivity vector that can be obtained from (8.23) is $\partial J / \partial\tilde{\omega}_B$. Since the manipulation of (8.23) is rather involved, such sensitivities are determined in an alternative way. First, observe that from (8.14):

$$\frac{\partial J(t)}{\partial\varepsilon_B(t)} = \breve{\varepsilon}_B^\dagger(t)\,\mathbf{U}^{-1}\;. \tag{8.27}$$

Then, from (8.5), one obtains:

$$\breve{\varepsilon}_B^\dagger(t)\,\mathbf{U}^{-1} = \breve{\mu}_B^\dagger(t)\,\mathbf{B}_{BB}\;, \tag{8.28}$$

which descends from the well-known property $(\mathbf{A}^\dagger\mathbf{B})^\dagger = \mathbf{B}^\dagger\mathbf{A}$. Then, multiplying (8.17) by $\breve{\mu}_B^\dagger$ and observing from (8.4) that $\breve{\mu}_B^\dagger\,\mathbf{B}_{BG} = \mathbf{0}_{1,n_G}$, one has:

$$\frac{\partial\varepsilon_B(t)}{\partial\tilde{\omega}_B(t)} = -\mathbf{1}_{n_B}\;. \tag{8.29}$$

Finally, multiplying together (8.27) and (8.29), imposing (8.28) and applying the chain rule, the sought expression is obtained:

$$\frac{\partial J(t)}{\partial\tilde{\omega}_B(t)} = \frac{\partial J(t)}{\partial\varepsilon_B(t)}\frac{\partial\varepsilon_B(t)}{\partial\tilde{\omega}_B(t)} = -\breve{\varepsilon}_B^\dagger(t)\,\mathbf{U}^{-1} = -\breve{\mu}_B^\dagger(t)\,\mathbf{B}_{BB}\;. \tag{8.30}$$

Such sensitivities can also be utilized to identify the buses whose frequency measurements improve the robustness of the rotor speed estimation.

8.1.2. Examples

The FDF-based state estimation of machine rotor speeds is illustrated below through a variety of examples based on the WSCC system in Section 8.1.2.1 and the AIITS in Section 8.1.2.2.

8.1.2.1. WSCC System

One PMU is assumed to be installed at every bus of the WSCC system described in Appendix A.2. All PMU devices are also assumed to be based on the same technology and have same measurement standard deviations, namely, $\sigma_{B,h} = 10^{-3}$, $\forall h = 1, \ldots, 9$. Hence, $\mathbf{U} = 10^{-6}\mathbf{I}_9$.

From the structure of $\mathbf{B}_{BG}$, it descends that the dual variables μ_B are always null at the generation buses. Matrix $\mathbf{B}_{BG}$, in fact, has nonnull elements only in

Table 8.1 Sensitivities $\partial\breve{\omega}_{\mathrm{G}}/\partial\tilde{\omega}_{\mathrm{B}} = \mathbf{D}^{+}$ for the WSCC system

$\partial\breve{\omega}_{\mathrm{G},k}$	$\partial\tilde{\omega}_{\mathrm{B},h}$								
	1	2	3	4	5	6	7	8	9
1	2.369	0	0	−1.369	0	0	0	0	0
2	0	3.534	0	0	0	0	−2.534	0	0
3	0	0	4.680	0	0	0	0	0	−3.680

the columns corresponding to the buses where generators are connected (see Table 5.1). The optimality condition (8.4) for the WSCC system gives:

$$0 = 12.682\,\breve{\mu}_{\mathrm{B},1} , \quad 0 = 6.315\,\breve{\mu}_{\mathrm{B},2} , \quad 0 = 4.637\,\breve{\mu}_{\mathrm{B},3} . \tag{8.31}$$

Then, dual variables $\boldsymbol{\mu}_{\mathrm{B}}$ are nonnull only at load or transition buses.

The sensitivities $\partial\breve{\omega}_{\mathrm{G}}/\partial\tilde{\omega}_{\mathrm{B}}$, obtained from (8.24) are shown in Table 8.1. Since $\mathbf{U}$ is diagonal and all its diagonal elements are equal, the sensitivities do not depend on its values and, in this case, one has:

$$\frac{\partial\breve{\omega}_{\mathrm{G}}(t)}{\partial\tilde{\omega}_{\mathrm{B}}(t)} = \mathbf{D}^{+} . \tag{8.32}$$

These sensitivities are constant if there is no change of the topology of the network. Note also that (8.32) can be obtained directly from differentiating (8.9).

As discussed in Section 8.1.1.1, $\mathbf{D}^{+}$ is very sparse. In this example, the three machines are connected in antenna and, hence, at least 6 PMU measurements at buses 1–4, 7, and 9 are needed to estimate all rotor speeds.

The sensitivities $\partial J/\partial\tilde{\omega}_{\mathrm{B}}$ which are given by (8.30) are studied next. For the sake of example, consider the sensitivity at the generator bus 1 and at the load bus 8, namely $\partial J/\partial\tilde{\omega}_{\mathrm{B},1}$ and $\partial J/\partial\tilde{\omega}_{\mathrm{B},8}$, respectively. From (8.30), one has:

$$\frac{\partial J(t)}{\partial\tilde{\omega}_{\mathrm{B},1}(t)} = 30.04\,\breve{\mu}_{\mathrm{B},1}(t) - 17.36\,\breve{\mu}_{\mathrm{B},4}(t) = -17.36\,\breve{\mu}_{\mathrm{B},4}(t) ,$$

$$\frac{\partial J(t)}{\partial\tilde{\omega}_{\mathrm{B},8}(t)} = -13.7\,\breve{\mu}_{\mathrm{B},7}(t) + 23.3\,\breve{\mu}_{\mathrm{B},8}(t) - 9.784\,\breve{\mu}_{\mathrm{B},9}(t) ,$$

$$\tag{8.33}$$

where the condition $\breve{\mu}_{\mathrm{B},1} = 0$ obtained from (8.31) is imposed in the first equation of (8.33). From (8.1) and again from (8.30), the sensitivities in (8.33) are equivalent to the expressions:

$$\frac{\partial J(t)}{\partial \tilde{\omega}_{\mathrm{B},1}(t)} = -\frac{\breve{\varepsilon}_{\mathrm{B},1}(t)}{\sigma_{\mathrm{B},1}^2(t)} = -10^6 \, \breve{\varepsilon}_{\mathrm{B},1}(t) \, ,$$

$$\frac{\partial J(t)}{\partial \tilde{\omega}_{\mathrm{B},8}(t)} = -\frac{\breve{\varepsilon}_{\mathrm{B},8}(t)}{\sigma_{\mathrm{B},8}^2(t)} = -10^6 \, \breve{\varepsilon}_{\mathrm{B},8}(t) \, ,$$

(8.34)

which indicate that the sensitivities of the objective function with respect to the bus frequency measurements are inversely proportional to the variance of the measurements themselves.

Consider the loss of load connected to bus 5 occurring at $t = 1$ s. Figure 8.3 shows the trajectory of the actual rotor speed $\omega_{\mathrm{G},1}$ of the synchronous machine connected to bus 1, i.e. the state variable of the machine swing equation; the estimated rotor speed $\breve{\omega}_{\mathrm{G},1}$ obtained by solving (8.4)–(8.6); and the error $|\omega_{\mathrm{G},1} - \breve{\omega}_{\mathrm{G},1}|$. Note that, in real-life applications, $\omega_{\mathrm{G},1}$ is actually not available to system operators (see also the scheme of Figure 8.1). Along the time domain simulation, the bus measurements $\tilde{\omega}_{\mathrm{B}}$ are obtained with SRF-PLLs (see Section 3.3.2).

No noise is considered in this example. The error between the actual and the estimated rotor speeds is due exclusively to the PMU tracking. Apart from the first instants after the contingency, such an error is below 0.0002 pu, i.e. 0.012 Hz for a 60-Hz system and 0.01 Hz for a 50-Hz one.

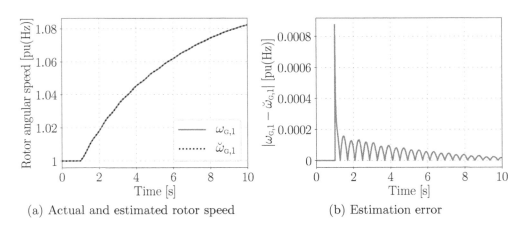

(a) Actual and estimated rotor speed   (b) Estimation error

Figure 8.3 WSCC system – Rotor speed of synchronous machine at bus 1 after a loss of load.

The average error decreases as the rotor speed stabilizes after the contingency, thus leading to the conclusion that the average error is proportional to the rate of change of frequency. This is a consequence of the dynamic behavior of the swing equations of synchronous machines. Following a contingency, the power imbalance in the grid accelerates or decelerates the machines, thus creating frequency variations, which are different from bus to bus and maximum in the first few seconds after the contingency. Then, the primary frequency controllers act on the turbine governors of the machines and help recover a synchronous condition, i.e. a condition in which the rotor speeds of the machines converge to a common value. In the first seconds after a contingency, thus, machine rotor speeds show the largest rate of change and, consequently, the highest estimation errors.

Figure 8.4 shows the trajectories of the sensitivities $\partial J/\partial \tilde{\omega}_{B,1}$ and $\partial J/\partial \tilde{\omega}_{B,8}$ for the contingency discussed above. Similarly to the error between actual and estimated rotor speeds, the corresponding sensitivities decrease as the rate of change of frequency decreases. In steady-state all sensitivities $\partial J/\partial \tilde{\omega}_{B,h}$ are null, since the frequency is the same at every bus of the network.

Finally, it is relevant to illustrate the formulation of the decoupled state estimation problem to determine a single rotor speed. Considering for example $\Delta \omega_{G,1}$, equation (8.9) and Table 8.1 indicate that the minimum set of measurements required to estimate this rotor speed are $\Delta \tilde{\omega}_{B,1}$ and $\Delta \tilde{\omega}_{B,4}$.

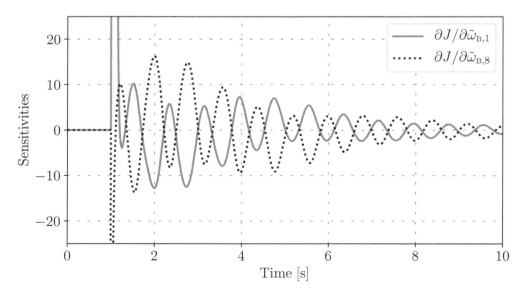

Figure 8.4 WSCC system – Sensitivities $\partial J/\partial \tilde{\omega}_{B,1}$ and $\partial J/\partial \tilde{\omega}_{B,8}$ after a loss of load.

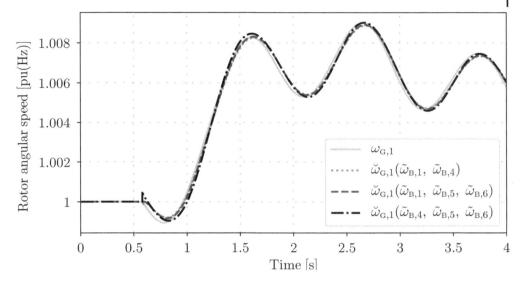

Figure 8.5 Actual rotor speed $\omega_{G,1}$ of the synchronous machine 1 and the estimated quantity $\breve{\omega}_{G,1}$ considering different sets of PMU measurements.

The resulting state estimation problem is:

$$\min_{(\Delta\omega_{G,1},\varepsilon_{B,1},\varepsilon_{B,4})} \quad J(t) = \frac{1}{2}\left(10^6\,\varepsilon_{B,1}^2(t) + 10^6\,\varepsilon_{B,4}^2(t)\right)$$

$$\text{s.t.} \quad 0 = 12.682\,\Delta\omega_{G,1}(t) - 30.04\left(\Delta\tilde{\omega}_{B,1}(t) + \varepsilon_{B,1}(t)\right)$$
$$+ 17.36\left(\Delta\tilde{\omega}_{B,4}(t) + \varepsilon_{B,4}(t)\right) \qquad : \mu_B(t) .$$

The KKT optimality conditions lead to $\breve{\varepsilon}_{B,1} = \breve{\varepsilon}_{B,4} = \breve{\mu}_B = 0$, and:

$$\Delta\breve{\omega}_{G,1}(t) = 2.369\,\Delta\tilde{\omega}_{B,1}(t) - 1.369\,\Delta\tilde{\omega}_{B,4}(t) .$$

The latter can be directly obtained from (8.9).

The full optimization problem (8.1)–(8.2) provides significantly more redundancy than (8.9), which is to be expected as (8.1)–(8.2) include a higher number of measurements. For example, if $\tilde{\omega}_{B,4}$ is missing, $\tilde{\omega}_{B,4}$ can still be estimated from $\tilde{\omega}_{B,1}$, $\tilde{\omega}_{B,5}$, and $\tilde{\omega}_{B,6}$ (see the fourth row of matrix $\mathbf{B}_{BB}$ in Table 5.1). Figure 8.5 shows the estimation of $\breve{\omega}_{G,1}$ considering different sets of frequency measurements.

8.1.2.2. All-Island Irish Transmission System

Four scenarios based on a dynamic model of the AIITS (see Appendix B) are studied: (i) sensitivities $\partial\breve{\omega}_G/\partial\tilde{\omega}_B$; (ii) the robustness of the state estimation problem (8.1)–(8.2), hereinafter OPT, and the weighted-least-square problem (8.7), hereinafter WLS, against the loss of some key frequency measurements; (iii) effect of measurement noise; and (iv) effect of measurement latency.

The contingency considered in all scenarios is a three-phase fault occurring at $t = 1$ s, and cleared after 150 ms by disconnecting the corresponding line. It is important to emphasize that the FDF-based frequency estimation is particularly relevant for the very first few seconds following severe events that trigger electromechanical oscillations of the machines. Only in this time frame, in fact, frequencies are different from bus to bus.

Due to the linear formulation, the high sparsity of all matrices involved, and the fact that no matrix factorization is required, the computational burden of the two state estimation problems is negligible. Solving the full OPT and the WLS problems for the whole AIITS requires about 100 µs and 5 µs, respectively, per time step. These times are orders of magnitude smaller than the phenomenon to be analyzed, i.e. synchronous machine electromechanical oscillations, and thus these problems are suited for implementation in real-life applications.

Example 1: Sensitivity Analysis

The sensitivities $\partial \breve{\omega}_G / \partial \tilde{\omega}_B = \mathbf{D}^+$ are relevant to determine the weight of bus frequency measurements in the estimation of rotor speeds. The sub-matrix of $\mathbf{D}^+$ for the AIITS including generator buses (columns) and neighboring buses (rows) is shown in Figure 8.6.

Matrix $\mathbf{D}^+$ has only 44 nonzero elements over a total of 32,538, thus showing that, if measurement redundancy is neglected, one only needs a very reduced set of bus frequency measurements to estimate the rotor speeds of all synchronous machines, even for large systems. For the AIITS, in fact, the minimum set includes 42 measurements, as two pairs of machines are connected with the topology depicted in Figure 8.2.c.

Example 2: Measurement Redundancy and Robustness

The robustness against the loss of one of the measurement signals of the two proposed estimation approaches is studied next. In this scenario, the accuracy of each approach is compared when estimating the rotor speed of a synchronous machine connected to the system as in Figure 8.2.a when a large transient occurs in the system, and one of the two PMUs fails to send the measurement $\tilde{\omega}_{B,h}$. The accuracy of the OPT and WLS problems assuming that all PMU measurements are available is shown in Figure 8.7.

Both techniques are able to capture the machine rotor speed variations with better accuracy than the measurement provided by the PMU connected

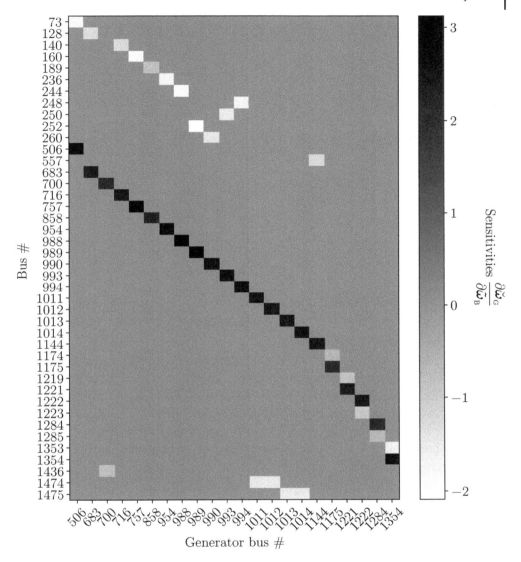

Figure 8.6 AIITS – Sensitivities $\partial\breve{\omega}_{\mathrm{G}}/\partial\tilde{\omega}_{\mathrm{B}}$.

to the machine bus. The OPT estimation is slightly better than that obtained with the WLS. The total amplitude of the characteristic spikes of the PMU during the discontinuous events (fault and line outage), and the consequent spikes of the estimation techniques, is not shown in order to better capture the more relevant electromechanical oscillations.

To study the impact of a PMU malfunction, it is assumed that $\Delta\tilde{\omega}_{\mathrm{B},h} = 0$ from the beginning of the simulation, and results are shown in Figure 8.8.

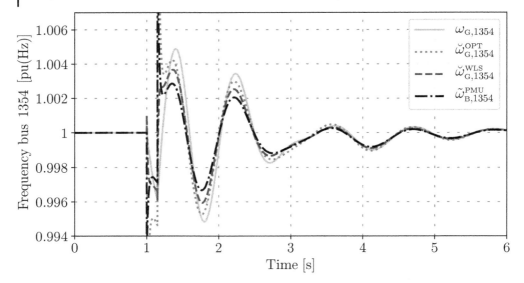

Figure 8.7 AIITS – Actual and estimated rotor speed of the synchronous machine at bus 1354, and PMU frequency measurement at the machine bus.

In the two cases simulated, namely the failure of the PMU at the machine neighboring bus (Figure 8.8.a), and at the machine bus (Figure 8.8.b), the measurement redundancy of the OPT problem enables the correction of the erroneous $\tilde{\omega}_{\mathrm{B},h}$, thus providing a highly accurate estimation $\breve{\omega}_{\mathrm{G},k}$. Bad data considerably impact on the WLS-based estimation, in particular if the bad signal is that of the PMU at the machine bus, where the estimation is in counter-phase with the actual $\omega_{\mathrm{G},k}$.

Example 3: Measurement Noise

The sensitivity of the two estimation approaches OPT and WLS to measurement noise is the focus of this example. The noise is modeled through the Ornstein-Uhlenbeck process given in equation (4.31) and is applied to all bus voltage phase angles [134]. The noisy time-varying voltages are then utilized as input signals of the PLL of the PMUs that measure bus frequencies. Results are depicted in Figure 8.9. As expected, the OPT problem shows a lower sensitivity to noise than the WLS. The OPT problem, in fact, can include more measurements than the WLS problem. The higher the number of measurements, the lower the impact of measurement errors on the estimation of rotor speeds.

Example 4: Measurement Latency

Latency is modeled as a constant delay that is introduced in the PLL output signals of the PMUs that provide bus frequency measurements. Each delay is

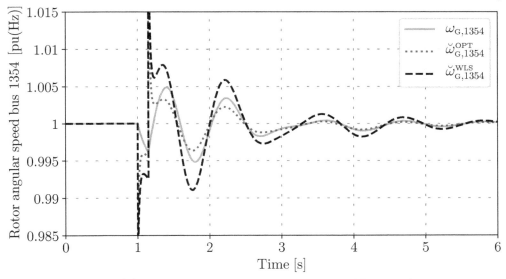

(a) Failure of PMU at machine neighboring bus

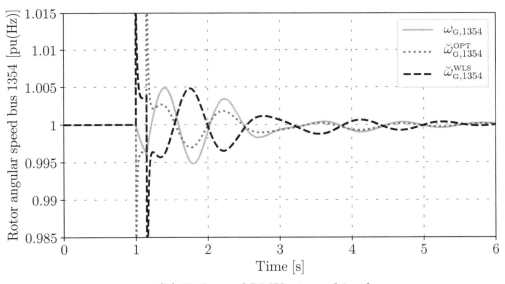

(b) Failure of PMU at machine bus

Figure 8.8 AIITS – Actual and estimated rotor speed of the synchronous machine at bus 1354 considering a PMU malfunction.

composed of a term due to the measurement process plus a communication delay in transmitting the signal to the control center. The simulations assume only a difference in communication delays, τ, between PMUs and, hence, the latency is applied to all but one PMUs involved in the estimation of each

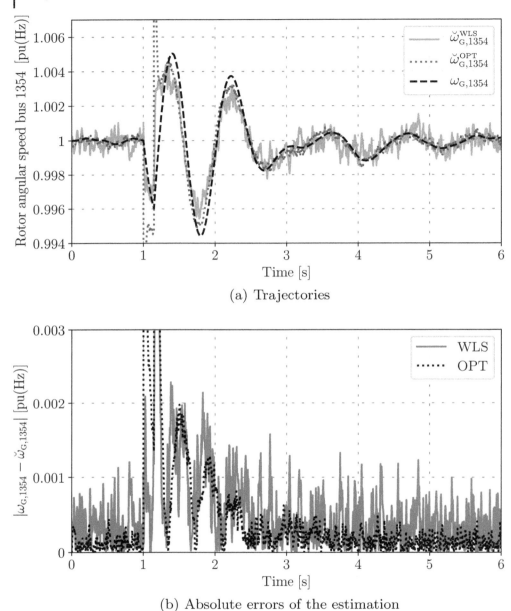

(a) Trajectories

(b) Absolute errors of the estimation

Figure 8.9 AIITS – Actual and estimated rotor speed of the synchronous machine at bus 1354 under the presence of noise.

rotor speed. Two cases are considered, namely $\tau = 25$ ms (short distance) and $\tau = 100$ ms (long distance). Results are shown in Figures 8.10 (τ in the signal of the machine bus) and 8.11 (τ in the signal of the machine neighboring bus).

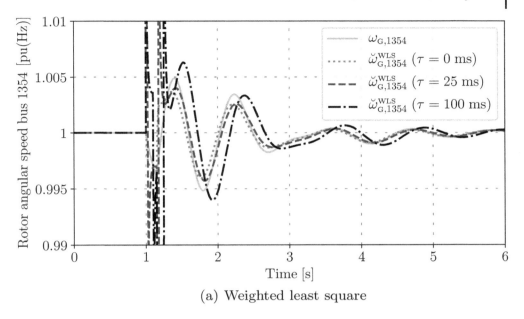

(a) Weighted least square

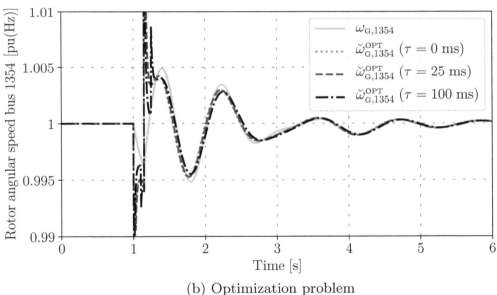

(b) Optimization problem

Figure 8.10 AIITS – Actual and estimated rotor speed of the synchronous machine at bus 1354 with latency in the PMU signal of the synchronous machine bus.

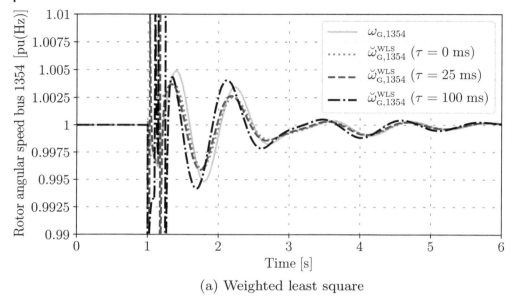

(a) Weighted least square

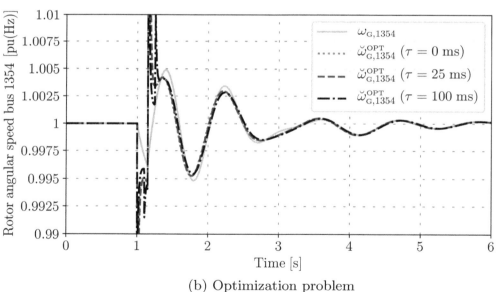

(b) Optimization problem

Figure 8.11 AIITS – Actual and estimated rotor speed of the synchronous machine at bus 1354 with latency in the PMU signal of the machine neighboring bus.

The OPT problem shows good robustness with respect to measurement latency, regardless of the affected PMU and the value of τ. The WLS is more sensitive to latency. WLS accuracy is compromised if $\tau = 100$ ms and shows significantly more and larger spikes than the OPT. Note, however, that the impact of noise and latency can be reduced through appropriate filtering and/or compensation techniques (see, for example, [175]).

8.2. Center of Inertia

The CoI is as a well-known concept utilized in transient stability analysis as the reference for synchronous machine rotor speeds (see Section 2.4.1.4). In practice, however, the frequency of the CoI cannot be estimated because its calculation requires the availability of the measures of the rotor speeds of all synchronous machines connected to the system.

For this reason, system operators do not estimate the frequency of the CoI online but, rather, measure the frequency at some relevant *pilot bus* of the system. This is acceptable for long term analysis, e.g. when the frequency of the system can be effectively assumed to be the same everywhere. However, in the first seconds after a large contingency, the frequency of no bus, and hence also that of the pilot bus, does accurately represent the average frequency of the system.

The WLS problem (8.7) described in Section 8.1 makes actually possible to estimate the frequency of the CoI without the need of measuring the rotor speed of synchronous machines.

8.2.1. Derivation

Consider an interconnected AC grid with n_B buses and n_G synchronous machines. The starting equations are the definition of the angular speed of the CoI (2.59) discussed in Section 2.4.1.4, and the FDF (5.13) presented in Section 5.2. The expression (2.59) of the speed of the CoI can be reformulated as:

$$\omega_{\text{CoI}}(t) = \boldsymbol{m}^\dagger \boldsymbol{\omega}_{\text{G}}(t) , \tag{8.35}$$

where $\boldsymbol{\omega}_{\text{G}}$ is a $n_G \times 1$ vector of synchronous machine rotor speeds and $\boldsymbol{m}$ is a $n_G \times 1$ vector of normalized inertia constants, i.e. the h-th element of $\boldsymbol{m}$ is:

$$m_h = M_h / M_{\text{tot}} , \tag{8.36}$$

where M_h is the inertia of the h-th machine and $M_{\text{tot}} = \sum_{h=1}^{n_G} M_h$.

Equation (8.35) can be rewritten as a function of angular speed variations, as follows:

$$\Delta\omega_{\mathrm{CoI}}(t) = \boldsymbol{m}^\dagger \Delta\boldsymbol{\omega}_{\mathrm{G}}(t) \; . \tag{8.37}$$

In fact, (8.35) derives directly from (8.36), thus $\boldsymbol{m}^\dagger \mathbf{1}_{n_{\mathrm{G}}} = \sum_{h=1}^{m} m_h = 1$. Then, recalling the solution (8.9) of the WLS problem (8.7) and the definition of the pseudo-inverse $\mathbf{D}^+$ given in (8.11), one has:

$$\Delta\boldsymbol{\omega}_{\mathrm{G}}(t) = \mathbf{D}^+ \Delta\boldsymbol{\omega}_{\mathrm{B}}(t) \; . \tag{8.38}$$

Merging together (8.37) and (8.38) leads to:

$$\Delta\omega_{\mathrm{CoI}}(t) = \boldsymbol{m}^\dagger \mathbf{D}^+ \Delta\boldsymbol{\omega}_{\mathrm{B}}(t) = \boldsymbol{w}^\dagger \Delta\boldsymbol{\omega}_{\mathrm{B}}(t) \; , \tag{8.39}$$

where $\boldsymbol{w}^\dagger = \boldsymbol{m}^\dagger \mathbf{D}^+$ is the sought vector of *weights* that enables the calculation of the frequency variations of the CoI from the measurements of the frequency variations of the buses $\Delta\tilde{\boldsymbol{\omega}}_{\mathrm{B}}$.

The following remarks are relevant.

- Equation (8.39) only involves sparse matrix-vector products, where matrix $\mathbf{D}^+$ tends to be extremely sparse, as discussed in the examples of Section 8.1.2.1.

- The elements of $\boldsymbol{w}$ can be viewed as the weights of the measurements $\Delta\tilde{\boldsymbol{\omega}}_{\mathrm{B}}$ for the evaluation of $\breve{\omega}_{\mathrm{CoI}}$. Note also that $\boldsymbol{w}^\dagger \mathbf{1}_{n_{\mathrm{B}}} \approx 1$, which derives from the definition of $\boldsymbol{m}$ and the properties of the rows of matrix $\mathbf{D}$, as discussed in Section 5.5.

- As a consequence of $\mathbf{D}^+$ being very sparse, a large number of elements of $\boldsymbol{w}$ is null. This fact has a relevant practical consequence: only a reduced number of measurements of bus frequencies are needed to estimate ω_{CoI}. This result is consistent with the discussion of the WLS problem for the estimation of the rotor angular speed of synchronous machines provided in Section 8.1.1.1.

Finally, the expression to estimate the frequency of the CoI is the following:

$$\boxed{\breve{\omega}_{\mathrm{CoI}}(t) = \boldsymbol{w}^\dagger \tilde{\boldsymbol{\omega}}_{\mathrm{B}}(t) + c_o} \; , \tag{8.40}$$

where $c_o = 1 - \boldsymbol{w}^\dagger \mathbf{1}_{n_{\mathrm{B}}}$ is an offset, with $|c_o| \ll 1$. As for the FDF expression, (8.40) is valid in time. $\boldsymbol{w}$ and c_o, are piece-wise constant and need to be recomputed only when a topological change occurs, e.g. a line outage, or a synchronous machine is connected to or disconnected from the grid.

8.2.1.1. Example

The AIITS described in Appendix B serves to carry out a numerical appraisal of the formula (8.40). As already discussed in the example of Section 8.1.2.2, only 42 bus frequency measurements are needed to define all machine rotor speeds of the AIITS. This leads to a vector $\boldsymbol{w}$ with 42 nonzero elements or, said in another way, only 2.8% of the total number of buses (1,479) of the system are needed to estimate the frequency of the CoI. Of these 42 measurements, 22 correspond to the terminal buses of the synchronous machines connected to the AIITS. The rationale behind the requirement of these 42 measurements is given in Section 8.1.

Observe that it is not accurate to simply use the 22 frequencies $\boldsymbol{\omega}_{\mathrm{B,G}}$ measured at the terminal buses where synchronous machines are connected, i.e.:

$$\breve{\omega}'_{\mathrm{CoI}}(t) \approx \boldsymbol{m}^{\dagger}\boldsymbol{\omega}_{\mathrm{B,G}}(t) . \tag{8.41}$$

The accuracy of the estimation provided by (8.40) using 42 and 22 measurements is illustrated in Figure 8.12, which shows the frequency response following a three-phase fault that occurs at $t = 1$ s and is cleared after 50 ms. The frequencies at the buses are estimated by means of the FDF.

The rationale behind this result, as discussed in Chapter 5, is that the frequencies at the terminal buses of the synchronous machines are not exactly equal to the rotor speeds because of the internal reactances of the machines themselves and the topology of the network. In other words, the term $\mathbf{D}^{+}$ in (8.39) cannot be neglected without introducing a significant estimation error.

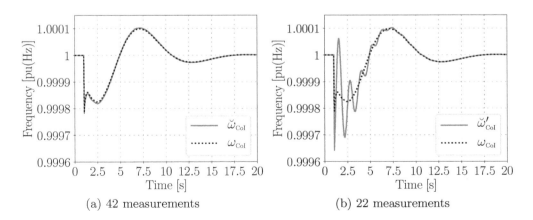

(a) 42 measurements   (b) 22 measurements

Figure 8.12 AIITS – Trajectories of the actual and estimated frequency of the CoI after a three-phase fault.

Table 8.2 Number of bus frequency measurements as a function of a threshold of the CoI weight vector $\boldsymbol{w}$.

Threshold $w^{\max}$	0	0.01	0.02	0.03	0.04	0.05	0.06	0.07
Length of $\hat{\boldsymbol{w}}$	42	41	40	39	31	26	25	20
# gen. buses	22	22	21	21	17	14	14	13

It is also relevant to note that the proposed expression (8.40) enables the estimation of the error when some signals are neglected, missing or erroneous. As previously discussed, in fact, the elements of $\boldsymbol{w}$ are the weights of bus frequencies to estimate the frequency of the CoI. This property can be utilized to identify the most important nodes – or areas – of the system with respect to frequency dynamics. For example, Table 8.2 shows the number of measurements as well as the number of measurements taken at generator terminal buses as a function of a threshold $w^{\max}$ used to filter the elements of $\boldsymbol{w}$, as follows:

$$\hat{\boldsymbol{w}} = \{w_h \in \boldsymbol{w}, \quad |w_h| \geq w^{\max}, \quad h = 1, \ldots, n_{\mathrm{B}}\} . \tag{8.42}$$

Interestingly, as $w^{\max}$ increases, the relevant frequency bus measurements that are retained are not always those at generator terminal buses.

The robustness of equation (8.40) considering real-world disturbances [118] is studied next. All 42 PMUs that are modeled in the AIITS include an SRF-PLL model (see Section 3.3.2.1). The parameters of SRF-PLL, namely K_{P}, T_{I} and T_{f}, are tuned by trial-and-error in order to obtain the *best* dynamic performance of the PLLs. Stochastic wind speed fluctuations modeled using the Ornstein-Uhlenbeck process given in equation (4.31) are considered. PMU measurement delays are modeled as a composite of pseudo-periodic, constant and stochastic delays [117].

The trajectory of the time-varying delay is shown in Figure 8.13 for 1 s of simulation. Noise modeled as an Ornstein-Uhlenbeck process are also applied to all bus voltage phase angles. The contingency considered is a three-phase fault occurring at $t = 1$ s and cleared after 150 ms by disconnecting the corresponding line.

The three-phase fault is simulated in the North-Ireland Transmission system, which is one of the 20 sub-area systems of the AIITS, and includes 5 synchronous power plants and 10 PMUs. A critical scenario is considered where the fault results in the loss of all PMUs of the North-Ireland sub-area system due to protection tripping. Figure 8.14 compares the estimation results of this scenario, $\breve{\omega}_{\mathrm{CoI}}^*$, with the scenario where all measurements are available, $\breve{\omega}_{\mathrm{CoI}}$, and with the actual frequency of the CoI, ω_{CoI}, computed using the machine rotor speeds from the simulation. Figure 8.14 also includes the frequency measured

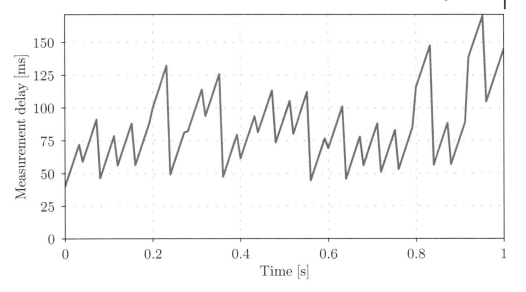

Figure 8.13 AIITS – Time-varying delay.

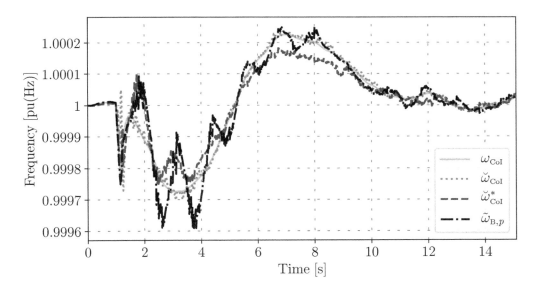

Figure 8.14 AIITS – Estimation of the system frequency of the CoI after a three-phase fault, and PMU measurement at the pilot bus.

at a pilot bus, $\tilde{\omega}_{B,p}$, to account for the common practice of TSOs to estimate the frequency of the CoI.

The estimation approach (8.40) with 42 PMUs accurately tracks the dynamic behavior of the frequency of the CoI following the fault clearance.

The accuracy of this estimation approach decreases when losing the 10 PMUs of the North-Ireland system, but it nevertheless shows a good robustness against the loss of about 25% of the total number of measurements. The worst estimation is obtained when measuring the frequency at the pilot bus, as it captures not only the overall trend of the frequency of the CoI, but also some local oscillatory modes that are naturally filtered out by the CoI.

The last example presented in this section considers the estimation of the frequency of the CoI of a sub-area system. With this aim, the North-Ireland sub-area transmission system of the AIITS is considered. In this sub-area system, the pilot bus chosen is that of a synchronous power plant that represents 34.54% of the total inertia of the region.

The results of the estimation comparison of the sub-area CoI frequency are shown in Figure 8.15.a. The estimation based on (8.40), $\breve{\omega}_{\mathrm{CoI}}^{\mathrm{NI}}$ and the pilot bus measurement $\tilde{\omega}_{\mathrm{B},p}$ have similar accuracy. The scenario where a PMU cannot be installed at the pilot bus is also studied, and results are shown in Figure 8.15.b, where the measurement at the neighboring bus, $\tilde{\omega}_{\mathrm{B},n}$, is included. Results indicate that the approach of (8.40) is still accurate even if the measurement at the bus where the machine with the highest inertia of the sub-area system is not available, despite the phase-shift introduced in the estimation.

8.2.2. Equivalent Radial Network

Figure 8.16 shows a radial system composed of a synchronous machine and an equivalent radial grid, similar to a Thévenin equivalent, connected to the terminal bus of the machine. X_h^{I} represents the internal reactance of the machine and the reactance of the step-up transformer and, if present, of the transmission line that connects the machine to bus h, whereas X_h^{E} is the equivalent reactance of the rest of the network as seen from bus h.

Note that, for the sake of the determination of the frequency of the CoI, X_h^{E} cannot be obtained using the well-known procedure of the Thévenin equivalent because the effect of the different frequencies of the generators of the external region of the grid connected to bus h must be preserved. However, one can always assume that the frequency at bus E, ω_{E} is the weighted average of the rotor speeds of the machines of region E such that:

$$\omega_{\mathrm{CoI}}(t) = \frac{M_{\mathrm{E}}\,\omega_{\mathrm{E}}(t) + M_{\mathrm{I}}\,\omega_{\mathrm{I}}(t)}{M_{\mathrm{tot}}}, \tag{8.43}$$

where M_{I} is the inertia of the synchronous machine; M_{E} is the combined inertia of the generators connected to the rest grid; and $M_{\mathrm{tot}} = M_{\mathrm{I}} + M_{\mathrm{E}}$.

The example presented in Section 5.2.1.1 and equation (5.20) shows that the CoI of this system can be found at a point along the connection EI such

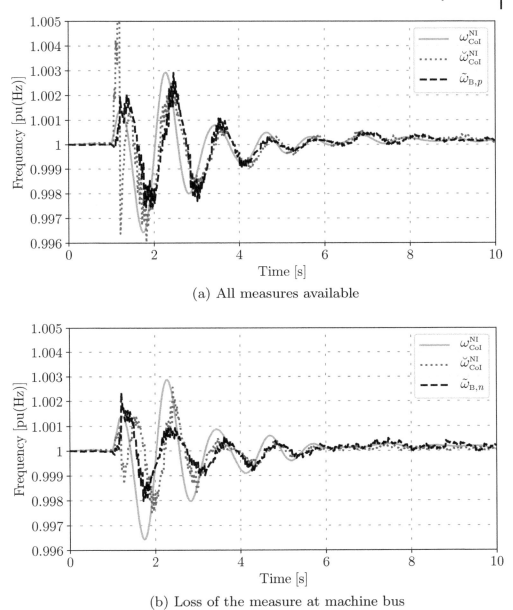

(a) All measures available

(b) Loss of the measure at machine bus

Figure 8.15 AIITS – Estimation of the frequency of the CoI of the North-Ireland sub-area system after a three-phase fault, and PMU measurements at the pilot bus, $\tilde{\omega}_{\mathrm{B},p}$, and its neighboring bus, $\tilde{\omega}_{\mathrm{B},n}$.

that:

$$X_{\mathrm{CoI}}^{\mathrm{I}} = \frac{M_{\mathrm{E}}}{M_{\mathrm{tot}}} X_{\mathrm{EI}} , \qquad X_{\mathrm{CoI}}^{\mathrm{E}} = \frac{M_{\mathrm{I}}}{M_{\mathrm{tot}}} X_{\mathrm{EI}} , \qquad (8.44)$$

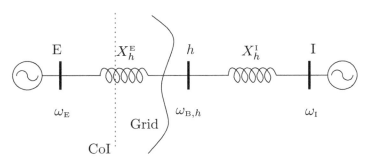

Figure 8.16 Radial equivalent system of the grid (E) connected to a synchronous machine (I) through an interface bus h.

where $X_{\mathrm{EI}} = X_h^{\mathrm{E}} + X_h^{\mathrm{I}}$. If the grid is a large interconnected system, then $M_{\mathrm{E}} \gg M_{\mathrm{I}}$ and the CoI almost coincides with the fictitious bus E of the Thévenin equivalent. Using the equation (5.18), one can determine the frequency ω_{E} as:

$$\omega_{\mathrm{E}}(t) = \frac{X_{\mathrm{EI}}}{X_h^{\mathrm{I}}} \omega_{\mathrm{B},h}(t) - \frac{X_h^{\mathrm{E}}}{X_h^{\mathrm{I}}} \omega_{\mathrm{I}}(t) \ , \tag{8.45}$$

and then estimate the frequency of the CoI as:

$$\breve{\omega}_{\mathrm{CoI}}(t) = \frac{X_{\mathrm{CoI}}^{\mathrm{I}}}{X_{\mathrm{EI}}} \omega_{\mathrm{E}}(t) + \frac{X_{\mathrm{CoI}}^{\mathrm{E}}}{X_{\mathrm{EI}}} \omega_{\mathrm{I}}(t) \ , \tag{8.46}$$

or, after some algebraic manipulations:

$$\breve{\omega}_{\mathrm{CoI}}(t) = \omega_{\mathrm{I}}(t) + \frac{M_{\mathrm{E}} X_{\mathrm{EI}}}{M_{\mathrm{tot}} X_h^{\mathrm{I}}} \left(\omega_{\mathrm{B},h}(t) - \omega_{\mathrm{I}}(t) \right) \ . \tag{8.47}$$

Assuming that X_{EI} can be estimated, the expression (8.47) can be utilized by any conventional power plant, where M_{I} and X_h^{I} are known and $\omega_{\mathrm{I}} = \tilde{\omega}_{\mathrm{G},h}$ and $\tilde{\omega}_{\mathrm{B},h}$ can be measured. Finally, M_{E} can be deduced from the knowledge of dispatched conventional generators or provided by the TSO.

8.2.2.1. Example

It is assumed that one PMU is installed at the terminal buses 1, 2, and 3 for the generators of the WSCC system described in Appendix A.2, and a three-phase fault is applied at bus 7 at $t = 1$ s and clear it after 70 ms. Figure 8.17 shows the estimation of the frequency of the CoI from the generators using (8.47). The values of X_{EI} for each power plant have been determined by trial-and-error, as follows:

$$X_{\mathrm{EI},1} \approx 0.36 \ , \quad X_{\mathrm{EI},2} \approx 0.59 \ , \quad X_{\mathrm{EI},3} \approx 1.10 \ .$$

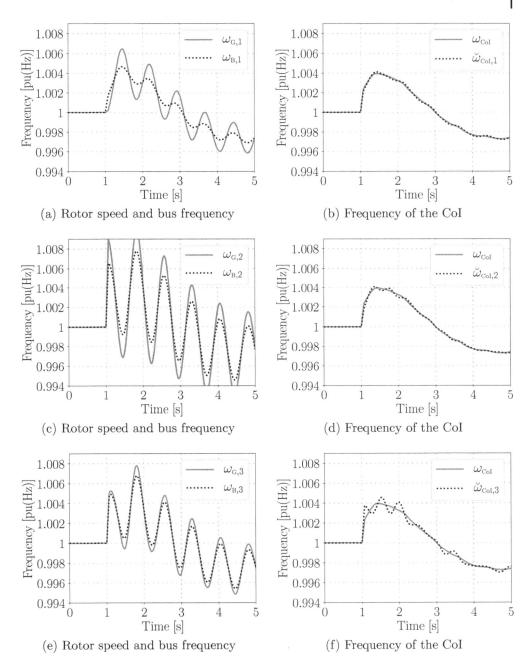

Figure 8.17 WSCC – Estimation of the frequency of the CoI through the generator rotor speeds and the radial network equivalents at the generator terminal buses.

The bigger the inertia of the machine with respect to the total inertia of the system, the easier is to estimate the frequency of the CoI. In fact $M_1 = 47.28$ s, $M_2 = 12.8$ s, and $M_3 = 6.02$ s and, hence $M_{\text{tot}} = 66.1$ s. Results also suggest that X_{EI} is not exactly constant but can be approximated as such if M_{I} is of the same order of M_{E}. In practice, expression (8.47) is more accurate if both regions E and I represent an equivalent of a large network.

8.3. Applications of the RoCoP

This section presents some relevant applications of the alternative formulation of the FDF discussed in Chapter 6 for the dynamic state estimation of bus frequencies (Section 8.3.1) and machine rotor speeds (Section 8.3.2), as well as an approximated expression to estimate the equivalent inertia of a device, valid in transient conditions, i.e. in the first instants after a contingency (Section 8.3.3).

8.3.1. Bus Frequencies

A direct application of the definition of the RoCoP given in (6.11) is that the knowledge of frequency variations at network buses enables the estimation of the variations of the regulating power as:

$$\Delta \breve{p}'_{\text{B},h}(t) = \int_t \ddot{p}'_{\text{B},h}(r)dr = \int_t \sum_{k \in \mathcal{B}_i} \hat{B}_{hk}\left(\Delta\tilde{\omega}_{\text{B},h}(r) - \Delta\tilde{\omega}_{\text{B},k}(r)\right)dr \ . \quad (8.48)$$

Equation (8.48) has the drawback that frequency variations have to be estimated with high accuracy. The technology of PMUs, however, has greatly improved in the last decade, and the required precision is already available, e.g. [47].

Vice versa, expression (6.3) suggests that active power measurements can be utilized to estimate frequency variations at network buses:

$$\Delta\boldsymbol{\omega}_{\text{B}}(t) = \hat{\mathbf{B}}_{\text{bus}}^{-1}\, \dot{\boldsymbol{p}}'_{\text{B}}(t) \ , \quad (8.49)$$

where $\dot{\boldsymbol{p}}'_{\text{B}}$ can also be approximated with a measurement of $\Delta\tilde{\boldsymbol{p}}'_{\text{B}}/\Delta t$ in a given finite time Δt.

The main difficulty in applying (8.49) is the estimation of $\Delta\tilde{\boldsymbol{p}}'_{\text{B}}$ which, as discussed in Section 6.2, is not trivially the vector of the total power injection/consumption variation at buses, say $\Delta\tilde{\boldsymbol{p}}_{\text{B}}$, but only the variation of the "regulated" power. For generators, storage devices and controlled loads,

$\Delta \tilde{p}_{\mathrm{B}} \approx \Delta \tilde{p}'_{\mathrm{B}}$. For conventional loads, however, one has first to determine the quota of power consumption that behaves as a constant admittance, e.g. using the expression (6.33).

8.3.2. Machine Rotor Speeds

The noteworthy result of Section 8.1 is that the estimation of the rotor angular speeds of synchronous machines can be achieved with a reduced set of bus frequency estimations and a linear optimization problem. Considering the simple example of Figure 6.1, the rotor speed of the machine can be obtained as:

$$\Delta \breve{\omega}_{\mathrm{G}}(t) = \frac{B_{12} + B_{10} + B_{\mathrm{G}}}{B_{\mathrm{G}}} \Delta \tilde{\omega}_{\mathrm{B},1}(t) - \frac{B_{12}}{B_{\mathrm{G}}} \Delta \tilde{\omega}_{\mathrm{B},2}(t) . \tag{8.50}$$

Equation (6.6) leads to an alternative expression to estimate the rotor speed of the machine based only on measurements at the terminal bus of the machine itself, thus also eliminating the issue of measurement delays discussed in Section 8.1.2.2:

$$\Delta \breve{\omega}_{\mathrm{G}}(t) = \Delta \tilde{\omega}_{\mathrm{B},1}(t) - \hat{X}_{\mathrm{G}} \dot{\tilde{p}}_{\mathrm{G}}(t) , \tag{8.51}$$

where $\hat{X}_{\mathrm{G}} = 1/\hat{B}_{\mathrm{G}}$ and, for a synchronous machine, $p_{\mathrm{G}} \approx p'_{\mathrm{G}}$. The former expression can be also used for any radial connection, such as a long transmission line, to estimate the frequency at one end by measuring the frequency and the regulating active power injection into the other end.

8.3.3. Inertia

Elaborating on (6.18), one can derive an expression to estimate the inertia of a synchronous machine or, more interestingly, to estimate the *equivalent* inertia of any nonsynchronous device in the transient following a contingency.

The time scale of the inertial response of the machine is faster than that of its PFC. Thus, in the first seconds after an event that causes a power unbalance in the system, it can be assumed that:

$$\frac{d}{dt} p_{\mathrm{PFC}}(t) \ll M_{\mathrm{G}} \frac{d^2}{dt^2} \omega_{\mathrm{G}}(t) . \tag{8.52}$$

Using (8.52), equation (6.18) can be rewritten as:

$$M_{\mathrm{G}} \frac{d^2}{dt^2} \omega_{\mathrm{G}}(t) \approx -\dot{p}'_{\mathrm{G}}(t) \approx -\dot{p}_{\mathrm{G}}(t), \quad \text{for } 0 < t < t_i , \tag{8.53}$$

with $t_i \approx 1$ s. Finally, substituting $d^2/dt^2\,(\omega_\mathrm{G})$ with the second time derivative of the expression obtained from (8.51), (8.53) becomes:

$$\check{M}_\mathrm{G}(t) \approx \frac{-\dot{\hat{p}}_\mathrm{G}(t)}{d^2/dt^2\left[\Delta\tilde{\omega}_{\mathrm{B},\mathrm{G}}(t) - \hat{X}_\mathrm{G}\dot{\hat{p}}_\mathrm{G}(t)\right]}, \quad \text{for } t < t_i \,. \tag{8.54}$$

Equation (8.54) determines the physical inertia of synchronous machines if applied in the proper time scale. Interestingly, however, the device to be monitored does not have to be a synchronous machine. In that case, (8.54) can be rewritten as an estimation of the *equivalent* inertia that such a device shows after a contingency, as follows:

$$\boxed{\check{M}_{\blacksquare,h}(t) \approx \frac{-\dot{\hat{p}}'_{\mathrm{B},h}(t)}{d^2/dt^2\left[\Delta\tilde{\omega}_{\mathrm{B},h}(t) - \hat{X}_{\blacksquare,h}\,\dot{\hat{p}}'_{\mathrm{B},h}(t)\right]}, \quad \text{for } t < t_i\,,} \tag{8.55}$$

where the parameter $\hat{X}_{\blacksquare,h}$ can be defined for nonsynchronous devices based on the nominal power capacity of the device and using typical values for synchronous machines of the same size. Clearly, the device does not need to have a physical inertia. To show $\check{M}_{\blacksquare,h} \neq 0$, in fact, the device can provide an inertia-like response through proper fast frequency response (FFR). This is possible if using power electronic converters, which can provide PFC much faster than synchronous machines, thus entering in the time scale of the inertial response. Expression (8.55) can thus be utilized by system operators to evaluate and possibly reward as an ancillary service the equivalent "inertial response" of nonsynchronous devices.

In steady state, both terms of the denominator of equation (8.55) are null, i.e. $\Delta\tilde{\omega}_{\mathrm{B},h} = \dot{\hat{p}}'_{\mathrm{B},h} = 0$, which leads to a singularity. This is not a problem, however, as, in steady-state conditions, the inertial response of synchronous machines and of all other devices is null. During transients, there exists the possibility that, at a given instant, $\Delta\tilde{\omega}_{\mathrm{B},h} = \hat{X}_{\blacksquare,h}\dot{\hat{p}}'_{\mathrm{B},h}$, thus also leading to a singularity. This point is further discussed in the example below.

8.3.3.1. Example

This example illustrates the dynamic behavior of the expression (8.54) to estimate the equivalent inertia of a device/subsystem. To test the accuracy of such an expression, the estimated inertia of the three synchronous machines of the system in Section 6.4.1.1 for the case with only PFC is shown in Figure 8.18. The estimations are compared with their actual inertia constant, represented by the dashed horizontal lines, and whose values are $M_{\mathrm{G},1} = 47.28$, $M_{\mathrm{G},2} = 12.8$ and $M_{\mathrm{G},3} = 6.02$ MW s/MVA. LPFs with time constant of 1 s have been added

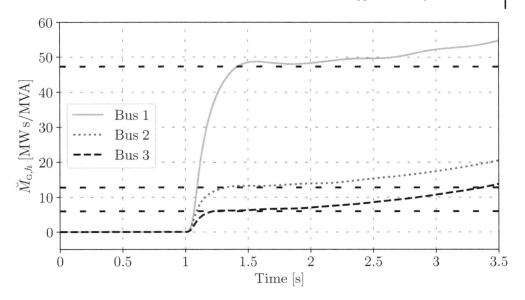

Figure 8.18 WSCC system – Estimated inertia of the three synchronous machines. Horizontal dashed lines represent the actual inertia constant of each machine.

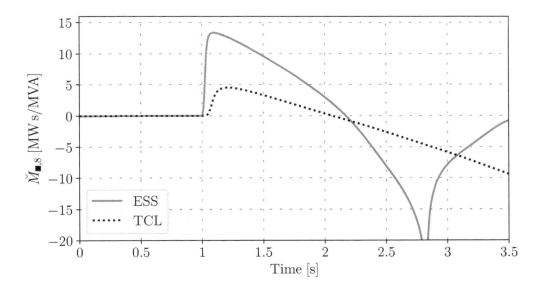

Figure 8.19 WSCC system – Estimated inertia of an ESS and a TCL connected to bus 8.

to the SRF-PLLs utilized to estimate bus frequencies for the three machines to clean the signals from numerical issues due to the sudden jumps of $\dot{p}'_{B,h}$. The value of the time constant has been chosen based on a trial-and-error process such that the estimations of the synchronous machine inertias show a first-order, critically damped response, based on the knowledge of the actual inertia of the machines.

Simulation results show that, in the first second after the contingency, the estimation of the machine inertias is highly accurate. Then, the trajectories start drifting away from their respective actual values due to the overlapping of the inertial response of the machines with the PFC.

The equivalent inertia of both the ESS and the TCL used in the test system of Section 6.4.1.2 is estimated by means of equation (8.55), and results are shown in Figure 8.19. LPFs are again required to filter out the signals from the numerical issues due to the sudden jumps observed in Figure 6.7.a when the loss of load occurs. An equivalent internal reactance of 1 pu(Ω) is considered in both cases. Same time constant for the LPFs as in the example above has been used in the estimation.

As expected, the ESS shows a considerably higher equivalent inertia than that of the TCL right after the loss of the load. The response of the FFR of the ESS rapidly leads to cancel the two terms in the denominator of equation (8.55), thus causing the singularity observed at $t \approx 2.8$ s. The slower response of the TCL delays the occurrence of such a singularity.

Chapter 9

Power System Model

9.1. Introduction

As discussed in Chapter 2, the conventional QSS model for transient stability analysis of power systems, namely the transient stability (TS) model, assumes that the frequency is constant and equal to the nominal one for the definition of network parameters such as transmission line series reactances and shunt susceptances. This approximation is widely adopted in simulation software tools for TS analysis. However, Chapters 4, 5, 6, and 7 discuss how, during a transient triggered by a large disturbance, e.g. the outage of a large in-feed, synchronous machine rotor speeds can deviate significantly from their nominal value. These variations can be properly captured by EMT models, which include detailed three-phase AC dynamic models of all elements of the grid. But EMT models are computationally too heavy to be used for the stability analysis of large power systems.

Some proprietary software tools, such as Eurostag and PSS®E, enable the utilization of a modified version of the TS model with inclusion of a variable frequency for the reactances of transmission lines and loads. Eurostag utilizes a *reference frequency*, common to all devices [211]. Such a frequency is the CoI, i.e. the weighted average of synchronous machine rotor speeds connected to the network (see Section 2.4.1.4). PSS®E implements another approach, i.e. calculates the numerical time derivative of bus voltage phase angles similar to the WF described in Section 3.3.1.2 using a filter time constant of 0.04 s [190].

As thoroughly discussed in Chapter 5, both techniques above show relevant drawbacks. On one hand, the frequency of the CoI is unique for the whole system and cannot take into account local variations of the frequency. The Eurostag model is thus not fully accurate in the first seconds after a large contingency. On the other hand, the numerical derivatives of bus voltage phase angles (as those calculated by PSS®E) are always affected by numerical issues and/or delays which can even lead to extraneous instabilities [161].

Frequency Variations in Power Systems: Modeling, State Estimation, and Control, First Edition.
Federico Milano and Álvaro Ortega Manjavacas.
© 2020 John Wiley & Sons Ltd. Published 2020 by John Wiley & Sons Ltd.

This chapter presents a frequency dependent (FD) model for TS analysis based on the FDF that overcomes both issues above. The FDF indicates that the value of the frequency varies as a *continuum* along the branches of the grid the boundary conditions are constituted by synchronous machine rotor speeds and measurements at the interconnection buses. Hence, during a transient, the frequency not only varies from bus to bus, but also along the length of series reactances of transmission lines and transformers. This observation leads to the following question: *what frequency should be used to compute correctly the parameters of network branches?* The remainder of this chapter provides an answer to this question.

9.2. Frequency Dependent Model

The vector of bus frequencies $\boldsymbol{\omega}_{\mathrm{B}}$ that can be calculated with the FDF through the expression (5.13) enables the straightforward definition of FD models of shunt devices, such as loads and capacitor banks. For example, for an inductive series impedance load:

$$
\begin{aligned}
\bar{s}_{\mathrm{D}}(t) &= \bar{v}_h(t)\,\bar{i}_{\mathrm{D}}^*(t) \\
&= \bar{v}_h(t)\left(\frac{\bar{v}_h(t)}{R_{\mathrm{D}} + j\omega_h(t)L_{\mathrm{D}}}\right)^* \\
&= \frac{v_h^2(t)}{R_{\mathrm{D}} - j\omega_h(t)L_{\mathrm{D}}}\,,
\end{aligned}
\tag{9.1}
$$

and for a capacitor bank:

$$
q_{\mathrm{C}}(t) = \omega_h(t)\,C_{\mathrm{C}}\,v_h^2(t)\ .
\tag{9.2}
$$

Then, as discussed in Section 2.4.2, empirical parameters for voltage and frequency dependent load models (2.61) are given, for example, in [14].

Less straightforward is to determine the frequency of series connections. Equation (5.13) is derived assuming that, for a series connection, the frequency variation between one point to another is linear, as previously illustrated in Figure 5.4 that, for convenience, is recalled below in Figure 9.1.

Recalling the first example of Section 5.2.1, the frequencies at buses h and k are imposed by the generators, namely ω_h and ω_k. Then, from (5.18), the frequency ω_i at the intermediate point i can be obtained from (5.13) as:

$$
\omega_i(t) = \frac{X_{ik}}{X_{hk}}\,\omega_h(t) + \frac{X_{hi}}{X_{hk}}\,\omega_k(t)\ ,
\tag{9.3}
$$

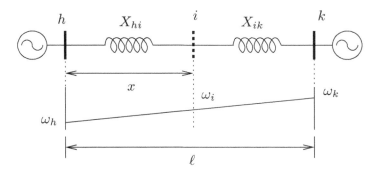

Figure 9.1 Two-bus system.

where $X_{hk} = X_{hi} + X_{ik}$. This expression can be utilized to derive the average frequency of any series connection, including the relevant cases of transmission lines, transformers and, surprisingly enough, synchronous machines, which are described below.

Transmission Line

Consider a uniform transmission line with total length ℓ from node h to k and let x the distance of point i from bus h. Then, $(\ell - x)$ is the distance of point i from bus k. Then, according to (9.3), the frequency along the connection between buses h and k can be rewritten as function of the position x:

$$\omega(t, x) = \frac{(\ell - x)\,\omega_h(t) + x\,\omega_k(t)}{\ell}, \quad x \in [0, \ell], \tag{9.4}$$

which states that the frequency varies linearly in a series connection as represented in Figure 9.1.

To determine the value of the frequency of frequency-dependent transmission line parameters, it is convenient to start from the infinitesimal length dx of a line, as shown in Figure 9.2 where R_ℓ, L_ℓ, G_ℓ, and C_ℓ are the resistance, inductance, susceptance, and capacitance, respectively, in per unit and per unit

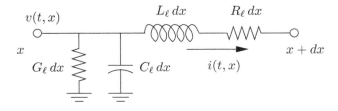

Figure 9.2 Transmission line section of length dx.

length of the line, e.g. R_ℓ is expressed in $pu(\Omega)/km$ (see also Section 2.4.3.1 the conventional lumped QSS model of overhead transmission lines). For uniform conductors, the series impedance of the line is given by:

$$\int_0^\ell (R_\ell + j\omega(t, x) L_\ell) dx = R_\ell \ell + j\frac{\omega_h(t) + \omega_k(t)}{2} L_\ell \ell$$
$$= R_L + j\langle\omega\rangle_{hk}(t) L_L$$
$$= R_L + j\langle\omega\rangle_{hk}(t) X_L \, ,$$

(9.5)

where $\langle\omega\rangle_{hk} = \frac{1}{2}(\omega_h + \omega_k)$ is the average frequency between nodes h and k; and R_L and L_L are the total resistance and inductance of the line, respectively, and X_L is the total series reactance of the line, which, in per unit, coincides with the inductance.[1]

Equation (9.5) can be easily extended to the case of a nonuniform transmission line, e.g. a connection obtained as a series of sections with different materials or topologies. Similarly, the shunt susceptance of the line is obtained as:

$$\int_0^\ell (G_\ell + j\omega(t, x)C_\ell) dx = G_\ell \ell + j\frac{\omega_h + \omega_k}{2} C_\ell \ell$$
$$= G_L + j\langle\omega\rangle_{hk}(t) C_L$$
$$= G_L + j\langle\omega\rangle_{hk}(t) B_L \, ,$$

(9.6)

where G_L and C_L are the total conductance and capacitance, respectively, of the line, and B_L is the total susceptance of the line, which, in per unit, coincides with the capacitance.

Transformer

Equation (9.5) applies also to transformers. In fact, the FDF expresses the variation of the frequency as a continuum of series reactance but does not impose any assumption on the physical dimension of the branch. For transformers,

---

[1]This assertion can be readily shown, as follows. Let ω_o and L_b be the reference angular frequency in rad/s and the inductance base in henries, respectively. Then the reactance base is $X_b = \omega_o L_b$ in ohms. In absolute values (av):

$$\omega_{av} L_{av} = X_{av} \, ,$$

and in per unit (pu):

$$\omega_{pu} L_{pu} = \frac{\omega_{av} L_{av}}{X_b} = \frac{\omega_o}{\omega_o} \frac{\omega_{av} L_{av}}{X_b} = \frac{\omega_{av}}{\omega_o} \frac{\omega_o L_{av}}{X_b} = \omega_{pu} X_{pu} \, ,$$

where $X_{pu} = \omega_o L_{av}/X_b$. Hence, $L_{pu} = X_{pu}$.

one has:

$$
\int_0^{R_T} dr + j \int_0^{L_T} \omega(t,l)\, dl = R_T + j\langle\omega\rangle_{hk}(t)\, L_T \tag{9.7}
$$
$$
= R_T + j\langle\omega\rangle_{hk}(t)\, X_T \ ,
$$

where dr and dl are the infinitesimal variations of the resistance and leakage inductance, respectively, along the transformer windings; R_T and L_T are the per unit total series resistance and leakage inductance, respectively, of the transformer; and X_T is the total series reactance, which, in per unit, coincides with the inductance.

The dependence of ω on the inductance l along the transformer windings is formally analogous to (9.4):

$$
\omega(t,l) = \frac{(L_T - l)\,\omega_h(t) + l\,\omega_k(t)}{L_T}, \quad l \in [0, L_T]\,. \tag{9.8}
$$

Synchronous Machine

The frequency of the internal impedances of synchronous machines can be studied in the same way as transmission lines and transformers. Each machine imposes the frequency, i.e. the rotor speed, at its EMF behind the internal reactances, not at the bus where the machine is connected. The internal reactances of the machine thus constitute an *antenna*, i.e. a series connection between the internal EMF and the terminal bus of the machine. The frequency at the EMF is, by definition, the rotor speed of the machine ω_G, while the frequency at the bus ω_h is that determined through (5.14).

Assuming a continuous linear variation of the frequency from ω_G to ω_h and proceeding in a similar manner as discussed for the transmission line and transformer, the average frequency along the internal stator reactances of the machine is $\langle\omega\rangle_{G,h} = \frac{1}{2}(\omega_G + \omega_h)$. Thus, the dependency of the machine stator voltage equations on the frequency of equation (2.31) can be written as follows:

$$
0 = R_a i_{s,d}(t) + \langle\omega\rangle_{G,h}(t)\, \psi_{s,q}(t) + v_{s,d}(t) \ ,
$$
$$
0 = R_a i_{s,q}(t) - \langle\omega\rangle_{G,h}(t)\, \psi_{s,d}(t) + v_{s,q}(t) \ . \tag{9.9}
$$

9.3. Example

The dynamic behavior of the FD model of power system devices described in the previous section is compared with that of the conventional TS model as well as of a dynamic phasor (DP) model through the 1,479-bus dynamic model of the AIITS described in Appendix B.

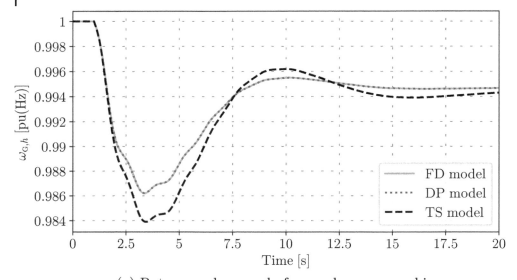

(a) Rotor angular speed of a synchronous machine

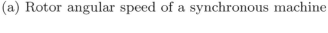

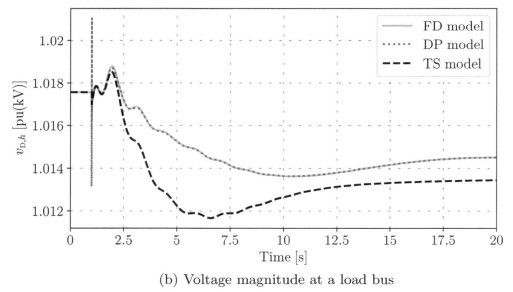

(b) Voltage magnitude at a load bus

Figure 9.3 AIITS – Response of the system following the outage of the largest in-feed.

Figure 9.3 shows the transient response of the AIITS following the outage of the largest in-feed, namely, the HVDC interconnection with UK. This is one of the most severe contingencies that can happen in the AIITS, and is considered here to dramatize frequency variations.

In Figure 9.3, the following three models are compared:

- The conventional TS model that considers constant reactances and susceptances everywhere in the grid (see Chapter 2);

- The FD model of loads, branches, and synchronous machines proposed in this chapter. Loads are assumed to be full-load induction motors with $\beta_p = 2.8$, $\beta_q = 1.8$, $\gamma_p = 0.1$, and $\gamma_q = 0.6$ [14]; and

- A DP model that includes machine flux and line dynamics (see Section 1.4.3).

Note that the DP model is the Park transform of the fully fledged EMT model with the following approximations: (i) load is assumed perfectly balanced; (ii) no harmonics are considered; and (iii) the fundamental frequency is shifted by means of the Park transform.

Simulation results show that the TS model is conservative, as the frequency nadir is about 0.002 pu (100 mHz) lower than that obtained with the more precise FD model. Voltage variations are also, in general, bigger for the standard model than for the FD model. The computational burden of the FD model is only about the double than the conventional one (14 s versus 7.5 s, using an integration time step of 0.01 s), despite the significant nonlinearity introduced by the dependency of network parameters on the frequency.

On the other hand, the FD and DP models give basically same results, except for some spikes in the voltages right after the occurrence of the contingency. However, the DP model requires a much smaller time step to properly integrate flux and line dynamics. To complete the DP model simulation shown in Figure 9.3, in fact, required about 165 s with a time step of 0.002 s.

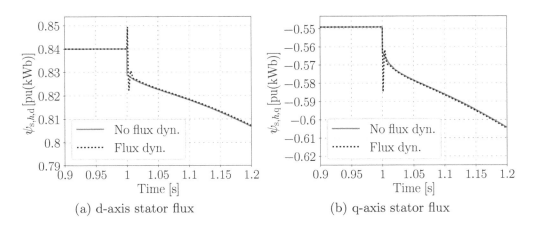

(a) d-axis stator flux

(b) q-axis stator flux

Figure 9.4 AIITS – Stator dq-frame fluxes of a synchronous machine of the system modeled with the FD model following the outage of the largest in-feed.

Figure 9.4 shows a further comparison of the trajectories of stator fluxes of a synchronous generator of the Irish system following the same contingency discussed above and considering the FD model with two scenarios, namely, with and without stator flux dynamics. The results indicate that the effect of the dynamics of the fluxes is visible only in the very first instants after the contingency and they do not modify the overall behavior of the machine. This is consistent with the discussion provided in Section 2.4.4.

Simulation results indicate that the FD model is a compromise model between the transient stability and EMT approaches. With respect to the TS model, the FD model is able to retains the actual local variations of the frequency at every bus of the system. This enables the consideration, during the first seconds after a contingency, of the electromechanical oscillations of the synchronous machines. With respect to the DP approach, the FD model neglects fast electromagnetic dynamics, which anyway damp in the first tens of milliseconds after a contingency and, thus, have little impact on the electromechanical dynamics. As a result, the FD model enhances the accuracy of the conventional TS model while avoiding the computational burden of the fully fledged EMT simulations. Overall, the FD model appears as a useful improvement with respect to conventional models as it is less conservative and can thus lead to better estimations of the transient behavior of power systems, especially those with low inertia, which are prone to high-frequency variations.

Chapter 10
Frequency in Power Systems

10.1. Definitions

The discussion given in Section 6.2 leads to the following expression:

$$\boxed{\Delta\boldsymbol{\omega}_{\mathrm{B}}(t) = \omega_o^{-1}\,\mathbf{B}_{\mathrm{bus}}^{-1}\,\dot{\boldsymbol{p}}_{\mathrm{B}}'(t)}\;, \tag{10.1}$$

which suggests a general definition of bus frequency variations in high-voltage transmission systems as *linear combinations of the rates of change of active power at network buses.*

A relevant special case of the general definition given in (10.1) is when the only time-variant power injections are given by synchronous machines. Recalling (6.4) and (6.5):

$$\begin{aligned}
\mathbf{B}_{\mathrm{BB}} &= \mathbf{B}_{\mathrm{bus}} + \mathbf{B}_{\mathrm{G}}\;, \\
\dot{\boldsymbol{p}}_{\mathrm{B}}'(t) &= -\omega_o\left[\mathbf{B}_{\mathrm{BG}}\Delta\boldsymbol{\omega}_{\mathrm{G}}(t) + \mathbf{B}_{\mathrm{G}}\Delta\boldsymbol{\omega}_{\mathrm{B}}(t)\right]\;,
\end{aligned} \tag{10.2}$$

equation (10.1) can be rewritten as:

$$\boxed{\Delta\boldsymbol{\omega}_{\mathrm{B}}(t) = -\mathbf{B}_{\mathrm{BB}}^{-1}\mathbf{B}_{\mathrm{BG}}\,\Delta\boldsymbol{\omega}_{\mathrm{G}}(t) = \mathbf{D}\,\Delta\boldsymbol{\omega}_{\mathrm{G}}(t)}\;. \tag{10.3}$$

The latter expression is the FDF defined in Section 5.2 and indicates that, if the network includes exclusively conventional devices, namely synchronous machines with their controllers, constant admittance loads, transmission lines and transformers, the bus frequency variations at network buses are *linear combinations of the rotor angular speeds of the synchronous machines.*

10.2. Final Remarks

The definition (10.1) has a fundamental feature in common with the standard definitions given in Section 1.1, namely the fact that frequency can be defined

Frequency Variations in Power Systems: Modeling, State Estimation, and Control, First Edition.
Federico Milano and Álvaro Ortega Manjavacas.
© 2020 John Wiley & Sons Ltd. Published 2020 by John Wiley & Sons Ltd.

only indirectly as a measurement of something else, in this case, the first time derivative of the *regulated* active power. However, electric power is a very tangible quantity. This makes also frequency a very tangible quantity, in spite of what is stated in some references.

In particular, (10.1) appears better suited than the other definitions discussed in Section 1.1 to be measured with a representationalist measurement approach, i.e. to be represented through an unambiguous number and thus less prone to the elusive idiosyncrasy of the operationalist measurement approach. This, of course, if one assumes that it is possible to properly define and measure the component p'_{B} of the instantaneous active power and then solve the issues related to the calculation of the numerical time derivative. But, at least, (10.1) does not require the definition *a priori* of a model of the signal whose frequency is to be estimated.

It is also worth noticing that (10.1) defines variations of the frequency that are relative to a reference value. This indicates that the reference frequency *per se* is a "potential" and, as such, is similar to the ground voltage or the phase reference angle in electric circuits.

Finally, (10.1) states that frequency variations in an AC circuit are intrinsically related to power and, in turn, to energy. This link is well-known for synchronous machines (see equation (1.9)) but it is not obvious that it applies in general, for any device connected to the grid. This result, of course, is just a consequence of the fact that (10.1) has been derived from the power flow equations and, in turn, from Kirchhoff's current law. A relevant consequence, however, is that no synchronous machine has to be connected to the grid for (10.1) to hold. This seems to indicate that frequency will remain a relevant quantity because it is intrinsically related to energy and, thus, independent from any technological change that AC power systems may undergo in the future.

Appendices

Appendix A
Data

A.1. Three-Bus System

The single line diagram of the three-bus, two-machine radial system utilized in Chapter 5 is shown in Figure A.1. The power and frequency bases are 100 MVA and 50 Hz, respectively. Bus 1 is the slack.

A.1.1. Static Data

The grid consists of three transmission lines with nominal voltage 6.9 kV, two synchronous generators at buses 1 and 2, and one load at bus 3. Transmission line and transformer data are shown in Table A.1.

The power flow data and solution for the base-case operating point are shown in Tables A.2 and A.3. The base-case total generation, demand and losses are shown in Table A.4.

A.1.2. Dynamic Data

Synchronous machines are represented with the sixth-order model described in Section 2.4.1. The parameters of the synchronous machines are shown in Table A.5. The data of the turbine governors and the AVRs of the WSCC system are shown in Tables A.6 and A.7, respectively.

A.2. WSCC System

The single line diagram of the WSCC nine-bus three-machine system (hereinafter, *WSCC system*) is shown in Figure A.2. The data given below are based on [8] and [186]. The power and frequency bases are 100 MVA and 60 Hz, respectively. Bus 1 is the slack.

Frequency Variations in Power Systems: Modeling, State Estimation, and Control, First Edition.
Federico Milano and Álvaro Ortega Manjavacas.
© 2020 John Wiley & Sons Ltd. Published 2020 by John Wiley & Sons Ltd.

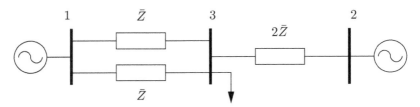

Figure A.1 Three-bus system.

Table A.1 Branch data of the three-bus system.

Branch #	From h	To k	R_L (R_T) [pu(Ω)]	X_L (X_T) [pu(Ω)]	B_L [pu(Ω^{-1})]	a $\left[\frac{pu(kV)}{pu(kV)}\right]$	ϕ [rad]	Type
1	1	3	0.025	0.075	0	–	–	line
2	1	3	0.025	0.075	0	–	–	line
3	2	3	0.050	0.150	0	–	–	line

Table A.2 Base-case power flow solution of the three-bus system.

Bus #	v_b [kV]	v [pu(kV)]	θ [rad]	p_G [pu(MW)]	q_G [pu(MVAr)]	p_D [pu(MW)]	q_D [pu(MVAr)]
1	6.9	1.0000	0	0.3386	0.3222	0	0
2	6.9	0.9900	0.0269	0.2222	-0.0285	0	0
3	6.9	0.9837	-0.0088	0	0	0.5556	0.2778

Table A.3 Base-case power flows and losses in transmission lines and transformers of the three-bus system.

Line h-k	p_{hk} [pu(MW)]	q_{hk} [pu(MVAr)]	p_{loss} [pu(MW)]	q_{loss} [pu(MVAr)]
1-3	0.1693	0.1611	0.0014	0.0041
1-3	0.1693	0.1611	0.0014	0.0041
2-3	0.2222	-0.0285	0.0026	0.0077

Table A.4 Total generation, demand and losses for the base-case operating point of the three-bus system.

Type	Active power [pu(MW)]	Reactive power [pu(MVAr)]
Generation	0.5609	0.2937
Demand	0.5556	0.2778
Losses	0.0053	0.0159

Table A.5 Data of the synchronous machines of the three-bus system.

Parameter	Unit	Machine 1	Machine 2
s_n	[MVA]	100.0	100.0
v_n	[kV]	6.9	6.9
D	[pu(MW)]	1.0	1.0
M	[s pu(MW)]	13.3617	13.3617
R_a	[pu(Ω)]	0.002	0.002
T'_{do}	[s]	4.2	4.2
T''_{do}	[s]	0.04	0.04
T'_{qo}	[s]	3.0	3.0
T''_{qo}	[s]	0.02	0.02
X_d	[pu(Ω)]	0.9111	0.9111
X'_d	[pu(Ω)]	0.408	0.408
X''_d	[pu(Ω)]	0.204	0.204
X_ℓ	[pu(Ω)]	0	0
X_q	[pu(Ω)]	0.58	0.58
X'_q	[pu(Ω)]	0.58	0.58
X''_q	[pu(Ω)]	0.3	0.3

Table A.6 Data of the turbines and turbine governors of the three-bus system.

Machine	$p^{\max}$ [pu(MW)]	$p^{\min}$ [pu(MW)]	$\mathcal{R}$ [pu(Hz/MW)]	T_g [s]	T_{rh} [s]	T_{sm} [s]	T_t [s]	κ_{rh} –
all	5.5	−1.5	0.05	0.1	50.0	0.45	0	0.25

Table A.7 Data of the AVRs of the three-bus system.

Machine	A_{ef} $-$	B_{ef} $-$	K_a $-$	K_{ef} $-$	K_f $-$	T_a [s]	T_b [s]	T_c [s]
	0.0006	0.9	0.5	1.0	1.0	20.0	0	0
all	T_{ef} [s]	T_f [s]	T_R [s]	v_a^{max} [pu(kV)]	v_a^{min} [pu(kV)]			
	1.0	0.35	1.0	9.99	0.0			

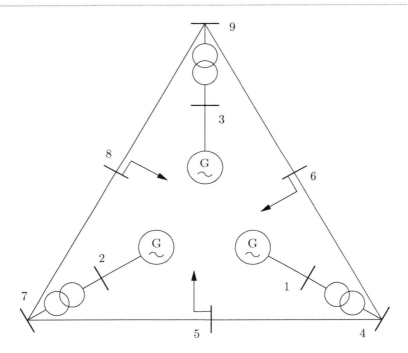

Figure A.2 WSCC nine-bus system.

A.2.1. Static Data

The grid consists of nine branches, three step-up transformers that connect the synchronous machines to the transmission grid and six transmission lines with nominal voltage 230 kV. The generators at buses 1, 2, and 3 have nominal voltages 16.5, 18 and 13.8 kV, respectively. Transmission line and transformer data are shown in Table A.8.

The power flow data and solution for the base-case operating point are shown in Tables A.9 and A.10. The base-case total generation, demand, and losses are shown in Table A.11.

Table A.8 Branch data of the WSCC system.

Branch #	From h	To k	R_L (R_T) [pu(Ω)]	X_L (X_T) [pu(Ω)]	B_L [pu(Ω^{-1})]	a $\left[\frac{\text{pu(kV)}}{\text{pu(kV)}}\right]$	ϕ [rad]	Type
1	1	4	0	0.0576	0	1.0	0	transf.
2	2	7	0	0.0625	0	1.0	0	transf.
3	3	9	0	0.0586	0	1.0	0	transf.
4	6	4	0.0170	0.0920	0.1580	–	–	line
5	5	4	0.0100	0.0850	0.1760	–	–	line
6	7	5	0.0320	0.1610	0.3060	–	–	line
7	9	6	0.0390	0.1700	0.3580	–	–	line
8	7	8	0.0085	0.0720	0.1490	–	–	line
9	9	8	0.0119	0.1008	0.2090	–	–	line

Table A.9 Base-case power flow solution of the WSCC system.

Bus #	v_b [kV]	v [pu(kV)]	θ [rad]	p_G [pu(MW)]	q_G [pu(MVAr)]	p_D [pu(MW)]	q_D [pu(MVAr)]
1	16.5	1.0400	0	0.7164	0.2705	0	0
2	18	1.0250	0.1620	1.6300	0.0665	0	0
3	13.8	1.0250	0.0814	0.8500	−0.1086	0	0
4	230	1.0258	−0.0387	0	0	0	0
5	230	0.9956	−0.0696	0	0	1.25	0.50
6	230	1.0127	−0.0644	0	0	0.90	0.30
7	230	1.0258	0.0649	0	0	0	0
8	230	1.0159	0.0127	0	0	1.00	0.35
9	230	1.0324	0.0343	0	0	0	0

A.2.2. Dynamic Data

Synchronous machines are represented with the simplified fourth-order models described in Section 2.4.1. The parameters of the synchronous machines are shown in Table A.12.

The data of the turbine governors and the AVRs of the WSCC system are shown in Tables A.13 and A.14, respectively.

Table A.10 Base-case power flows and losses in transmission lines and transformers of the WSCC system.

Line h-k	p_{hk} [pu(MW)]	q_{hk} [pu(MVAr)]	p_{loss} [pu(MW)]	q_{loss} [pu(MVAr)]
1-4	0.7164	0.2705	0	0.0312
2-7	1.6300	0.0665	0	0.1583
3-9	0.8500	−0.1086	0	0.0410
6-4	−0.3054	−0.1654	0.0017	−0.1551
5-4	−0.4068	−0.3869	0.0026	−0.1579
7-5	0.8662	−0.0838	0.0230	−0.1969
9-6	0.6082	−0.1807	0.0135	−0.3153
7-8	0.7638	−0.0080	0.0048	−0.1150
9-8	0.2418	0.0312	0.0009	−0.2118

Table A.11 Total generation, demand, and losses for the base-case operating point of the WSCC system.

Type	Active power [pu(MW)]	Reactive power [pu(MVAr)]
Generation	3.1964	0.2284
Demand	3.15	1.15
Losses	0.04641	−0.9216

Table A.12 Data of the synchronous machines of the WSCC system.

Machine	v_n [kV]	M [s pu(MW)]	T'_{do} [s]	T'_{qo} [s]	X_d [pu(Ω)]	X'_d [pu(Ω)]	X_q [pu(Ω)]	X'_q [pu(Ω)]
1	16.5	47.28	8.96	0.310	0.1460	0.0608	0.0969	0.0969
2	18.0	12.80	6.00	0.535	0.8958	0.1198	0.8645	0.1969
3	13.8	6.02	5.89	0.600	1.3125	0.1813	1.2578	0.2500

A.3. IEEE 14-Bus System

Figure A.3 depicts the single-line diagram of the IEEE 14-bus test system. This benchmark network consists of two synchronous machines and three synchronous compensators, two two-winding, and one three-winding transformers, 15 transmission lines and 11 loads. The system also includes AVRs, TGs, and

Table A.13 Data of the turbines and turbine governors of the WSCC system.

Machine	p^{max} [pu(MW)]	p^{min} [pu(MW)]	$\mathcal{R}$ [pu(Hz/MW)]	T_g [s]	T_{rh} [s]	T_{sm} [s]	T_t [s]	κ_{rh} –
1	1.6	0	0.05	0	50.0	0.45	0.1	0.25
2	3.2	0	0.05	0	50.0	0.45	0.1	0.25
3	1.7	0	0.05	0	50.0	0.45	0.1	0.25

Table A.14 Data of the AVRs of the WSCC system.

Machine	A_{ef} –	B_{ef} –	K_a –	K_{ef} –	K_f –	T_a [s]	T_b [s]	T_c [s]
	0.0039	1.555	20.0	1.0	0.063	0.2	0	0

all	T_{ef} [s]	T_f [s]	T_R [s]	v_a^{max} [pu(kV)]	v_a^{min} [pu(kV)]
	0.314	0.35	0.001	5.0	−5.0

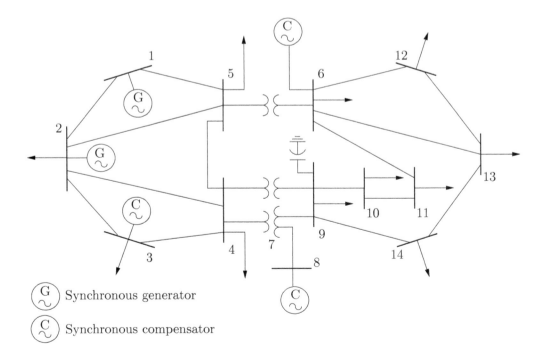

G ⁓ Synchronous generator

C ⁓ Synchronous compensator

Figure A.3 IEEE 14-bus system.

Table A.15 Branch data of the IEEE 14-bus system.

Branch #	From h	To k	R_L (R_T) [pu(Ω)]	X_L (X_T) [pu(Ω)]	B_L [pu(Ω^{-1})]	a $\left[\frac{\text{pu(kV)}}{\text{pu(kV)}}\right]$	ϕ [rad]	Type
1	1	2	0.01938	0.05917	0.0528	–	–	line
2	1	5	0.05403	0.22304	0.0492	–	–	line
3	2	3	0.04699	0.19797	0.0438	–	–	line
4	2	4	0.05811	0.17632	0.0374	–	–	line
5	2	5	0.05695	0.17388	0.0340	–	–	line
6	3	4	0.06701	0.17103	0.0346	–	–	line
7	4	5	0.01335	0.04211	0.0128	–	–	line
8	4	7	0	0.20912	0	0.978	0	transf.
9	4	9	0	0.55618	0	0.969	0	transf.
10	5	6	0	0.25202	0	0.932	0	transf.
11	6	11	0.09498	0.19890	0	–	–	line
12	6	12	0.12291	0.25581	0	–	–	line
13	6	13	0.06615	0.13027	0	–	–	line
14	7	8	0	0.17615	0	1.000	0	transf.
15	7	9	0	0.11001	0	–	–	line
16	9	10	0.03181	0.08450	0	–	–	line
17	9	14	0.12711	0.27038	0	–	–	line
18	10	11	0.08205	0.19207	0	–	–	line
19	12	13	0.22092	0.19988	0	–	–	line
20	13	14	0.17093	0.34802	0	–	–	line

an AGC. the dynamic data of the IEEE 14-bus system as well as a detailed discussion of its transient behavior can be found in [107] and [127].

A.3.1. Static Data

The transmission system consists of three voltage levels, 69 kV for buses 1-5, 13.8 kV for buses 6, 7, and 9–14, and 18 kV for bus 8. The power and frequency bases are 100 MVA and 60 Hz, respectively. Bus 1 is the slack. Table A.15 shows transmission line and transformer data. The fixed shunt compensator at bus 9 has a capacitive susceptance of 0.19 pu(S).

Table A.16 Base-case power flow solution of the IEEE 14-bus system.

Bus #	v_b [kV]	v [pu(kV)]	θ [rad]	p_G [pu(MW)]	q_G [pu(MVAr)]	p_D [pu(MW)]	q_D [pu(MVAr)]	b_{sh} [pu(S)]
1	69.0	1.0600	0	2.324	−0.1689	0	0	0
2	69.0	1.0450	−0.0869	0.400	0.4240	0.217	0.127	0
3	69.0	1.0100	−0.2220	0	0.2339	0.942	0.190	0
4	69.0	1.0186	−0.1802	0	0	0.478	−0.039	0
5	69.0	1.0203	−0.1533	0	0	0.076	0.016	0
6	13.8	1.0700	−0.2482	0	0.1224	0.112	0.075	0
7	13.8	1.0620	−0.2333	0	0	0	0	0
8	18.0	1.0900	−0.2333	0	0.1736	0	0	0
9	13.8	1.0563	−0.2609	0	0	0.295	0.166	0.19
10	13.8	1.0513	−0.2636	0	0	0.090	0.058	0
11	13.8	1.0571	−0.2582	0	0	0.035	0.018	0
12	13.8	1.0552	−0.2632	0	0	0.061	0.016	0
13	13.8	1.0504	−0.2646	0	0	0.135	0.058	0
14	13.8	1.0358	−0.2799	0	0	0.149	0.050	0

The power flow data and solution for the base-case operating point are shown in Tables A.16 and A.17. The total generation, demand, and losses at the base-case operating point are shown in Table A.18.

A.3.2. Dynamic Data

Synchronous machines are represented by the sixth-order model described in Section 2.4.1. The parameters of the five machines are shown in Table A.19.

The dynamic data of the AVRs and the PSS of the IEEE 14-bus system are shown in Tables A.20 and A.21, respectively.

A.4. New England System

The single-line diagram of the New England 39-bus 10-machine system (hereinafter, *New England system*) is shown in Figure A.4. The power and frequency bases are 100 MVA and 60 Hz, respectively. Bus 31 is the slack.

Table A.17 Base-case power flows and losses in the branches of the IEEE 14-bus system.

Line h-k	p_{hk} [pu(MW)]	q_{hk} [pu(MVAr)]	p_{loss} [pu(MW)]	q_{loss} [pu(MVAr)]
1-2	1.5683	−0.2039	0.0429	0.0726
1-5	0.7555	0.0350	0.0276	0.0608
2-3	0.7319	0.0357	0.0232	0.0515
2-4	0.5614	−0.0229	0.0168	0.0111
2-5	0.4251	0.0076	0.0090	−0.0087
3-4	−0.2333	0.0281	0.0037	−0.0261
4-5	−0.6122	0.1567	0.0052	0.0030
4-7	0.2809	−0.0942	0	0.0169
4-9	0.1609	−0.0032	0	0.0130
5-6	0.4406	0.1282	0	0.0443
6-11	0.0734	0.0347	0.0005	0.0011
6-12	0.0778	0.0249	0.0007	0.0015
6-13	0.1774	0.0717	0.0021	0.0042
7-8	0	−0.1691	0	0.0045
7-9	0.2809	0.0580	0	0.0080
9-10	0.0524	0.0431	0.0001	0.0003
9-14	0.0944	0.0367	0.0012	0.0025
10-11	−0.0377	−0.0153	0.0001	0.0003
12-13	0.0161	0.0074	0.0001	0.0001
13-14	0.0563	0.0169	0.0005	0.0011

Table A.18 Total generation, demand and losses for the base-case operating point of the IEEE 14-bus system.

Type	Active power [pu(MW)]	Reactive power [pu(MVAr)]
Generation	2.724	0.785
Demand	2.590	0.523
Losses	0.134	0.262

Table A.19 Data of the synchronous machines of the IEEE 14-bus system.

Parameter	Unit	Machine				
		1	2	3	6	8
s_n	[MVA]	615.0	60.0	60.0	25.0	25.0
v_n	[kV]	69.0	69.0	69.0	13.8	18.0
D	[pu(MW)]	2.0	2.0	2.0	2.0	2.0
M	[s pu(MW)]	10.296	13.08	13.08	10.12	10.12
R_a	[pu(Ω)]	0	0.0031	0.0031	0.0041	0.0041
T'_{do}	[s]	7.4	6.1	6.1	4.75	4.75
T''_{do}	[s]	0.03	0.04	0.04	0.06	0.06
T'_{qo}	[s]	0	0.3	0.3	1.5	1.5
T''_{qo}	[s]	0.033	0.099	0.099	0.21	0.21
X_d	[pu(Ω)]	0.8979	1.05	1.05	1.25	1.25
X'_d	[pu(Ω)]	0.2995	0.185	0.185	0.232	0.232
X''_d	[pu(Ω)]	0.23	0.13	0.13	0.12	0.12
X_ℓ	[pu(Ω)]	0.2396	0	0	0.134	0.134
X_q	[pu(Ω)]	0.646	0.98	0.98	1.22	1.22
X'_q	[pu(Ω)]	0.646	0.36	0.36	0.715	0.715
X''_q	[pu(Ω)]	0.4	0.13	0.13	0.12	0.12

A.4.1. Static Data

The transmission system consists of 34 transmission lines, 12 transformers, 15 loads, and 10 synchronous machines. The voltage levels of the grid are 18 kV at the generator terminal bus and 230 kV for the transmission system. Table A.22 shows transmission line and transformer data.

The power flow data and solution for the base-case operating point are shown in Tables A.23 and A.24. The total generation, demand, and losses at the base-case operating point are shown in Table A.25.

Table A.20 Data of the AVRs of the IEEE 14-bus system.

Parameter	Unit	Machine				
		1	2	3	6	8
A_{ef}	–	0.0006	0.0006	0.0006	0.0006	0.0006
B_{ef}	–	0.9	0.9	0.9	0.9	0.9
K_a	–	200.0	20.0	20.0	20.0	20.0
K_{ef}	–	1.0	1.0	1.0	1.0	1.0
K_f	–	0.0012	0.001	0.001	0.001	0.001
T_a	[s]	0.02	0.02	0.02	0.02	0.02
T_b	[s]	0	0	0	0	0
T_c	[s]	0	0	0	0	0
T_{ef}	[s]	0.19	1.98	1.98	0.7	0.7
T_f	[s]	1.0	1.0	1.0	1.0	1.0
T_R	[s]	0.001	0.001	0.001	0.001	0.001
v_a^{max}	[pu(kV)]	9.99	2.05	1.7	2.2	2.2
v_a^{min}	[pu(kV)]	0.0	0.0	0.0	1.0	1.0

Table A.21 Data of the PSS of the IEEE 14-bus system.

Machine	K_w –	T_w [s]	T_1 [s]	T_2 [s]	T_3 [s]	T_4 [s]	v_s^{max} [pu(kV)]	v_s^{min} [pu(kV)]
1	5.0	10.0	0.28	0.02	0.28	0.02	0.1	−0.1

A.4.2. Dynamic Data

In the New England system, synchronous machines are represented by the simplified 4th order models described in Section 2.4.1. The parameters of the ten machines are shown in Table A.26.

The dynamic data of the turbine governors, the automatic voltage regulators and PSSs of the New England system are shown in Tables A.27, A.28, and A.29, respectively. The gain of the integrator of the AGC is $K_o = 0.01$.

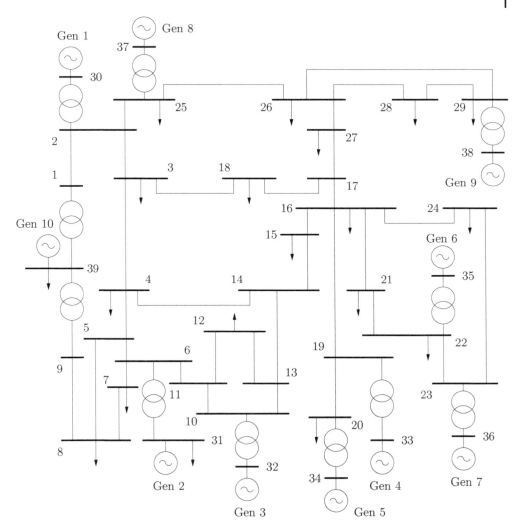

Figure A.4 New England system.

Table A.22 Branch data of the New England system.

Branch #	From h	To k	R_L (R_T) [pu(Ω)]	X_L (X_T) [pu(Ω)]	B_L [pu(Ω^{-1})]	a $\left[\frac{pu(kV)}{pu(kV)}\right]$	ϕ [rad]	Type
1	1	2	0.0035	0.0411	0.6987	–	–	line
2	1	39	0.0010	0.0250	0	1.000	0	transf.
3	2	3	0.0013	0.0151	0.2572	–	–	line
4	2	25	0.0070	0.0086	0.1460	–	–	line
5	2	30	0	0.0181	0	1.025	0	transf.
6	3	4	0.0013	0.0213	0.2214	–	–	line
7	3	18	0.0011	0.0133	0.2138	–	–	line
8	4	5	0.0008	0.0128	0.1342	–	–	line
9	4	14	0.0008	0.0129	0.1382	–	–	line
10	5	8	0.0008	0.0112	0.1476	–	–	line
11	6	5	0.0002	0.0026	0.0434	–	–	line
12	6	7	0.0006	0.0092	0.1130	–	–	line
13	6	11	0.0007	0.0082	0.1389	–	–	line
14	6	31	0	0.0250	0	1.070	0	transf.
15	7	8	0.0004	0.0046	0.0780	–	–	line
16	8	9	0.0023	0.0363	0.3804	–	–	line
17	9	39	0.0010	0.0250	0	1.0	0	transf.
18	10	11	0.0004	0.0043	0.0729	–	–	line
19	10	13	0.0004	0.0043	0.0729	–	–	line
20	10	32	0	0.0200	0	1.070	0	transf.
21	12	11	0.0016	0.0435	0	–	–	line
22	12	13	0.0016	0.0435	0	–	–	line
23	13	14	0.0009	0.0101	0.1723	–	–	line
24	14	15	0.0018	0.0217	0.3660	–	–	line
25	15	16	0.0009	0.0094	0.1710	–	–	line
26	16	17	0.0007	0.0089	0.1342	–	–	line
27	16	19	0.0016	0.0195	0.3040	–	–	line
28	16	21	0.0008	0.0135	0.2548	–	–	line
29	16	24	0.0003	0.0059	0.0680	–	–	line
30	17	18	0.0007	0.0082	0.1319	–	–	line
31	17	27	0.0013	0.0173	0.3216	–	–	line
32	19	20	0.0007	0.0138	0	–	–	line
33	19	33	0.0007	0.0142	0	1.070	0	transf.
34	20	34	0.0009	0.0180	0	1.009	0	transf.
35	21	22	0.0008	0.0140	0.2565	–	–	line
36	22	23	0.0006	0.0096	0.1846	–	–	line
37	22	35	0	0.0143	0	1.025	0	transf.
38	23	24	0.0022	0.0350	0.3610	–	–	line
39	23	36	0.0005	0.0272	0	1.000	0	transf.
40	25	26	0.0032	0.0323	0.5130	–	–	line
41	25	37	0.0006	0.0232	0	1.025	0	transf.
42	26	27	0.0014	0.0147	0.2396	–	–	line
43	26	28	0.0043	0.0474	0.7802	–	–	line
44	26	29	0.0057	0.0625	1.0290	–	–	line
45	28	29	0.0014	0.0151	0.2490	–	–	line
46	29	38	0.0008	0.0156	0	1.025	0	transf.

Table A.23 Base-case power flow solution of the New England system.

Bus #	v_b [kV]	v [pu(kV)]	θ [rad]	p_G [pu(MW)]	q_G [pu(MVAr)]	p_D [pu(MW)]	q_D [pu(MVAr)]
1	230	1.0406	−0.1474	0	0	0	0
2	230	1.0472	−0.1006	0	0	0	0
3	230	1.0286	−0.1505	0	0	3.2200	0.0240
4	230	1.0018	−0.1682	0	0	5.000	1.8400
5	230	1.0026	−0.1507	0	0	0	0
6	230	1.0051	−0.1391	0	0	0	0
7	230	0.9936	−0.1772	0	0	2.3380	0.8400
8	230	0.9921	−0.1858	0	0	5.2200	1.7600
9	230	1.0175	−0.1802	0	0	0	0
10	230	1.0155	−0.0950	0	0	0	0
11	230	1.0107	−0.1100	0	0	0	0
12	230	0.9983	−0.1093	0	0	0.0750	0.8800
13	230	1.0126	−0.1067	0	0	0	0
14	230	1.0100	−0.1340	0	0	0	0
15	230	1.0143	−0.1355	0	0	3.2000	1.5300
16	230	1.0310	−0.1084	0	0	3.2900	0.3230
17	230	1.0326	−0.1279	0	0	0	0
18	230	1.0297	−0.1440	0	0	1.5800	0.3000
19	230	1.0496	−0.0182	0	0	0	0
20	230	0.9910	−0.0356	0	0	6.2800	1.0300
21	230	1.0312	−0.0664	0	0	2.7400	1.1500
22	230	1.0495	0.0113	0	0	0	0
23	230	1.0445	0.0079	0	0	2.4750	0.8460
24	230	1.0366	−0.1063	0	0	3.0860	−0.9200
25	230	1.0565	−0.0766	0	0	2.2400	0.4720
26	230	1.0513	−0.0969	0	0	1.3900	0.1700
27	230	1.0369	−0.1313	0	0	2.8100	0.7550
28	230	1.0497	−0.0356	0	0	2.0600	0.2760
29	230	1.0497	0.0126	0	0	2.8350	0.2690
30	18	1.0475	−0.0583	2.5000	1.5466	0	0
31	18	0.9820	0	5.2082	2.0767	0.0920	0.0460
32	18	0.9831	0.0448	6.5000	2.1294	0	0
33	18	0.9972	0.0728	6.3200	1.1181	0	0
34	18	1.0123	0.0550	5.0800	1.6665	0	0
35	18	1.0493	0.0979	6.5000	2.1456	0	0
36	18	1.0635	0.1450	5.6000	1.0240	0	0
37	18	1.0278	0.0419	5.4000	0.0497	0	0
38	18	1.0265	0.1359	8.3000	0.2463	0	0
39	18	1.0300	−0.1760	10.000	2.6370	11.040	−2.500

Table A.24 Base-case power flows and losses in the branches of the New England system.

Line h-k	p_{hk} [pu(MW)]	q_{hk} [pu(MVAr)]	p_{loss} [pu(MW)]	q_{loss} [pu(MVAr)]
1-2	−1.2416	−0.4106	0.0050	−0.7029
1-39	1.2416	0.4106	0.0016	0.0395
2-3	3.6468	0.9228	0.0171	−0.0784
2-25	−2.3933	0.7735	0.0412	−0.1109
2-30	−2.5000	−1.4040	0	0.1426
3-4	0.9311	1.1304	0.0030	−0.1794
3-18	−0.5214	−0.1532	0.0003	−0.2230
4-5	−1.3677	−0.0347	0.0015	−0.1109
4-14	−2.7042	−0.4954	0.0060	−0.0435
5-8	3.1737	0.6914	0.0085	−0.0281
6-5	4.5471	0.6258	0.0042	0.0106
6-7	4.2068	1.0118	0.0112	0.0587
6-11	−3.6378	−0.3923	0.0092	−0.0329
6-31	−5.1162	−1.2452	0	0.7855
7-8	1.8576	0.1131	0.0014	−0.0607
8-9	−0.1986	−0.8667	0.0012	−0.3657
9-39	−0.1997	−0.5011	0.0003	0.0070
10-11	3.6520	0.7707	0.0054	−0.0165
10-13	2.8480	0.3906	0.0032	−0.0404
10-32	−6.5000	−1.1613	0	0.9681
12-11	0.0007	−0.4200	0.0003	0.0078
12-13	−0.0757	−0.4600	0.0004	0.0096
13-14	2.7687	−0.0387	0.0067	−0.1007
14-15	0.0518	−0.3900	0.0001	−0.3740
15-16	−3.1483	−1.5459	0.0105	−0.0689
16-17	2.3004	−0.4144	0.0036	−0.0976
16-19	−5.0266	−0.5062	0.0382	0.1367
16-21	−3.2959	0.1127	0.0082	−0.1321
16-24	−0.4267	−0.9921	0.0003	−0.0666
17-18	2.1046	0.1243	0.0029	−0.1059
17-27	0.1922	−0.4411	0.0001	−0.3425
19-33	−6.2910	−0.5299	0.0290	0.5882
19-20	1.2262	−0.1130	0.0011	0.0213
20-34	−5.0549	−1.1644	0.0251	0.5021
21-22	−6.0441	−0.9051	0.0279	0.2111
22-23	0.4280	0.4208	0.0002	−0.1984
22-35	−6.5000	−1.5371	0	0.6085
23-24	3.5384	0.0178	0.0253	0.0123
23-36	−5.5857	−0.2446	0.0143	0.7794
25-26	0.7089	−0.1784	0.0015	−0.5549
25-37	−5.3834	0.5908	0.0166	0.6405
26-27	2.6275	0.6933	0.0096	−0.1603
26-28	−1.4083	−0.2247	0.0079	−0.7741
26-29	−1.9017	−0.2620	0.0191	−0.9257
28-29	−3.4762	0.2734	0.0156	−0.1065
29-38	−8.2477	0.7745	0.0523	1.0208

Table A.25 Total generation, demand and losses for the base-case operating condition of the New England system.

Type	Active power [pu(MW)]	Reactive power [pu(MVAr)]
Generation	61.408	14.640
Demand	60.971	14.091
Losses	0.4372	0.5488

Table A.26 Data of the synchronous machines of the New England system.

Machine Gen. #	v_n [kV]	M [s pu(MW)]	T'_{do} [s]	T'_{qo} [s]	X_d [pu(Ω)]	X'_d [pu(Ω)]	X_q [pu(Ω)]	X'_q [pu(Ω)]
1	18.0	84.0	10.2	0.10	0.1000	0.0310	0.0690	0.0080
2	18.0	60.6	6.56	1.50	0.2950	0.0697	0.2820	0.0697
3	18.0	71.6	5.70	1.50	0.2495	0.0531	0.2370	0.0876
4	18.0	57.2	5.69	1.50	0.2620	0.0436	0.2580	0.1660
5	18.0	52.0	5.40	0.44	0.6700	0.1320	0.6200	0.1660
6	18.0	69.6	7.30	0.40	0.2540	0.0500	0.2410	0.0814
7	18.0	52.8	5.66	1.50	0.2950	0.0490	0.2920	0.1860
8	18.0	48.6	6.70	0.41	0.2900	0.0570	0.2800	0.0911
9	18.0	69.0	4.79	1.96	0.2106	0.0570	0.2050	0.0587
10	18.0	1,000.0	7.00	0.70	0.0200	0.0060	0.0190	0.0080

Table A.27 Data of the turbines and turbine governors of the New England system.

Machine Gen. #	p^{max} [pu(MW)]	p^{min} [pu(MW)]	$\mathcal{R}$ [pu(Hz/MW)]	T_g [s]	T_{rh} [s]	T_{sm} [s]	T_t [s]	κ_{rh} −
1	4.0	0	0.0355	1.82	50.0	0.45	0.0	0.25
2	7.0	0	0.0385	6.67	50.0	0.45	0.0	0.25
3	8.0	0	0.0316	5.00	50.0	0.45	0.0	0.25
4	8.0	0	0.0316	5.00	50.0	0.45	0.0	0.25
5	7.0	0	0.0232	20.00	50.0	0.45	0.0	0.25
6	8.0	0	0.0316	5.00	50.0	0.45	0.0	0.25
7	7.0	0	0.0303	2.00	50.0	0.45	0.0	0.25
8	7.0	0	0.0303	2.00	50.0	0.45	0.0	0.25
9	10.0	0	0.0350	10.00	50.0	0.45	0.0	0.25
10	13.0	0	0.0539	25.00	50.0	0.45	0.0	0.25

Table A.28 Data of the AVRs of the New England system.

Machine	A_{ef} —	B_{ef} —	K_{a} —	K_{ef} —	K_{f} —	T_{a} [s]	T_{b} [s]	T_{c} [s]
	0	0	200.0	0	0	0.015	10.0	1.0

all	T_{ef} [s]	T_{f} [s]	T_{R} [s]	$v_{\text{a}}^{\max}$ [pu(kV)]	$v_{\text{a}}^{\min}$ [pu(kV)]
	0	0	0.01	5.0	−5.0

Table A.29 Data of the PSSs of the New England system.

Machine Gen. #	K_{w} —	T_{w} [s]	T_1 [s]	T_2 [s]	T_3 [s]	T_4 [s]	$v_{\text{s}}^{\max}$ [pu(kV)]	$v_{\text{s}}^{\min}$ [pu(kV)]
1	1.0	10.0	1.0	0.05	3.0	0.50	0.2	−0.2
2	0.5	10.0	5.0	0.40	1.0	0.10	0.2	−0.2
3	0.5	10.0	3.0	0.20	2.0	0.20	0.2	−0.2
4	2.0	10.0	1.0	0.10	1.0	0.30	0.2	−0.2
5	1.0	10.0	1.5	0.20	1.0	0.10	0.2	−0.2
6	4.0	10.0	0.5	0.10	0.5	0.05	0.2	−0.2
7	7.5	10.0	0.2	0.02	0.5	0.10	0.2	−0.2
8	2.0	10.0	1.0	0.20	1.0	0.10	0.2	−0.2
9	2.0	10.0	1.0	0.50	2.0	0.10	0.2	−0.2
10	1.0	10.0	5.0	0.60	3.0	0.50	0.2	−0.2

Appendix B
Irish Transmission System

A detailed transient-stability model of the All-Island Irish Transmission System (AIITS) is utilized in the examples discussed in Chapters 5–9. The high penetration of wind power plants makes this system an excellent test-bed to study the behavior of nonsynchronous generation and the frequency estimation techniques discussed in this book.

The topology and the static data of the system are available at the webpage of the EirGrid Group, the Irish TSO. Dynamic data are guessed based on the technologies and the capacities of conventional and renewable power plants. Although dynamic data are realistic, simulation results do not represent any operating condition of the actual Irish grid.

The model includes 1,479 buses, 1,851 transmission lines and transformers, 245 loads, 22 conventional power plants modeled with sixth-order synchronous machine models with AVRs and TGs, six PSSs, and 176 WECSs, of which 142 are equipped with DFIGs and 34 with CSWTs.

The map of the AIITS is shown in Figure B.1 [57].

Frequency Variations in Power Systems: Modeling, State Estimation, and Control, First Edition.
Federico Milano and Álvaro Ortega Manjavacas.

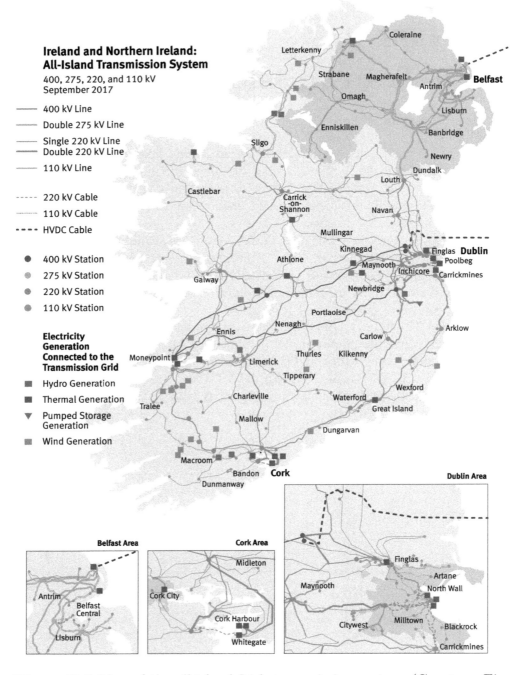

Figure B.1 Map of the all-island Irish transmission system. (Courtesy: Eir-Grid Group, www.eirgridgroup.com).

Bibliography

[1] S. M. Abdelkader, D. J. Morrow, and A. J. Conejo, 'Network usage determination using a transformer analogy,' *IET Generation, Transmission Distribution*, **8** (*1*), 81–90, Jan. 2014.

[2] A. Abur and A. Gómez Expósito, *Power System State Estimation: Theory and Applications*, ser. Power Engineering. New York, NY: CRC Press, 2004.

[3] E. Acha and M. Madrigal, *Power Systems Harmonics: Computer Modelling and Analysis*. Chichester, UK: John Wiley & Sons, 2001.

[4] V. Akhmatov, H. Knudsen, and A. H. Nielsen, 'Advanced simulation of windmills in the electric power supply,' *Electrical Power & Energy Systems*, **22** (*6*), 421–434, Aug. 2000.

[5] A. Ametani, Ed., *Numerical Analysis of Power System Transient and Dynamics*. London, UK: IET, 2015.

[6] O. Anaya Lara, F. Hughes, N. Jenkins, and G. Strbac, 'Contribution of DFIG-based wind farms to power system short-term frequency regulation,' *IEE Proceedings on Generation, Transmission and Distribution*, **153** (*2*), 164–170, Mar. 2006.

[7] P. M. Anderson, *Analysis of Faulted Power Systems*. New York, NY: Wiley-IEEE Press, 1995.

[8] P. M. Anderson and A. A. Fouad, *Power System Control and Stability*, 2nd ed. New York, NY: Wiley-IEEE Press, 2002.

[9] A. E. Angueth de Araújo and D. A. Vieira Tonidandel, 'Steinmetz and the concept of phasor: A forgotten story,' *Journal of Control, Automation and Electrical Systems*, **24** (*3*), 388–395, June 2013.

[10] J. Arrillaga and N. R. Watson, *Power System Harmonics*, 2nd ed. Chichester, UK: John Wiley & Sons, 2003.

[11] F. C. Aschmoneit and J. F. Verstege, 'An external system equivalent for on-line steady-state generator outage simulation,' *IEEE Transactions on Power Apparatus and Systems*, **PAS-98** (*3*), 770–779, May 1979.

Frequency Variations in Power Systems: Modeling, State Estimation, and Control, First Edition.
Federico Milano and Álvaro Ortega Manjavacas.
© 2020 John Wiley & Sons Ltd. Published 2020 by John Wiley & Sons Ltd.

[12] J. A. Barnes, A. R. Chi, L. S. Cutler *et al.*, 'Characterization of frequency stability,' *IEEE Transactions on Instrumentation and Measurement*, **IM-20** (*2*), 105–120, May 1971.

[13] H. M. Beides and G. T. Heydt, 'Dynamic state estimation of power system harmonics using Kalman filter methodology,' *IEEE Transactions on Power Delivery*, **6** (*4*), 1663–1670, Oct. 1991.

[14] G. L. Berg, 'Power system load representation,' *Proceedings of the IEEE*, **120** (*3*), 344–348, 1973.

[15] H. Bevrani, A. Ghosh, and G. Ledwich, 'Renewable energy sources and frequency regulation: Survey and new perspectives,' *IET Renewable Power Generation, IET*, **4** (*5*), 438–457, Sep. 2010.

[16] P. Bhui, N. Senroy, A. K. Singh, and B. C. Pal, 'Estimation of inherent governor dead-band and regulation using unscented Kalman filter,' *IEEE Transactions on Power Systems*, **33** (*4*), 3546–3558, July 2018.

[17] J. Bialek, 'Allocation of transmission supplementary charge to real and reactive loads,' *IEEE Transactions on Power Systems*, **13** (*3*), 749–754, Aug. 1998.

[18] H. Bilil and H. Gharavi, 'MMSE-based analytical estimator for uncertain power system with limited number of measurements,' *IEEE Transactions on Power Systems*, **33** (*5*), 5236–5247, Sep. 2018.

[19] F. Bizzarri, A. Brambilla, and F. Milano, 'Analytic and numerical study of TCSC devices: Unveiling the crucial role of phase-locked loops,' *IEEE Transactions on Circuits and Systems I: Regular Papers*, **65** (*6*), 1840–1849, June 2018.

[20] B. Boashash, 'Estimating and interpreting the instantaneous frequency of a signal. I. Fundamentals,' *Proceedings of the IEEE*, **80** (*4*), 520–538, Apr. 1992.

[21] B. Boashash, 'Estimating and interpreting the instantaneous frequency of a signal. II. Algorithms and applications,' *Proceedings of the IEEE*, **80** (*4*), 540–568, Apr. 1992.

[22] S. Boyd and L. Vandenberghe, *Convex Optimization*. Cambridge, UK: Cambridge University Press, 2004.

[23] K. E. Brenan, S. L. Campbell, and L. R. Petzold, *Numerical Solution of Initial-Value Problems in Differential-Algebraic Equations*. Philadelphia, PA: SIAM, 1996.

[24] X. Cai, C. Wang, and R. Kennel, 'A fast and precise grid synchronization method based on fixed-gain filter,' *IEEE Transactions on Industrial Electronics*, **65** (*9*), 7119–7128, Sep. 2018.

[25] E. Castillo, A. J. Conejo, C. Castillo, and R. Mínguez, 'Closed formulas in local sensitivity analysis for some classes of linear and non-linear problems,' *TOP*, **15** (*2*), 355–371, Dec 2007.

[26] E. Castillo, A. J. Conejo, C. Castillo *et al.*, 'Perturbation approach to sensitivity analysis in mathematical programming,' *Journal of Optimization Theory and Applications*, **128** (*1*), 49–74, Jan 2006.

[27] E. Castillo, A. S. Hadi, A. Conejo, and A. Fernández-Canteli, 'A general method for local sensitivity analysis with application to regression models and other optimization problems,' *Technometrics*, **46** (*4*), 430–444, 2004.

[28] L. M. Castro and E. Acha, 'On the provision of frequency regulation in low inertia AC grids using HVDC systems,' *IEEE Transactions on Smart Grid*, **7** (*6*), 2680–2690, Nov. 2016.

[29] A. Cataliotti, V. Cosentino, and S. Nuccio, 'A phase-locked loop for the synchronization of power quality instruments in the presence of stationary and transient disturbances,' *IEEE Transactions on Instrumentation and Measurement*, **56** (*6*), 2232–2239, Dec. 2007.

[30] F. E. Cellier and E. Kofman, *Continuous System Simulation*. London, UK: Springer, 2006.

[31] J. Cerqueira, F. Bruzzone, C. Castro *et al.*, 'Comparison of the dynamic response of wind turbine primary frequency controllers,' in *Proceedings of the IEEE PES General Meeting*, Chicago, IL, July 2017, 1–5.

[32] N. R. Chaudhuri, B. Chaudhuri, R. Majumder, and A. Yazdani, *Multi-terminal Direct-current Grids: Modeling, Analysis, and Control*. Hoboken, NJ: John Wiley & Sons, 2014.

[33] H. Chen, J. Liu, J. Giri *et al.*, 'Standard handbook for electrical engineers,' in *Standard Handbook for Electrical Engineers*, 17th ed., S. Santoso and H. Wayne Beaty, Eds. New York, NY: McGraw Hill Education, 2018, ch. 18, 1097–1168.

[34] Y. C. Chen and S. V. Dhople, 'Power divider,' *IEEE Transactions on Power Systems*, **31** (*6*), 5135–5143, Nov. 2016.

[35] M. Cheng, S. S. Sami, and J. Wu, 'Benefits of using virtual energy storage system for power system frequency response,' *Applied Energy*, **194**, 376–385, 2017.

[36] M. Ciobotaru, R. Teodorescu, and F. Blaabjerg, 'A new single-phase PLL structure based on second order generalized integrator,' in *Proceedings of the 37th IEEE Power Electronics Specialists Conference*, Jeju, South Korea, June 2006, 1–6.

[37] E. Clarke, *Circuit Analysis of AC Power Systems – Volume I: Symmetrical and Related Components*, ser. General Electric Series. New York, NY: J. Wiley & Sons, 1943.

[38] L. Cohen, 'Time-frequency distributions – A review,' *Proceedings of the IEEE*, **77** (*7*), 941–981, July 1989.

[39] A. J. Conejo, E. Castillo, R. Mínguez, and F. Milano, 'Locational marginal price sensitivities,' *IEEE Transactions on Power Systems*, **20** (*4*), 2026–2033, Nov. 2005.

[40] A. J. Conejo, J. Contreras, D. A. Lima, and A. Padilha-Feltrin, 'Z-bus transmission network cost allocation,' *IEEE Transactions on Power Systems*, **22** (*1*), 342–349, Feb. 2007.

[41] A. J. Conejo, F. D. Galiana, and I. Kockar, 'Z-bus loss allocation,' *IEEE Transactions on Power Systems*, **16** (*1*), 105–110, Feb. 2001.

[42] P. M. Costa and M. A. Matos, 'Loss allocation in distribution networks with embedded generation,' *IEEE Transactions on Power Systems*, **19** (*1*), 384–389, Feb. 2004.

[43] A. M. L. da Silva, J. G. de Carvalho Costa, and L. H. Lopes Lima, 'A new methodology for cost allocation of transmission systems in interconnected energy markets,' *IEEE Transactions on Power Systems*, **28** (*2*), 740–748, May 2013.

[44] J. C. Das, *Power System Harmonics and Passive Filter Designs*. Hoboken, NJ: IEEE Press – John Wiley & Sons, 2015.

[45] I. K. Dassios, P. Cuffe, and A. Keane, 'Visualizing voltage relationships using the unity row summation and real valued properties of the FLG matrix,' *Electric Power Systems Research*, **140**, 611–618, 2016.

[46] S. Deckmann, A. Pizzolante, A. Monticelli *et al.*, 'Studies on power system load flow equivalencing,' *IEEE Transactions on Power Apparatus and Systems*, **PAS-99** (*6*), 2301–2310, November/December 1980.

[47] A. Derviškadić, P. Romano, and M. Paolone, 'Iterative-interpolated DFT for synchrophasor estimation: A single algorithm for P- and M-class compliant PMUs,' *IEEE Transactions on Instrumentation and Measurement*, **67** (*3*), 547–558, Mar. 2018.

[48] DIgSILENT, *PowerFactory Technical Reference Ver. 15*, Gomaringen, Germany, 2015.

[49] P. Dimo, *Nodal Analysis of Power Systems*. Kent, UK: Abacus Press, 1975.

[50] Q. Ding and A. Abur, 'Transmission loss allocation based on a new quadratic loss expression,' *IEEE Transactions on Power Systems*, **21** (*3*), 1227–1233, Aug. 2006.

[51] H. W. Dommel, *Electro-Magnetic Transient Program (EMTP) – Theory Book*. Portland, OR: Booneville Power Administration, 1994.

[52] P. Du and Y. Makarov, 'Using disturbance data to monitor primary frequency response for power system interconnections,' *IEEE Transactions on Power Systems*, **29** (*3*), 1431–1432, May 2014.

[53] P. Du and J. Matevosyan, 'Forecast system inertia condition and its impact to integrate more renewables,' *IEEE Transactions on Smart Grid*, **9** (*2*), 1531–1533, Mar. 2018.

[54] W. C. Duesterhoeft, M. W. Schulz, and E. Clarke, 'Determination of instantaneous currents and voltages by means of alpha, beta, and zero components,' *Transactions of the American Institute of Electrical Engineers*, **70** (*2*), 1248–1255, July 1951.

[55] T. E. Dy Liacco, S. C. Suvulesco, and K. A. Ramarao, 'An on-line topological equivalent of a power system,' *IEEE Transactions on Power Apparatus and Systems*, **PAS-97** (*5*), Sep. 1978.

[56] S. Eftekharnejad, V. Vittal, G. T. Heydt *et al.*, 'Impact of increased penetration of photovoltaic generation on power systems,' *IEEE Transactions on Power Systems*, **28** (*2*), 893–901, May 2013.

[57] EirGrid Group, 'All island Irish transmission system map.' [Online]. Available:
`http://smartgriddashboard.eirgrid.com/#all/transmission-map`

[58] EirGrid, SONI, 'Quarterly wind dispatch down report user guide.' [Online]. Available: `www.eirgridgroup.com`

[59] J. Ekanayake and N. Jenkins, 'Comparison of the response of doubly fed and fixed-speed induction generator wind turbines to changes in network frequency,' *IEEE Transactions on Energy Conversion*, **19** (*4*), 800–802, Dec. 2004.

[60] K. Emami, T. Fernando, H. H. C. Iu *et al.*, 'Application of unscented transform in frequency control of a complex power system using noisy PMU data,' *IEEE Transactions on Industrial Informatics*, **12** (*2*), 853–863, Apr. 2016.

[61] ENTSO-E AISBL, 'Continental Europe significant frequency deviations – January 2019,' ENTSO-E, Tech. Rep., Apr. 2019. [Online]. Available: `https://bit.ly/2Wh4Hvt`

[62] ERCOT Concept Paper, 'Future ancillary services in ERCOT,' 2013.

[63] A. M. Ersdal, L. Imsland, and K. Uhlen, 'Model predictive load-frequency control,' *IEEE Transactions on Power Systems*, **31** (*1*), 777–785, Jan. 2016.

[64] N. Etherden, V. Vyatkin, and M. H. J. Bollen, 'Virtual power plant for grid services using IEC 61850,' *IEEE Transactions on Industrial Informatics*, **12** (*1*), 437–447, Feb. 2016.

[65] J. H. Eto, J. Undrill, P. Mackin *et al.*, 'Use of frequency response metrics to assess the planning and operating requirements for reliable integration of variable renewable generation,' Lawrence Berkeley National Laboratory, Berkeley, Tech. Rep. LBNL-4145E, 2010.

[66] D. Fabozzi and T. Van Cutsem, 'On angle references in long-term time-domain simulations,' *IEEE Transactions on Power Systems*, **26** (*1*), 483–484, Feb. 2011.

[67] E. Farantatos, G. K. Stefopoulos, G. J. Cokkinides, and A. P. Meliopoulos, 'PMU-based dynamic state estimation for electric power systems,' in *Proceedings of the IEEE PES General Meeting*, Calgary, AB, July 2009, 1–8.

[68] J. A. Fleming, *The Alternate Current Transformer in Theory and Practice – Volume I: The Induction of Electric Currents*. London, UK: 'The Electrician' Printing and Publishing Company, 1896.

[69] C. L. Fortescue, 'Method of symmetrical co-ordinates applied to the solution of polyphase networks,' *Transactions of the American Institute of Electrical Engineers*, **37** (*2*), 1027–1140, June 1918.

[70] A. Fradi, S. Brignone, and B. E. Wollenberg, 'Calculation of energy transaction allocation factors,' *IEEE Transactions on Power Systems*, **16** (*2*), 266–272, May 2001.

[71] G. Frigo, A. Derviškadić, Y. Zuo, and M. Paolone, 'PMU-based RO-COF measurements: Uncertainty limits and metrological significance in power system applications,' *IEEE Transactions on Instrumentation and Measurement*, 1–1, 2019, preprint available at ieeexplore.org.

[72] M. Frigo and S. G. Johnson, 'The design and implementation of FFTW3,' *Proceedings of the IEEE*, **93** (*2*), 216–231, 2005, special issue on "Program Generation, Optimization, and Platform Adaptation."

[73] B. Gao, G. K. Morison, and P. Kundur, 'Voltage stability evaluation using modal analysis,' *IEEE Transactions on Power Systems*, **7** (*4*), 1529–1542, Nov. 1992.

[74] E. Ghahremani and I. Kamwa, 'Dynamic state estimation in power system by applying the extended Kalman filter with unknown inputs to phasor measurements,' *IEEE Transactions on Power Systems*, **26** (*4*), 2556–2566, Nov. 2011.

[75] E. Ghahremani and I. Kamwa, 'Online state estimation of a synchronous generator using unscented Kalman filter from phasor measurements units,' *IEEE Transactions on Energy Conversion*, **26** (*4*), 1099–1108, Dec. 2011.

[76] E. Ghahremani and I. Kamwa, 'Local and wide-area PMU-based decentralized dynamic state estimation in multi-machine power systems,' *IEEE Transactions on Power Systems*, **31** (*1*), 547–562, Jan. 2016.

[77] C. K. Gharban and B. J. Cory, 'Non-linear dynamic power system state estimation,' *IEEE Transactions on Power Systems*, **1** (*3*), 276–283, Aug. 1986.

[78] M. J. Gibbard, O. Pourbeik, and D. J. Vowles, *Small-signal Stability, Control and Dynamic Performance of Power Systems*. Adelaide, Australia: University of Adelaide Press, 2015.

[79] D. T. Gillespie, 'Exact numerical simulation of the Ornstein-Uhlenbeck process and its integral,' *Physical Review E*, **54** (*2*), 2084–2091, 1996.

[80] J. D. Glover, M. S. Sarma, and T. Overbye, *Power System Analysis and Design*. Boston, MA: Cengage Learning, 2011.

[81] Ö. Göksu, R. Teodorescu, C. L. Bak *et al.*, 'Instability of wind turbine converters during current injection to low voltage grid faults and PLL frequency based stability solution,' *IEEE Transactions on Power Systems*, **29** (*4*), 1683–1691, July 2014.

[82] M. Göl and A. Abur, 'A fast decoupled state estimator for systems measured by PMUs,' *IEEE Transactions on Power Systems*, **30** (*5*), 2766–2771, Sep. 2015.

[83] D. J. Griffiths and D. F. Schroeter, *Introduction to Quantum Mechanics*, 3rd ed. Cambridge, UK: Cambridge University Press, 2018.

[84] S. Heier, *Grid Integration of Wind Energy Conversion Systems*. Chichester, UK: John Wiley & Sons, 1998.

[85] P. Hirsch, *Extended Transient-Midterm Stability Program (ETMSP) Ver. 3.1: User's Manual*, EPRI, TR-102004-V2R1, May 1994.

[86] I. A. Hiskens, 'Power system modeling for inverse problems,' *IEEE Transactions on Circuits and Systems - I: Regular Papers*, **51** (*3*), 539–551, Mar. 2004.

[87] I. A. Hiskens, 'Dynamics of Type-3 wind turbine generator models,' *IEEE Transactions on Power Systems*, **27** (*1*), 465–474, Feb. 2012.

[88] G.-C. Hsieh and J. C. Hung, 'Phase-locked loop techniques – a survey,' *IEEE Transactions on Industrial Electronics*, **43** (*6*), 609–615, Dec. 1996.

[89] C.-S. Hsu, M.-S. Chen, and W. Lee, 'Approach for bus frequency estimating in power system simulations,' *IEE Proceedings-Generation, Transmission and Distribution*, **145** (*4*), 431–435, July 1998.

[90] J. Hu, S. Wang, W. Tang, and X. Xiong, 'Full-capacity wind turbine with inertial support by adjusting phase-locked loop response,' *IET Renewable Power Generation*, **11** (*1*), 44–53, 2017.

[91] S. J. Huang and K. R. Shih, 'Dynamic-state-estimation scheme including nonlinear measurement-function considerations,' *IEE Proceedings on Generation, Transmission and Distribution*, **149** (*6*), 673–678, Nov. 2002.

[92] Z. Huang, P. Du, D. Kosterev, and S. Yang, 'Generator dynamic model validation and parameter calibration using phasor measurements at the point of connection,' *IEEE Transactions on Power Systems*, **28** (*2*), 1939–1949, May 2013.

[93] Z. Huang, K. Schneider, and J. Nieplocha, 'Feasibility studies of applying Kalman filter techniques to power system dynamic state estimation,' in *Proceedings of the International Power Engineering Conference (IPEC 2007)*, Singapore, Dec. 2007, 376–382.

[94] IEEE C37.118.1-2011, *IEEE Standard for Synchrophasor Measurements for Power Systems*, IEEE Std., 2011.

[95] IEEE Committee Report, 'Recommended phasor diagram for synchronous machines,' *IEEE Transactions on Power Apparatus and Systems*, **PAS-88** (*11*), 1593–1610, Nov. 1969.

[96] IEEE Power System Dynamic Performance Committee, 'Dynamic Models Turbine-Governors in Power System Studies,' IEEE, Tech. Rep., 2013, PES-TR1.

[97] IEEE Task Force on Load Representation for Dynamic Performance, 'Load representation for dynamic performance analysis [of power systems],' *IEEE Transactions on Power Systems*, **8** (*2*), 472–482, May 1993.

[98] IEEE Working Group on Computer Modelling of Excitation Systems, 'Excitation system models for power system stability studies,' *IEEE Transactions on Power Apparatus and Systems*, **100** (*2*), 494–509, Feb. 1981.

[99] M. Ilić and J. Zaborszky, *Dynamic and Control of Large Electric Power Systems*. New York, NY: Wiley-Interscience Publication, 2000.

[100] G. M. Jónsdóttir and F. Milano, 'Data-based continuous wind speed models with arbitrary probability distribution and autocorrelation,' *Renewable Energy*, **143**, 368–376, 2019.

[101] W. Juanjuan, F. Chuang, and Z. Yao, 'Design of WAMS-based multiple HVDC damping control system,' *IEEE Transactions on Smart Grid*, **2** (*2*), 363–374, June 2011.

[102] M. Karimi Ghartemani and M. R. Iravani, 'Robust and frequency-adaptive measurement of peak value,' *IEEE Transactions on Power Delivery*, **19** (*2*), 481–489, Apr. 2004.

[103] P. Kessel and H. Glavitsch, 'Estimating the voltage stability of a power system,' *IEEE Transactions on Power Delivery*, **1** (*3*), 346–354, July 1986.

[104] E. W. Kimbark, *Power System Stability – Vol. I – Elements of Stability Calculations*. New York, NY: John Wiley & Sons, 1948.

[105] H. Kirkham, W. Dickerson, and A. Phadke, 'Defining power system frequency,' in *Proceedings of the IEEE PES General Meeting*, Portland, OR, Aug. 2018, 1–5.

[106] D. Kirschen, R. Allan, and G. Strbac, 'Contributions of individual generators to loads and flows,' *IEEE Transactions on Power Systems*, **12** (*1*), 52–60, Feb. 1997.

[107] S. K. M. Kodsi and C. A. Cañizares, 'Modeling and simulation of IEEE 14-bus system with FACTS controllers,' University of Waterloo, Waterloo, Tech. Rep. 2003-3, Mar. 2003.

[108] D. Koraki and K. Strunz, 'Wind and solar power integration in electricity markets and distribution networks through service-centric virtual power plants,' *IEEE Transactions on Power Systems*, **33** (*1*), 473–485, Jan. 2018.

[109] E. Koutroulis and K. Kalaitzakis, 'Design of a maximum power tracking system for wind-energy-conversion applications,' *IEEE Transactions on Industrial Electronics*, **53** (*2*), 486–494, Apr. 2006.

[110] P. Krause, O. Wasynczuk, S. Sudhoff, and S. Pekarek, *Analysis of Electric Machinery and Drive Systems*, 3rd ed., ser. IEEE Press Series on Power Engineering. Hoboken, NJ: John Wiley & Sons, 2013.

[111] Y. H. Ku, 'Transient analysis of A-C. machinery,' *Transactions of the American Institute of Electrical Engineers*, **48** (*3*), 707–714, July 1929.

[112] S. Kundu, N. Sinitsyn, I. Hiskens, and S. Backhaus, 'Modeling and control of thermostatically controlled loads,' in *Proceedings of the Power Systems Computation Conference (PSCC)*, Stockholm, Sweden, Aug. 2011.

[113] P. Kundur, *Power System Stability and Control*. New York, NY: McGraw-Hill, 1994.

[114] W. W. Lewis, 'Short circuit currents on grounded neutral systems,' *General Electric Review*, 524–527, June 1917.

[115] G. Li, S. Cheng, J. Wen *et al.*, 'Power system stability enhancement by a double-fed induction machine with a flywheel energy storage system,' in *Proceedings of the IEEE PES General Meeting*, Montreal, Quebec, 2006.

[116] C. Liu, A. Bose, and P. Tian, 'Modeling and analysis of HVDC converter by three-phase dynamic phasor,' *IEEE Transactions on Power Delivery*, **29** (*1*), 3–12, Feb. 2014.

[117] M. Liu, I. Dassios, G. Tzounas, and F. Milano, 'Stability analysis of power systems with inclusion of realistic-modeling WAMS delays,' *IEEE Transactions on Power Systems*, **34** (*1*), 627–636, Jan. 2019.

[118] M. Liu, Á. Ortega, and F. Milano, 'PMU-based estimation of the frequency of the center of inertia and generator rotor speeds,' in *Proceedings of the IEEE PES General Meeting*, Atlanta, GA, Aug. 2019.

[119] Lu Miao, Jinyu Wen, Hailian Xie *et al.*, 'Coordinated control strategy of wind turbine generator and energy storage equipment for frequency support,' *IEEE Transactions on Industry Applications*, **51** (*4*), 2732–2742, July 2015.

[120] J. Machowski, J. W. Bialek, and J. R. Bumby, *Power System Dynamics and Stability*, 2nd ed. New York, NY: John Wiley & Sons, 2008.

[121] B. Mandelbrot, 'How long is the coast of Britain? Statistical self-similarity and fractional dimension,' *Science*, **156** (*3775*), 636–638, 1967.

[122] D. P. Mandic, S. Kanna, Y. Xia *et al.*, 'A data analytics perspective of power grid analysis – Part 1: The Clarke and related transforms [lecture notes],' *IEEE Signal Processing Magazine*, **36** (*2*), 110–116, March 2019.

[123] R. Marconato, *Electric Power Systems*. Milano, Italy: CEI, Italian Electrotechnical Committee, 2002, **2**.

[124] J. L. Mathieu, S. Koch, and D. S. Callaway, 'State estimation and control of electric loads to manage real-time energy imbalance,' *IEEE Transactions on Power Systems*, **28** (*1*), 430–440, 2013.

[125] J. M. Mauricio, A. Marano, A. Gómez Expósito, and J. L. Martínez Ramos, 'Frequency regulation contribution through variable-speed wind energy conversion systems,' *IEEE Transactions on Power Systems*, **24** (*1*), 173–180, Feb. 2009.

[126] S. Mehrotra, 'On the implementation of a primal-dual interior point method,' *SIAM Journal on Optimization*, **2** (*4*), 575–601, 1992.

[127] F. Milano, *Power System Modelling and Scripting*. London, UK: Springer, 2010.

[128] F. Milano, 'A Python-based software tool for power system analysis,' in *Proceedings of the IEEE PES General Meeting*, Vancouver, BC, July 2013.

[129] F. Milano, 'Semi-implicit formulation of differential-algebraic equations for transient stability analysis,' *IEEE Transactions on Power Systems*, **31** (*6*), 4534–4543, Nov. 2016.

[130] F. Milano, A. J. Conejo, and R. Zárate Miñano, 'General sensitivity formulas for maximum loading conditions in power systems,' *IET Proceedingss on Generation, Transmission and Distribution*, **1** (*3*), 516–526, May 2007.

[131] F. Milano, F. Dörfler, G. Hug *et al.*, 'Foundations and challenges of low-inertia systems,' in *Proceedings of the Power Systems Computation Conference (PSCC)*, Dublin, Ireland, June 2018, 1–22.

[132] F. Milano and Á. Ortega, *Converter-Interfaced Energy Storage Systems: Context, Modelling and Dynamic Analysis*. Cambridge, UK: Cambridge University Press, 2019.

[133] F. Milano, Á. Ortega, and A. J. Conejo, 'Model-agnostic linear estimation of generator rotor speeds based on phasor measurement units,' *IEEE Transactions on Power Systems*, **33** (*6*), 7258–7268, Nov. 2018.

[134] F. Milano and R. Zárate Miñano, 'A systematic method to model power systems as stochastic differential algebraic equations,' *IEEE Transactions on Power Systems*, **28** (*4*), 4537–4544, Nov. 2013.

[135] F. Milano, Ed., *Advances in Power System Modelling, Control and Stability Analysis*, ser. Energy Engineering. Institution of Engineering and Technology, 2016.

[136] F. Milano and Á. Ortega, 'Frequency makers vs frequency takers,' *IEEE Transactions on Power Systems*, 2019, under review.

[137] N. Miller, D. Lew, and R. Piwko, 'Technology capabilities for fast frequency response,' GE Energy Consulting, Tech. Rep., 2017.

[138] R. Mínguez and A. J. Conejo, 'State estimation sensitivity analysis,' *IEEE Transactions on Power Systems*, **22** (*3*), 1080–1091, Aug. 2007.

[139] Minyuan Guan, Wulue Pan, Jing Zhang *et al.*, 'Synchronous generator emulation control strategy for voltage source converter (VSC) stations,' *IEEE Transactions on Power Systems*, **30** (*6*), 3093–3101, Nov. 2015.

[140] S. Mohagheghi, R. H. Alaileh, G. Cokkinides, and A. P. S. Meliopoulos, 'Distributed state estimation based on the supercalibrator concept - laboratory implementation,' in *Proceedings of the IREP Symposium - Bulk Power System Dynamics and Control - VII. Revitalizing Operational Reliability*, Charleston, SC, Aug. 2007, 1–9.

[141] N. Mohan, T. M. Undeland, and W. P. Robbins, *Power Electronics: Converters, Applications and Design*, 3rd ed. New York, NY: John Wiley & Sons, 2003.

[142] A. Monti, F. Milano, E. Bompard, and X. Guillaud, Eds., *Converter-based dynamics and control of modern power systems*. London, UK: Academic Press, 2020.

[143] A. Monticelli, *State Estimation in Electric Power Systems: A Generalized Approach*, ser. Power Electronics and Power Systems. Norwell, MA: Springer, 1999.

[144] G. K. Morison, B. Gao, and P. Kundur, 'Voltage stability analysis using static and dynamic approaches,' *IEEE Transactions on Power Systems*, **8** (*3*), 1159–1171, Aug. 1993.

[145] P. Moutis, A. Vassilakis, A. Sampani, and N. Hatziargyriou, 'DC switch driven active power output control of photovoltaic inverters for the provision of frequency regulation,' *IEEE Transactions on Sustainable Energy*, **6** (*4*), 1485–1493, Oct. 2015.

[146] M. A. A. Murad and F. Milano, 'Modelling and simulation of PI-controllers limiters for the dynamic analysis of VSC-based devices,' *IEEE Transactions on Power Systems*, 1–1, 2019, preprint available at ieeexplore.org.

[147] M. A. A. Murad, B. Hayes, and F. Milano, 'Application of Filippov theory to the IEEE Standard 421.5-2016 anti-windup PI controller,' in *Proceedings of the IEEE PowerTech*, Milan, Italy, June 2019, 1–6.

[148] C. Murphy and A. Keane, 'Local and remote estimations using fitted polynomials in distribution systems,' *IEEE Transactions on Power Systems*, **32** (*4*), 3185–3194, July 2017.

[149] G. Neidhöfer, 'The evolution of the synchronous machine,' *Engineering Science & Education Journal*, **1**, 239–248(9), Oct. 1992.

[150] A. Nicastri and A. Nagliero, 'Comparison and evaluation of the PLL techniques for the design of the grid-connected inverter systems,' in *Proceedings of the IEEE International Symposium on Industrial Electronics*, July 2010, 3865–3870.

[151] M. V. A. Nunes, J. A. P. Lopes, H. H. Zurn *et al.*, 'Influence of the variable-speed wind generators in transient stability margin of the conventional generators integrated in electrical grids,' *IEEE Transactions on Energy Conversion*, **19** (*4*), 692–701, Dec. 2004.

[152] J. Nutaro and V. Protopopescu, 'Calculating frequency at loads in simulations of electro-mechanical transients,' *IEEE Transactions on Smart Grid*, **3** (*1*), 233–240, Mar. 2012.

[153] M. L. Oatts, S. R. Erwin, and J. L. Hart, 'Application of the REI equivalent for operations planning analysis of interchange schedules,' *IEEE Transactions on Power Systems*, **5** (*2*), 547–555, May 1990.

[154] D. Ochoa and S. Martínez, 'Fast-frequency response provided by DFIG-wind turbines and its impact on the grid,' *IEEE Transactions on Power Systems*, **32** (*5*), 4002–4011, Sep. 2017.

[155] A. V. Oppenheim and R. W. Schafer, *Discrete-time Signal Processing*. Englewood Cliffs NJ: Prentice-Hall, 1989.

[156] Á. Ortega and F. Milano, 'Modeling, simulation, and comparison of control techniques for energy storage systems,' *IEEE Transactions on Power Systems*, **32** (*3*), 2445–2454, May 2017.

[157] Á. Ortega and F. Milano, 'Comparison of different PLL implementations for frequency estimation and control,' in *Proceedings of the 18th International Conference on Harmonics and Power Quality (ICHPQ)*, Ljubljana, Slovenia, May 2018.

[158] Á. Ortega and F. Milano, 'Frequency participation factors,' *IEEE Transactions on Power Systems*, **33** (*5*), 5563–5571, Sep. 2018.

[159] Á. Ortega and F. Milano, 'Combined frequency and RoCoF control of converter-interfaced energy storage systems,' in *Proceedings of the IFAC Workshop on Control of Smart Grid and Renewable Energy Systems (CS-GRES)*, Jeju, South Korea, June 2019, 1–6.

[160] Á. Ortega and F. Milano, 'Design of a control limiter to improve the dynamic response of energy storage systems,' in *Proceedings of the IEEE PES General Meeting*, Denver, CO, July 2015.

[161] Á. Ortega and F. Milano, 'Comparison of bus frequency estimators for power system transient stability analysis,' in *Proceedings of the POWERCON Conference*, Wollongong, Australia, Sep. 2016.

[162] Á. Ortega and F. Milano, 'Generalized model of VSC-based energy storage systems for transient stability analysis,' *IEEE Transactions on Power Systems*, **31** (*5*), 3369–3380, Sep. 2016.

[163] Á. Ortega and F. Milano, 'Impact of frequency estimation for VSC-based devices with primary frequency control,' in *Proceedings of the IEEE PES Innovative Smart Grid Technologies Europe (ISGT Europe)*, Torino, Italy, Sep. 2017.

[164] Á. Ortega, A. Musa, A. Monti, and F. Milano, 'Hardware-in-the-loop validation of the frequency divider formula,' in *Proceedings of the IEEE PES General Meeting*, Portland, OR, Aug. 2018, 1–5.

[165] Á. Ortega and F. Milano, 'Frequency control of distributed energy resources in distribution networks,' in *Proceedings of the 10th IFAC Symposium on Control of Power and Energy Systems CPES 2018*, Tokyo, Japan, Sep. 2018.

[166] J. Pan, Y. Teklu, S. Rahman, and K. Jun, 'Review of usage-based transmission cost allocation methods under open access,' *IEEE Transactions on Power Systems*, **15** (*4*), 1218–1224, Nov. 2000.

[167] L. Papangelis, M. S. Debry, P. Panciatici, and T. V. Cutsem, 'Coordinated supervisory control of multi-terminal HVDC grids: A model predictive control approach,' *IEEE Transactions on Power Systems*, **32** (*6*), 4673–4683, Nov. 2017.

[168] R. H. Park, 'Two-reaction theory of synchronous machines generalized method of analysis – Part I,' *Transactions of the American Institute of Electrical Engineers*, **48** (*3*), 716–727, July 1929.

[169] M. Pavella, D. Ernst, and D. Ruiz Vega, *Transient Stability of Power Systems – A Unified approach to Assessment and Control*. Boston, MA: Kluwer Academic Publishers, 2000.

[170] M. Perninge, V. Knazkins, M. Amelin, and L. Söder, 'Risk estimation of critical time to voltage instability induced by saddle-node bifurcation,' *IEEE Transactions on Power Systems*, **25** (*3*), 1600–1610, Aug. 2010.

[171] A. G. Phadke and B. Kasztenny, 'Synchronized phasor and frequency measurement under transient conditions,' *IEEE Transactions on Power Delivery*, **24** (*1*), 89–95, Jan. 2009.

[172] A. G. Phadke, J. S. Thorp, and M. G. Adamiak, 'A new measurement technique for tracking voltage phasors, local system frequency, and rate of change of frequency,' *IEEE Transactions on Power Apparatus and Systems*, **PAS-102** (*5*), 1025–1038, May 1983.

[173] R. Preece, J. V. Milanović, A. M. Almutairi, and O. Marjanovic, 'Damping of inter-area oscillations in mixed AC/DC networks using WAMS based supplementary controller,' *IEEE Transactions on Power Systems*, **28** (*2*), 1160–1169, May 2013.

[174] T. J. Price, 'James Blyth – Britain's first modern wind power pioneer,' *Wind Engineering*, **29** (*3*), 191–200, 2005.

[175] J. Qiu, Y. Wei, H. R. Karimi, and H. Gao, 'Reliable control of discrete-time piecewise-affine time-delay systems via output feedback,' *IEEE Transactions on Reliability*, **67** (*1*), 79–91, Mar. 2018.

[176] G. Ramtharan, J. Ekanayake, and N. Jenkins, 'Frequency support from doubly fed induction generator wind turbines,' *IET Renewable Power Generation*, **1** (*1*), 3–9, Mar. 2007.

[177] P. Rodríguez, A. Luna, I. Candela *et al.*, 'Multiresonant frequency-locked loop for grid synchronization of power converters under distorted grid conditions,' *IEEE Transactions on Industrial Electronics*, **58** (*1*), 127–138, Jan. 2011.

[178] P. Rodríguez, A. Luna, M. Ciobotaru *et al.*, 'Advanced grid synchronization system for power converters under unbalanced and distorted operating conditions,' in *Proceedings of the 32nd Annual Conference on IEEE Industrial Electronics (IECON)*, Paris, France, Nov. 2006, 5173–5178.

[179] A. Rouhani and A. Abur, 'Linear phasor estimator assisted dynamic state estimation,' *IEEE Transactions on Smart Grid*, **9** (*1*), 211–219, Jan. 2018.

[180] S. Ruthe, C. Rehtanz, and S. Lehnhoff, 'Towards frequency control with large scale virtual power plants,' in *Proceedings of the IEEE PES Innovative Smart Grid Technologies Europe (ISGT Europe)*, Berlin, Germany, Oct. 2012, 1–6.

[181] L. Ruttledge and D. Flynn, 'Emulated inertial response from wind turbines: Gain scheduling and resource coordination,' *IEEE Transactions on Power Systems*, **31** (*5*), 3747–3755, Sep. 2016.

[182] L. Ruttledge, N. W. Miller, J. O'Sullivan, and D. Flynn, 'Frequency response of power systems with variable speed wind turbines,' *IEEE Transactions on Sustainable Energy*, **3** (*4*), 683–691, Oct. 2012.

[183] K. Samarakoon, J. Ekanayake, and N. Jenkins, 'Investigation of domestic load control to provide primary frequency response using smart meters,' *IEEE Transactions on Smart Grid*, **3** (*1*), 282–292, Mar. 2012.

[184] G. Sánchez Ayala, V. Centeno, and J. Thorp, 'Gain scheduling with classification trees for robust centralized control of PSSs,' *IEEE Transactions on Power Systems*, **31** (*3*), 1933–1942, May 2016.

[185] S. Sarri, L. Zanni, M. Popovic *et al.*, 'Performance assessment of linear state estimators using synchrophasor measurements,' *IEEE Transactions on Instrumentation and Measurement*, **65** (*3*), 535–548, Mar. 2016.

[186] P. W. Sauer and M. A. Pai, *Power System Dynamics and Stability*. Upper Saddle River, NJ: Prentice Hall, 1998.

[187] A. Semlyen, 'Analysis of disturbance propagation in power systems based on a homogeneous dynamic model,' *IEEE Transactions on Power Apparatus and Systems*, **PAS-93** (*2*), 676–684, Mar. 1974.

[188] A. Semlyen, 'Effect of nonuniformity on the continuous representation of electromechanical dynamics in large power systems,' *IEEE Power Engineering Review*, **18** (*5*), 60–61, May 1998.

[189] R. Shah, N. Mithulananthan, and K. Y. Lee, 'Large-scale PV plant with a robust controller considering power oscillation damping,' *IEEE Transactions on Energy Conversion*, **28** (*1*), 106–116, Mar. 2013.

[190] 'PSS®E 33.5 Program Application Guide,' Siemens Power Technologies International, Oct. 2013.

[191] A. M. Simoes, D. C. Savelli, P. C. Pellanda *et al.*, 'Robust design of a TCSC oscillation damping controller in a weak 500-kV interconnection considering multiple power flow scenarios and external disturbances,' *IEEE Transactions on Power Systems*, **24** (*1*), 226–236, Feb. 2009.

[192] A. K. Singh and B. C. Pal, 'Decentralized dynamic state estimation in power systems using unscented transformation,' *IEEE Transactions on Power Systems*, **29** (*2*), 794–804, Mar. 2014.

[193] J. G. Slootweg, H. Polinder, and W. L. Kling, 'Dynamic modelling of a wind turbine with doubly fed induction generator,' in *Proceedings of the IEEE PES Summer Meeting*, **1**, Vancouver, BC, July 2001, 644–649 vol.1.

[194] W. L. Snyder, 'Load flow equivalent circuits. An overview,' in *Proceedings of the IEEE PES Winter Meeting*, New York, NY, 1972.

[195] S. E. Z. Soudjani and A. Abate, 'Aggregation of thermostatically controlled loads by formal abstractions,' in *Proceedings of the Control Conference (ECC), 2013 European*, 2013, 4232–4237.

[196] A. M. Stanković and T. Aydin, 'Analysis of asymmetrical faults in power systems using dynamic phasors,' *IEEE Transactions on Power Systems*, **15** (*3*), 1062–1068, Aug. 2000.

[197] C. P. Steinmetz and E. J. Berg, *Theory and Calculation of Alternating Current Phenomena*. New York, US: W. J. Johnston Co., 1897.

[198] H. Stockman and G. Hok, 'A note on frequency modulation terminology,' *Proceedings of the Institute of Radio Engineers*, **32** (*3*), p. 181, Mar. 1944.

[199] B. Stott, J. Jardim, and O. Alsac, 'DC power flow revisited,' *IEEE Transactions on Power Systems*, **24** (*3*), 1290–1300, Aug. 2009.

[200] M. Swierczynski, D. Stroe, A.-I. Stan *et al.*, 'Selection and performance-degradation modeling of $LiMO_2/Li_4$ Ti_5 O_{12} and $LiFePO_4/C$ battery cells as suitable energy storage systems for grid integration with wind power plants: An example for the primary frequency regulation service,' *IEEE Transactions on Sustainable Energy*, **5** (*1*), 90–101, Jan. 2014.

[201] B. Tamimi, C. Cañizares, and K. Bhattacharya, 'Modeling and performance analysis of large solar photo-voltaic generation on voltage stability and inter-area oscillations,' in *Proceedings of the IEEE PES General Meeting*, July 2011, 1–6.

[202] B. Tamimi, C. Cañizares, and K. Bhattacharya, 'System stability impact of large-scale and distributed solar photovoltaic generation: The case of Ontario, Canada,' *IEEE Transactions on Sustainable Energy*, **4** (*3*), 680–688, July 2013.

[203] C. J. Tavora and O. J. M. Smith, 'Characterization of equilibrium and stability in power systems,' *IEEE Transactions on Power Apparatus and Systems*, **PAS-91** (*3*), 1127–1130, May 1972.

[204] C. Taylor and S. Lefebvre, 'HVDC controls for system dynamic performance,' *IEEE Transactions on Power Systems*, **6** (*2*), 743–752, May 1991.

[205] R. Teodorescu, M. Liserre, and P. Rodríguez, *Grid Converters for Photovoltaic and Wind Power Systems*. Chichester, UK: Wiley, 2011.

[206] D. Thukaram and C. Vyjayanthi, 'Relative electrical distance concept for evaluation of network reactive power and loss contributions in a deregulated system,' *IET Generation, Transmission Distribution*, **3** (*11*), 1000–1019, Nov. 2009.

[207] P. Tielens and D. Van Hertem, 'The relevance of inertia in power systems,' *Renewable and Sustainable Energy Reviews*, **55**, 999–1009, 2016.

[208] M. M. Tiller, *Introduction to Physical Modeling with Modelica*. Boston, MA: Kluwer Academic Publishers, 2001.

[209] W. F. Tinney and W. L. Powell, 'The REI approach to power network equivalents,' in *Proceedings of the Power Industry Computer Applications Conference*, Toronto, ON, May 1977, 314–320.

[210] L. Toma, M. Sanduleac, S. A. Baltac *et al.*, 'On the virtual inertia provision by BESS in low inertia power systems,' in *Proceedings of the IEEE International Energy Conference (ENERGYCON)*, Limassol, Chipre, June 2018, 1–6.

[211] 'Eurostag User Manual,' Tractebel Engineering GDF SUEZ – RTE. Available: at `www.eurostag.be`.

[212] A. Ulbig, T. S. Borsche, and G. Andersson, 'Impact of low rotational inertia on power system stability and operation,' in *Proceedings of the 19th IFAC World Conference*, **47**, Cape Town, South Africa, Aug. 2014, 7290–7297.

[213] T. Van Cutsem and C. Vournas, *Voltage Stability of Electric Power Systems*. New York, NY: Springer Science, 1998.

[214] L. Vanfretti, J. H. Chow, S. Sarawgi, and B. Fardanesh, 'A phasor-data-based state estimator incorporating phase bias correction,' *IEEE Transactions on Power Systems*, **26** (*1*), 111–119, Feb. 2011.

[215] V. A. Venikov, *Transient Processes in Electrical Power Systems*. Moscow, Russia: Mir Publishers, 1977.

[216] J. C. M. Vieira, W. Freitas, W. Xu, and A. Morelato, 'Efficient coordination of ROCOF and frequency relays for distributed generation protection by using the application region,' *IEEE Transactions on Power Delivery*, **21** (*4*), 1878–1884, Oct. 2006.

[217] J. Ville, 'Théorie et applications de la notion de signal analytique,' *Câbles et Transmissions*, **2** (*1*), 61–74, 1948.

[218] C. D. Vournas, M. A. Pai, and P. W. Sauer, 'The effect of automatic voltage regulation on the bifurcation evolution in power systems,' *IEEE Transactions on Power Systems*, **11** (*4*), 37–43, Dec. 1996.

[219] E. A. Wan and R. Van Der Merwe, 'The unscented Kalman filter for nonlinear estimation,' in *Proceedings of the IEEE 2000 Adaptive Systems for Signal Processing, Communications, and Control Symposium*, Lake Louise, Alberta, Canada, Oct. 2000, 153–158.

[220] S. Wang, W. Gao, and A. P. S. Meliopoulos, 'An alternative method for power system dynamic state estimation based on unscented transform,' *IEEE Transactions on Power Systems*, **27** (*2*), 942–950, May 2012.

[221] S. Wang, J. Hu, X. Yuan, and L. Sun, 'On inertial dynamics of virtual-synchronous-controlled DFIG-based wind turbines,' *IEEE Transactions on Energy Conversion*, **30** (*4*), 1691–1702, Dec. 2015.

[222] Y. Wang, G. Delille, H. Bayem *et al.*, 'High wind power penetration in isolated power systems – assessment of wind inertial and primary frequency responses,' *IEEE Transactions on Power Systems*, **28** (*3*), 2412–2420, Aug. 2013.

[223] J. B. Ward, 'Equivalent Circuits for Power Flow Studies,' *IEEE Transactions on Power Apparatus and Systems*, **PAS-68**, 373–382, 1974.

[224] J. Winkelman, J. Chow, B. Bowler *et al.*, 'An analysis of interarea dynamics of multi-machine systems,' *IEEE Transactions on Power Apparatus and Systems*, **PAS-100** (*2*), 754–763, Feb. 1981.

[225] K. Yang and A. Walid, 'Outage-storage tradeoff in frequency regulation for smart grid with renewables,' *IEEE Transactions on Smart Grid*, **4** (*1*), 245–252, Mar. 2013.

[226] T. Yang, H. Sun, and A. Bose, 'Transition to a two-level linear state estimato – Part I: Architecture,' *IEEE Transactions on Power Systems*, **26** (*1*), 46–53, Feb. 2011.

[227] T. Yang, H. Sun, and A. Bose, 'Transition to a two-level linear state estimator – Part II: Algorithm,' *IEEE Transactions on Power Systems*, **26** (*1*), 54–62, Feb. 2011.

[228] A. Yazdani and R. Iravani, *Voltage-Sourced Converters in Power Systems – Modeling, Control and Applications*. Hoboken, NJ: Wiley-IEEE Press, 2010.

[229] R. Zárate Miñano and F. Milano, 'Construction of SDE-based wind speed models with exponentially decaying autocorrelation,' *Renewable Energy*, **94**, 186–196, 2016.

[230] L. Zhao and A. Abur, 'Multi area state estimation using synchronized phasor measurements,' *IEEE Transactions on Power Systems*, **20** (*2*), 611–617, May 2005.

[231] N. Zhou, D. Meng, Z. Huang, and G. Welch, 'Dynamic state estimation of a synchronous machine using PMU data: A comparative study,' *IEEE Transactions on Smart Grid*, **6** (*1*), 450–460, Jan. 2015.

[232] N. Zhou, D. Meng, and S. Lu, 'Estimation of the dynamic states of synchronous machines using an extended particle filter,' *IEEE Transactions on Power Systems*, **28** (*4*), 4152–4161, Nov. 2013.

[233] P. Zhou, X. Yuan, J. Hu, and Y. Huang, 'Stability of DC-link voltage as affected by phase locked loop in VSC when attached to weak grid,' in *Proceedings of the IEEE PES General Meeting*, National Harbor, MD, 2014, 1–5.

Index

Frequency Variations in Power Systems: Modeling, State Estimation, and Control, First Edition.
Federico Milano and Álvaro Ortega Manjavacas.
© 2020 John Wiley & Sons Ltd. Published 2020 by John Wiley & Sons Ltd.